An Epidemiologic Study of Mortality and Radiation-Related Risk of Cancer Among Workers at the Idaho National Engineering and Environmental Laboratory, a U.S. Department of Energy Facility

NIOSH Occupational Energy Research Program Final Report

National Institute for Occupational Safety and Health

Division of Surveillance, Hazard Evaluations, and Field Studies

U.S. DEPARTMENT OF HEALTH AND HUMAN SERVICES
Centers for Disease Control and Prevention
National Institute for Occupational Safety and Health

January 2005

Disclaimer

Mention of any company or product does not constitute endorsement by NIOSH. In addition, citations to Web sites external to NIOSH do not constitute NIOSH endorsement of the sponsoring organizations or their programs or products. Furthermore, NIOSH is not responsible for the content of these Web sites.

Ordering Information

To receive documents or more information about occupational safety and health topics, contact the National Institute for Occupational Safety and Health (NIOSH) at

NIOSH
Publications Dissemination
4676 Columbia Parkway
Cincinnati, OH 45226-1998

Telephone: 1– 800–35–NIOSH **(1–800–356–4674)**
Fax: 513–533–8573
E-mail: Pubstaft@cdc.gov
or visit the NIOSH Web site at **www.cdc.gov/niosh**

HHS (NIOSH) Publication No. 2005-131

Authors

Schubauer-Berigan Mary K, National Institute for Occupational Safety and Health, Division of Surveillance, Hazard Evaluations, and Field Studies, Cincinnati, Ohio

Macievic Gregory V, National Institute for Occupational Safety and Health, Division of Surveillance, Hazard Evaluations, and Field Studies, Cincinnati, Ohio

Utterback David F, National Institute for Occupational Safety and Health, Division of Surveillance, Hazard Evaluations, and Field Studies, Cincinnati, Ohio

Tseng Chih-Yu, National Institute for Occupational Safety and Health, Division of Surveillance, Hazard Evaluations, and Field Studies, Cincinnati, Ohio

Flora Jason T, National Institute for Occupational Safety and Health, Division of Surveillance, Hazard Evaluations, and Field Studies, Cincinnati, OH; Current address: 3M Corporation, St. Paul, MN

Preface

Ionizing radiation and its sources are used every day in medical, industrial and governmental facilities around the world. Although some health risks from ionizing radiation exposures are widely recognized, the association of these exposures to specific diseases, especially various types of cancer, remains uncertain. Workers at U.S. Department of Energy (DOE) facilities have produced nuclear weapons, provided nuclear fuel materials for power reactors, and conducted a wide spectrum of research related to nuclear safety and other scientific issues. While completing this work, many of the employees have been exposed to ionizing radiation and other potentially hazardous materials.

Since 1991, the National Institute for Occupational Safety and Health (NIOSH) has conducted analytical epidemiologic studies of workers at DOE nuclear facilities, through a Memorandum of Understanding between the DOE and the U.S. Department of Health and Human Services (DHHS). The agreement occurred in response to recommendations to the Secretary of Energy in 1989 by the independent Secretarial Panel for the Evaluation of Epidemiologic Research Activities (SPEERA).

This technical report, entitled *An Epidemiologic Study of Mortality and Radiation Risk of Cancer Among Workers at the Idaho National Engineering and Environmental Laboratory, a U.S. Department of Energy Facility,* is one several products of the NIOSH Occupational Energy Research Program that are being published as a series. Most of these studies include detailed historical exposure assessments for radiation and other potentially hazardous agents so the health risks at different levels of exposure can be accurately estimated. Each of these studies contributes to the knowledge required to ensure that workers are adequately protected from chronic disease over their working lifetimes.

Distribution of this final report addresses the recommendation of the SPEERA panel to make reports of study results more readily available to workers and the interested public. Additional information about the NIOSH epidemiologic research program of occupational health studies involving the DOE nuclear weapons workforce may be obtained by contacting NIOSH toll free at 1-800-35-NIOSH (1-800-356-4674) or by visiting the NIOSH website for the Occupational Energy Research Program at **http://www.cdc.gov/niosh/2001-133.html**.

Executive Summary

Introduction

The Idaho National Engineering and Environmental Laboratory (INEEL) is a large U.S. Department of Energy (DOE) facility near Idaho Falls, Idaho. Since its construction in 1949 the INEEL has conducted a wide variety of activities, including engineering and basic scientific research, nuclear reactor design and testing, nuclear material chemical processing, and the construction, servicing and demolition of large-scale facilities. In addition, the U.S. Navy maintains its Naval Reactors Facility (NRF) at the INEEL, where research and testing of Navy ship reactors occurs, as well as training of military and civilian personnel involved in the naval nuclear surface ship and submarine program. An epidemiologic cohort mortality study was initiated to evaluate hazards associated with ionizing radiation and other exposures among civilian employees at the INEEL facility.

Methods

This cohort study included 63,561 civilian workers ever employed by the DOE, its contractors, or the NRF at the INEEL at any time between 1949 and 1991. Vital status (whether the worker was living or deceased) and causes of death if deceased were ascertained for each worker through 1999. Exposures were estimated to external ionizing radiation (gamma and neutron radiation) using site records available through 1998. Potential exposure to internal radiation (including beta radiation, fission products and transuranic radionuclides) was also categorized. These radiation exposures are described in the report.

The mortality experience of workers who were badged for ionizing radiation exposure was compared to that of unbadged workers. Workers receiving higher external radiation doses were compared to those receiving lower doses. In addition to these radiological exposures, the cohort was also divided into subcohorts for the evaluation of non-radiological hazards and other factors at the INEEL. Subcohorts that could be identified include construction and maintenance/service workers, asbestos workers, painters, reactor workers, chemists and chemical operators, security workers, and drivers. Mortality patterns were also evaluated for cohort members by the type of employer they worked for ("prime" contractors, subcontractors, or multiple types of contractors).

The statistical analysis of mortality patterns consisted of standardized mortality ratios (SMRs), standardized rate ratios (SRRs), and Poisson regression analysis. SMRs were calculated for the cohort by comparing mortality to both the U.S. population and to a regional population consisting of the states of Idaho, Montana and Wyoming, while standardizing on sex, race (white or non-white), age in five-year intervals and calendar year in five-year intervals. SRRs were calculated based on a comparison of the baseline categories to the regional population and on internal comparisons, for exposed subcohorts. Poisson regression was used to evaluate associations between external radiation and cancers, by estimating the risk of death per unit of dose for the following groups of interest:

- All solid cancers combined
- "Radiogenic" solid cancers (as defined in previous studies of radiation-exposed populations)

- All leukemia
- Leukemia excluding chronic lymphocytic
- Chronic lymphocytic leukemia (CLL)
- Any individual cancer found to be elevated in the SRR analysis

Although no smoking information was available for the cohort, mortality was evaluated for workers by two surrogates for smoking behavior: socioeconomic status (SES) as defined by the worker's first job title at the INEEL, and state of origin as Utah, Idaho, Montana or Wyoming, which was used as an indicator of possible membership in the Latter Day Saints religion (which is associated with lower rates of smoking and alcohol consumption). Smoking-related cancers were also analyzed separately as part of the Poisson regression analysis, to determine whether they were related differently to radiation than non-smoking-related cancers.

Results

The INEEL cohort was predominantly white (96%) and male (81%). About 18% of the cohort was white and female. The median year of birth was 1942, and was much earlier for white males (WM) than for non-whites and white females (WF). The median length of follow-up for the cohort was 21 years. The median year of hire for WM was 1973, about 6 years earlier than for WF and non-whites.

The workforce consisted of many short-term workers. The median duration of employment was just over three years. About 57% of the cohort was ever monitored for exposure to external radiation. The average cumulative external dose among the monitored workers was about 13 mSv. The highest average doses for workers, as well as collective doses across the site, occurred during the 1960s.

About 47% of the cohort was classified as ever having been a construction or maintenance service worker. The asbestos, painter, reactor and chemical worker subcohorts numbered 2741, 690, 1440 and 5332, respectively. There were 1276 security workers and 1947 truck or bus drivers identified at the facility. About half the cohort came from the states of Utah, Idaho, Montana and Wyoming; the rest were from elsewhere or unknown. The cohort consisted of many professional (16%) and administrative/technical (15%) workers, but also had a large percentage who were skilled manual or non-manual (33%) and partly skilled or unskilled (15%) workers. About 21% were of unknown SES.

Overall, about 17% of the cohort was deceased. A much greater percentage of WM than of other groups was deceased. INEEL workers had much lower mortality rates than the general U.S. population. When compared to the regional population, however, INEEL workers exhibited only slightly lower mortality rates [overall SMR: 0.96, 95% confidence interval (CI) 0.94-0.97, 10788 deaths], but cancer rates were elevated (all-cancer SMR: 1.07, 95% CI: 1.03-1.11, 2873 deaths). Workers monitored internally and externally for ionizing radiation exposures showed lower mortality than non-monitored workers for most causes of death. However, two of three bone cancer deaths in the cohort occurred among workers with some indication of positive internal dose, leading to an elevated bone cancer SRR for this group compared to other workers (SRR=7.33, 95% CI: 0.66-81.3).

Mortality rates for certain cancers were elevated among the INEEL cohort, or among individual subcohorts. Non-Hodgkin lymphoma (NHL) was elevated in the overall cohort (SMR=1.26, 95% CI: 1.05-1.50), particularly among painters (SRR=2.46, 95% CI: 0.89-6.80) and female construction workers (SRR=4.07, 95% CI: 1.08-15.3). Brain tumor mortality rates were elevated among male chemical workers (SRR=2.12, 95% CI: 0.82-5.49) and security workers (SRR=2.29, 95% CI: 0.78-6.71). Construction and maintenance service workers showed elevated mortality rates from asbestosis (SMR=4.92, 95% CI: 2.35-9.26) and cancers likely to be mesotheliomas (SRR=4.54, 95% CI: 1.01-20.4). Mortality rates for these causes were particularly high among those identified as asbestos workers (asbestosis SRR=25.6, 95% CI: 6.25-105; likely mesothelioma SRR=4.28, 95% CI=1.19-15.5). Bus and truck drivers showed elevated rates of death from transportation accidents (SRR=1.63, 95% CI: 1.07-2.48), and security workers exhibited higher mortality rates from accidental falls (SRR=18.9, 95% CI: 2.62-136) and other non-transportation accidents (SRR=2.34, 95% CI: 0.90-6.06) compared to other workers.

An inverse dose-response relation was observed for emphysema, heart disease, and smoking-related cancers, suggesting negative confounding by smoking (that is, those receiving higher doses may have smoked with lower frequency or at a lower rate than those receiving lower doses). Radiogenic non-smoking-related cancers showed a very weak negative association with radiation exposure, caused by a lower RR in workers receiving more than 100 mSv. The excess relative risk (ERR) per 10 mSv (1 rem) of cumulative exposure was -0.0023, with an upper 95% confidence limit (CL) of 0.0459 (the lower limit could not be calculated).

Positive, but non-significant, associations were detected for brain tumors, for leukemia, and for lymphatic cancers, particularly when off-site dose was included in the model. At a 20-year dose lag, the ERR per 10 mSv for all brain tumors combined was 0.087; (95% CI: -0.0037 to 0.338). At a 7-year lag, the ERR per 10 mSv for non-CLL leukemia was 0.0543 (95% CI: -0.0114 to 0.238). The ERR per 10 mSv for CLL was negative, even when a longer lag period was used. For NHL, the ERR per 10 mSv was 0.0199 (upper 95% CL: 0.100) and for multiple myeloma was 0.0638 (95% CI: -0.0150 to 0.345). These associations appear to have been driven by exposures in the highest dose groups ($\geq$100 mSv).

Conclusions

Overall cancer mortality in the INEEL cohort was somewhat higher than expected based on regional rates, but for most cancer types was unlikely to be related to ionizing radiation exposure. Cancers that did show some evidence of association with ionizing radiation exposure include leukemia (excluding chronic lymphocytic), NHL, brain cancer, and other "radiogenic" cancers. In addition, there were elevated rates of mortality for asbestos-related diseases and accidents and some cancers among other groups of workers at the INEEL.

Some strengths of the present study include the large size and well-characterized external radiation exposures of the cohort, and the relatively large population of female nuclear workers. Some limitations of the study include apparent confounding by smoking for many cancers (which may have reduced the ability to observe an association between external radiation and cancer), the fact that less than 20% of the cohort was deceased (making generalizations about mortality patterns difficult to discern for the entire cohort), the diversity of exposures across the facility, and the difficulty in assessing internal radiation and non-radiological exposures at the site. In particular, the INEEL cohort was relatively young (with a median age of 54.4) at the end of follow-up. Risk estimation for specific subcohorts and causes of death would likely be more precise with additional follow-up of this cohort. Other limitations may also be overcome through the continued follow-up of the INEEL cohort, and possible nested case-control studies within the cohort.

Contents

Disclaimer ii
Authors iii
Preface iv
Executive Summary vi
Contents x
List of Figures xiii
List of Tables xv
Acknowledgments xxii
1 Introduction and Background 1
   1.1   Introduction 1
   1.2   INEEL Overview 2
   1.3   Research Strategy and Objectives 7
         1.3.1   Specific Objectives 7
         1.3.2   Population Studied 8
         1.3.3   Outcomes Studied 8
         1.3.4   Workplace Exposures at INEEL 9
               1.3.4.1   Radiological Exposures 9
               1.3.4.2   Non-Radiological Exposures 10
2 Methods 11
   2.1   Cohort Definition and Identification 11
   2.2   INEEL Record Capture and Coding 14
         2.2.1   Demographic and Employment Records 16
         2.2.2   Radiological Record Sources 17
               2.2.2.1   Off-site Dose Adjustments 17
               2.2.2.2   The Updated RDS Database 18
               2.2.2.3   The NRF Database 18
               2.2.2.4   Internal Dose Data 19
         2.2.3   Record Linkage 19
   2.3   Exposure Assessment 20
         2.3.1   Radiological Exposures 20
               2.3.1.1   External Dosimetry 20
               2.3.1.2   Internal Dosimetry 25
         2.3.2   Non-radiological (Subcohort Identification) 27
               2.3.2.1   Contractor Code Collapse and Identification 28
               2.3.2.2   Job Title Collapse and Identification 30
               2.3.2.3   Identification of Exposure-based Subcohorts 31
   2.4   Confounder Assessment 36
         2.4.1   Demographic Variables 36
         2.4.2   Socioeconomic Status 36
   2.5   Vital Status and Cause of Death Ascertainment 41
   2.6   Statistical analysis 41
         2.6.1   Radiological Data 41

      2.6.2 Epidemiologic Data 42
            2.6.2.1 Analytic Variables 42
            2.6.2.2 SMR and SRR Analysis 42
            2.6.2.3 Multivariate regression modeling analysis........................46
3 Radiation Exposure Assessment Results...49
   3.1 External Dosimetry..49
   3.2 Internal Dose 54
   3.3 Missed Dose Estimate 56
   3.4 Doses for Naval Reactors Facility (NRF) and Construction/Service Workers...........57
      3.4.1 Naval Reactors Facility (NRF) Doses..57
      3.4.2 Construction and Service Worker Doses 59
4 Cohort Descriptive Results..62
   4.1 Demographic, Work History and Vital Status Information 62
   4.2 Cause of Death Information 63
   4.3 Covariates..72
   4.4 Non-radiological Exposures (Subcohorts)..72
   4.5 Radiological Exposures...73
5 Life Table Analysis Results...81
   5.1 Required Information for Analysis 81
   5.2 Full Cohort Analysis 83
      5.2.1 Comparison to General U.S. Population......................................83
            5.2.1.1 Combined Cohort, 1949-1999 83
            5.2.1.2 White Males, 1949-1999 83
            5.2.1.3 White Females, 1949-1999...84
            5.2.1.4 Non-white Males..84
            5.2.1.5 Non-white Females, 1949-1999.......................................84
            5.2.1.6 Comparison to U.S. Population, 1960-1999....................85
      5.2.2 Comparison to Regional Population (ID, WY, MT combined) 1960-1999......90
   5.3 Badged and Unbadged Worker Subcohorts..96
      5.3.1 White Males 96
      5.3.2 White Females 102
      5.3.3 Non-white Males and Females 103
   5.4 External Radiation Dose-response Analysis in the Full Cohort...............108
      5.4.1 White Males 108
      5.4.2 White Females 115
   5.5 Internal Exposure Category...119
      5.5.1 White Males 119
      5.5.2 White Females 121
   5.6 SES Subcohort...123
      5.6.1 White Males 123
      5.6.2 White Females 128
      5.6.3 Non-white Males 131
   5.7 Local vs. Migrant Subcohorts 132
   5.8 Construction and Service Worker Subcohort...136
      5.8.1 White Males 136
      5.8.2 White Females 137

5.9	Asbestos Worker Subcohort	140
5.10	Chemical Worker Subcohort	143
5.11	Drivers	145
5.12	Reactor Workers	147
5.13	Painter Subcohort	147
5.14	Security Worker Subcohort	149
5.15	Employer Type Subcohorts	152
6	Multivariable Modeling Results	157
6.1	All Solid Cancers	157
6.1.1	Development of Baseline Model	157
6.1.2	Dose-response Analysis with External Ionizing Radiation	158
6.2	Smoking-related and Non-smoking-related Cancers	164
6.3	Inclusion of Off-site Dose: Smoking-related and Non-smoking-Related Cancers	167
6.4	"Radiogenic" Non-smoking-related Cancers	167
6.5	Brain Tumor Dose-response	171
6.6	Leukemias	173
6.6.1	Baseline Model	173
6.6.2	Dose-response Analysis with External Ionizing Radiation	174
6.6.3	Sensitivity Analyses	175
6.7	Emphysema and Ischemic Heart Disease	184
6.8	Non-Hodgkin Lymphoma and Multiple Myeloma	186
7	Discussion	188
7.1	Exposures at INEEL	188
7.1.1	Radiological Exposures	188
7.1.2	Non-radiological Exposures	191
7.1.2.1	Construction and Service Workers	193
7.1.2.2	Painters	194
7.1.2.3	Asbestos Workers	195
7.1.2.4	Chemical Workers	196
7.1.2.5	Security Workers	198
7.1.2.6	Reactor Workers	199
7.1.2.7	Drivers	200
7.2	Epidemiological Findings	200
7.2.1	Life Table Analyses	200
7.2.2	Interpretation of Life Table Results	206
7.2.3	Radiation Dose-response Analyses	206
7.3	Strengths and Limitations of Study	208
8	Conclusions and Recommendations	211
Literature Cited		212
Appendix		A-1

List of Figures

Figure 1-1. INEEL site map 3
Figure 1-2. INEEL site contractor relationships..4
Figure 2-1. Distribution of hire years for INEEL workers in the cohort, and not in the cohort..14
Figure 2-2. Limit of detection and accuracy of the dosimeter types over time.....................21
Figure 2-3. Annual frequency of workers with external ionizing radiation doses at or below the detection limit 25
Figure 2-4. Educational level of engineer categories at NRF...38
Figure 2-5. Educational level of technician and other Intermediate SES categories at NRF...39
Figure 2-6. Distribution of highest educational levels within various SES categories assigned for 14,077 workers in the HRS file 39
Figure 3-1. Collective and average dose (in mSv) and number monitored by year at INEEL 51
Figure 3-2. Distribution of dosimeters by facility Area Code 52
Figure 3-3. Total off-site dose and worker number by year..52
Figure 3-4. Distribution of cumulative photon dose by number of employees.....................53
Figure 3-5. Distribution of cumulative neutron dose by number of employees 53
Figure 3-6. Variation of average neutron dose with worker number by year 54
Figure 3-7. Example (plutonium-238) for estimating internal dose to a maximally exposed organ..55
Figure 3-8. Annual frequency of externally monitored and non-monitored workers, and workers with potentially positive internal dose...56
Figure 3-9. Missed dose estimated from log probability regression of 1-5 mSv dose values from 1951-1958 57
Figure 3-10. Comparison of worker population and photon doses among NRF and non-NRF workers 58
Figure 3-11. Comparison of neutron doses among NRF and non-NRF site employees.........59
Figure 3-12. Number of externally monitored construction/service and non-construction/non-service worker-years by time period 60
Figure 3-13. Fractions of the construction and the non-construction worker groups with positive internal dose by time period and construction/service worker status....60
Figure 3-14. Average cumulative penetrating photon dose (by five-year time intervals) for construction and non-construction workers. 61
Figure 3-15. Collective penetrating photon dose (by five-year time intervals) for construction and non-construction workers. 61
Figure 4-1. Distribution of age at DLO by cumulative dose categories (217 missing). 79
Figure 4-2. Least-squares mean on-site external dose (mSv) for migrants by SES class......80

Figure 5-1. Standardized rate ratios for a) non-Hodgkin lymphoma and b) leukemia by dose category (including those with zero dose) among white males. 113

Figure 5-2. Standardized rate ratios for (a) malignant melanoma (includes zero dose group) and (b) brain cancer (includes only positive dose group), by 20-year-lagged dose category, white males. 114

Figure 5-3. Standardized rate ratios for all digestive cancers by dose category (including those with zero dose), white females. 115

Figure 5-4. Standardized rate ratios for breast cancer by dose category (including those with zero dose), white females. 116

Figure 6-1. Relative risks for smoking related (N=811) and non-smoking related (N=483) cancer mortality by dose category, after stratifying on sex, age, calendar time and duration of employment (<10 years and >10 years), and adjusting in the model for internal exposure category, SES, migrant status, and the interaction of SES and migrant status, analyzed with on-site (a) and total (b) dose. 166

Figure 6-2. Relative risks for radiogenic non-smoking related cancers, including just underlying (N=220), and non-underlying (N=20) in addition to underlying, cancers by dose category 171

Figure 6-3. Relative risks for brain tumors, including just underlying (N=66), and non-underlying (N=2) in addition to underlying, cancers by dose category 173

Figure 6-4. Relative risks for non-CLL (N=51 underlying and non-underlying) for two different dose categorizations 183

List of Tables

Table 1-1. Major contractors at the INEEL since 1949 with periods of operations. 4
Table 1-2. Reactors and critical assembly facilities operating or operable as of 1992. 5
Table 1-3. Dismantled, transferred, or standby status reactors and critical assembly facilities at the INEEL. 6
Table 1-4. Radiogenic cancers from BEIR V (NRC 1990). 8
Table 1-5. Agents with known industrial hygiene data at the INEEL. 11
Table 2-1. Application of exclusion criteria to INEEL cohort records. 13
Table 2-2. INEEL records sources used in cohort mortality study. 15
Table 2-3. Distribution of hire and termination dates by source file. 16
Table 2-4. Examples of errors in end date sequences in NRF file 19
Table 2-5. Summary of dosimeter characteristics over time (Perry 2002) 20
Table 2-6. Radionuclides/groups of interest for internal dosimetry and the commencement of monitoring methods at INEEL. 26
Table 2-7. Categories assigned to INEEL employers and the total number of individuals with work history records that identify an employer in the category (workers may appear in more than one category). 29
Table 2-8. Availability of job title fields by source file for INEEL cohort. 30
Table 2-9. Collapsed job titles and 1980 Bureau of Census occupational codes (indicated by BOC) used for each collapsed job title. 32
Table 2-10. Identification of exposure-based subcohorts of interest using contractor categories, collapsed job titles and 1980 Bureau of Census occupational codes (BOC_1980) 35
Table 2-11. 1980 Bureau of Census occupational codes assigned to SES groups for INEEL cohort analysis. 37
Table 2-12. Random sample of problematic job titles with respect to conflicts in SES determinations, and changes used to solve the problem. 40
Table 2-13. Additional subcohorts of interest in the INEEL cohort analysis. 44
Table 2-14. ICD codes used in the creation of special rate files for all brain tumors, all leukemia other than chronic lymphocytic leukemia (CLL) and pleura and peritoneum cancer combined. 44
Table 2-15. Cumulative dose category cut points for standardized rate ratio (SRR) analysis of INEEL cohort. 46
Table 3-1. Penetrating photon dose (mSv) percentiles by year for monitored workers 50
Table 3-2. Facilities that used greater than 5% of the dosimeters by Area Code 51
Table 4-1. Distribution of race/ethnicity by sex in the (a) INEEL demographic file and (b) final INEEL cohort after application of cohort eligibility criteria. 62
Table 4-2. INEEL cohort descriptive statistics. 64
Table 4-3. State of death distribution for 10,906 deceased INEEL study cohort members. ..65

Table 4-4.	Distribution of underlying COD (ICD-9 codes) in the INEEL cohort, by sex and race.	66
Table 4-5.	Distribution of covariates within the INEEL cohort: demographic variables and employer type.	75
Table 4-6.	Average duration employed (in years) among contractor types, of workers with non-missing DOH and DOT.	76
Table 4-7.	Distribution of covariates within the INEEL cohort: exposure-related variables.	77
Table 4-8.	Dosimetry characteristics of INEEL cohort.	78
Table 5-1.	Distribution of INEEL cohort members by presence of dates required for life table analysis.	81
Table 5-2.	Characteristics of workers ineligible for life table analysis, compared to eligible cohort.	82
Table 5-3.	Standardized mortality ratios (SMRs) for all workers, combined and by sex and race categories, compared to general U.S. population rates 1940-1999, adjusted for five-year age and calendar period intervals.	86
Table 5-4.	SMRs for selected causes of death for INEEL cohort, compared to U.S. rates 1960-1999, adjusted for age and calendar period.	89
Table 5-5.	Standardized mortality ratios (SMRs) for all workers, combined and by sex and race categories, compared to combined Idaho, Montana and Wyoming population 1960-1999, adjusted for age and calendar period.	92
Table 5-6.	Standardized mortality ratios by employment duration (compared to regional population of ID, MT, WY 1960-1999) for (a) all deaths combined and (b) cancer.	95
Table 5-7.	Standardized mortality ratio (compared to regional population) and standardized rate ratio (for employment duration categories) for combined tumors of brain, including malignant, benign and tumors of unspecified nature.	96
Table 5-8.	Standardized mortality ratios (SMRs) for white male radiation-badged and -unbadged workers, and standardized rate ratios (SRRs) for badged compared to unbadged workers.	98
Table 5-9.	SRRs for white males who were badged with zero dose and badged with positive dose, compared to unbadged workers.	100
Table 5-10.	Standardized mortality ratios (SMRs) and standardized rate ratios (SRRs) for white female workers	104
Table 5-11.	SRRs for white females who were badged with zero dose and badged with positive dose, compared to unbadged workers.	106
Table 5-12.	Slopes and standard errors (for badged white male workers, using on-site dose only), adjusted for age and calendar year.	110
Table 5-13.	SMR and dose-response trends (for badged white female workers, on-site dose only), adjusted for age and calendar year.	117
Table 5-14.	Standardized rate ratios for internal monitoring compared to unmonitored person-time, white males only	119

Table 5-15. SRRs for internal monitoring compared to unmonitored person-time, white females.. 122

Table 5-16. SRRs for white males in SES groups Intermediate, Skilled non-manual, Skilled manual, Partly skilled, Unskilled and unknown, compared to the Professional SES group.. 125

Table 5-17. Distribution of age (at DLO) among SES categories for white females. 128

Table 5-18. SRRs for white females in SES groups Intermediate, Skilled non-manual, Skilled manual, Partly skilled, Unskilled and Unknown, compared to the Professional SES group 129

Table 5-19. SMR results for non-white males, for SES groups Professional, Intermediate, Skilled non-manual, Skilled manual, Partly skilled, and Unskilled. 131

Table 5-20. Standardized rate ratios for migrants compared to local workers, for white males and white females 133

Table 5-21. SRR results for construction workers compared to non-construction workers, for white males and white females 137

Table 5-22. SRR results for asbestos workers compared to non-asbestos workers, for white males (only one death occurred among white female asbestos workers). 140

Table 5-23. SMRs (compared to combined ID, MT, UT and WY) and SRRs for combined pleura and peritoneal cancers, for asbestos workers and other workers............ 142

Table 5-24. SRR results for chemical workers compared to non-chemical workers, for white males and white females.. 143

Table 5-25. SRR results for drivers compared to non-drivers, for white males only. 146

Table 5-26. Standardized mortality ratios by time since hire for transportation accidents among white male drivers and non-drivers at INEEL 147

Table 5-27. SRR results for reactor workers compared to non-reactor workers, for white males only. 148

Table 5-28. SRR results for painters compared to non-painters, for white males only 149

Table 5-29. SRR results for security workers compared to non-security workers, for white males.. 151

Table 5-30. SRRs for white males by employer types (prime contractor, multiple contractors, unknown contractor) compared to subcontractor employee rates.. 153

Table 5-31. SRRs for white females by prime contractor and multiple contractor employer types, compared to subcontractor employer type rates, for causes of death with more than 5 expected or observed in the total cohort of white female workers. 156

Table 6-1. Comparison of all solid cancer (excluding lymphomas, myeloma and leukemias, and with 10-year lag) maximum-likelihood risk estimates and likelihood-based CIs produced with (N=34,916) and without (N=29,585) monitored employees who worked < 1 year, with DOB, date first monitored and migrant status available 159

Table 6-2. Comparison of all-solid-cancer (excluding lymphomas, myeloma and leukemias, and with 10-year lag) maximum-likelihood risk estimates and likelihood-based

	CIs produced with (N=34,916) monitored employees with DOB, date first monitored and migrant status available	160
Table 6-3.	Final baseline model: comparison of all-solid-cancer (excluding lymphomas, myeloma and leukemias, and with 10-year lag) maximum-likelihood risk estimates and likelihood-based CIs produced with (N=34,916) monitored employees with DOB, date first monitored and migrant status available	162
Table 6-4.	Results of risk estimation for smoking-related* (N=811) and non-smoking-related (N=483) cancer mortality	164
Table 6-5.	Non-smoking-related cancer risk coefficients (N=34,916 monitored employees with DOB, date of first monitoring and migrant status available) with inclusion of off-site dose	168
Table 6-6.	Smoking-related cancer risk coefficients by dose type (N=34,916 monitored employees with DOB, date of first monitoring and migrant status available).	170
Table 6-7.	Brain tumor (underlying only) risk coefficients by dose type (N=34,916 monitored employees with DOB, date of first monitoring and migrant status available)	172
Table 6-8.	Comparison of all leukemia (N=69 with 5-year lag) maximum-likelihood risk estimates and likelihood-based CIs produced with (N=34,916) monitored employees with DOB, date first monitored and migrant status available	176
Table 6-9.	Final baseline model for all leukemia (N=70 with 5-year lag) maximum-likelihood risk estimates and likelihood-based CIs produced with (N=36,169) monitored employees with DOB and date first monitored available	178
Table 6-10.	Leukemia (N=70) risk coefficients (ERR per mSv) by dose type (N=36,169 monitored employees with DOB and date of first monitoring available)	179
Table 6-11.	Leukemia risk coefficients (ERR per mSv) by dose type (N=36,169 monitored employees with DOB and date of first monitoring available), separated by chronic lymphocytic and non-CLL	180
Table 6-12.	Effect of varying lags on radiation related leukemia (N=70) risk, for 36,169 monitored employees with DOB and date of first monitoring available	181
Table 6-13.	Effect of varying lags on CLL and non-CLL leukemia risk, for 36,169 monitored employees with DOB and date of first monitoring available	182
Table 6-14.	Results of risk estimation for emphysema (N=69) and ischemic heart disease (N=1296) for 34916 workers with DOB, date of first monitoring and migrant status available	184
Table 6-15.	Radiation-related risk for non-Hodgkin lymphoma (NHL; N=82) and multiple myeloma (N=20) among 34916 workers with DOB, date of first monitoring and migrant status available	187
Table 7-1.	Comparison of monitored employees at four DOE facilities	190
Table 7-2.	Distribution of study cohort members by monitoring status, cumulative dose range and facility.	190
Table 7-3.	Dosimeter issuance and LOD by facility, for INEEL and Hanford.	191
Table 7-4.	Construction Worker Classifications from U.S. Census Bureau.	193

Acronyms and Abbreviations

AFSR	Argonne Fast Source Reactor
AIW-A	Large Ship Reactor Prototype "A"
AIW-B	Large Ship Reactor Prototype "B"
ALARA	As Low As Reasonably Achievable
ANL-W	Argonne National Laboratory-West (INEEL subdivision)
ARA	Auxiliary Reactor Area (INEEL subdivision)
ARMF-1	Advanced Reactivity Measurement Facility No. 1
ATLAS	Automatic Thermoluminescence Analyzer System
ATR	Advanced Test Reactor
ATRC	Advanced Test Reactor Critical Facility
BN	Benign Neoplasms
BORAX 1-5	Boiling Water Reactor Experiment 1-5 (INEEL subdivision)
B&W	Babcock and Wilcox
CDC	Centers for Disease Control and Prevention
CET	Critical Experiment Tank
CFA	Central Facilities Area (INEEL subdivision)
CFRMF	Coupled Fast Reactivity Measurement Facility
CI	Confidence Interval
CL	Confidence Limit
CLL	Chronic Lymphocytic Leukemia
CNS	Central Nervous System
COD	Cause of death
CRCE	Cavity Reactor Critical Experiment
DHHS	U.S. Department of Health and Human Services
DLO	Date Last Observed
DMF	Death Master File
DOB	Date of Birth
DOE	Department of Energy
DOELAP	Department of Energy Standard for the Performance Testing of Personnel Dosimetry Systems
DOH	Date of hire
DOT	Date of termination
EBOR	Experimental Beryllium Oxide Reactor
EBR I-II	Experimental Breeder Reactor I-II (INEEL subdivision)
ECF	Expended Core Facility
EOCR	Experimental Organic Cooled Reactor
ERR	Excess Relative Risk
ETR	Engineering Test Reactor
ETAC	Engineering Test Reactor Critical
FRAN	Nuclear Effects Reactor
GCRE	Gas Cooled Reactor Experiment Metals Reactor
HCFA	Health Care Finance Administration
HOTCE	Hot Critical Experiment
HRS	Human Resources System
HTRE 1-3	Heat Transfer Reactor Experiment 1-3
IARC	International Agency for Research on Cancer
ICD	International Classification of Diseases
ICPP	Idaho Chemical Processing Plant (INEEL subdivision)

IMBA	Integrated Modules for Bioassay Analysis
INEEL	Idaho National Engineering and Environmental Laboratory
LDS	Church of Jesus Christ of Latter Day Saints
LOD	Limit of detection
LOFT	Loss of Fluid Test Facility
LLD	Lower Limit of Detection
LTAS	Life Table Analysis System
MK-F	MK-Ferguson
ML	Mobile Low Power Reactor
MN	Malignant Neoplasms
MTR	Materials Test Reactor
MUD	Master Update Dump
NDI	National Death Index
NHL	Non-Hodgkin lymphoma
NIOSH	National Institute for Occupational Safety and Health
NRAD	Neutron Radiography Facility
NRC	Nuclear Regulatory Commission
NRF	Naval Reactors Facility (INEEL subdivision)
NS	Nervous System
NUB	Neoplasms of Uncertain Behavior
NUN	Neoplasms of Unspecified Nature
NWF	Non-White Females
NWM	Non-White Males
ODU	Operational Dosimetry Unit
OMP	Occupational Medicine Program
OMRE	Organic Moderated Reactor Experiment
ORNL	Oak Ridge National Laboratory
PBF	Power Burst Facility (INEEL subdivision)
PC	Pocket Chambers
PMB	Personnel Metering Branch
PTL	Protection Technologies Idaho
RALA	Radioactive Lanthanum Facility
RDS	Radiation Dosimetry System
RMF	Reactivity Measurement Facility
RR	Relative Risk
RWMC	Radioactive Waste Management Complex (INEEL subdivision)
SECIMS	Security Information Monitoring System
SES	Socioeconomic status
SL	Stationary Lower Power Reactor
SMC	Specific Manufacturing Capabilities
SMR	Standardized mortality ratio
SNAP 10A	System for Nuclear Auxiliary Power
SNAPTRAN 1-3	System for Nuclear Auxiliary Power Transient Program 1-3
SPERT	Special Power Excursion Reactor Tests
SRCE	Spherical Cavity Reactor Critical Experiment
SRR	Standardized rate ratio
SSA	Social Security Administration
SSN	Social Security number
STF	Security Training Facility (INEEL subdivision)
STR	Split Table Reactor

SUSIE	Shield Test Pool Facility
S1W (STR)	Submarine Thermal Reactor
S5G	Natural Circulation Submarine Prototype
TAN	Test Area North (INEEL subdivision)
TLD	Thermoluminescent dosimeter
TMI	Three Mile Island
TRA	Test Reactor Area (INEEL subdivision)
TREAT	Transient Reactor Test Facility
WBC	Whole Body Counting
WEC	Westinghouse Electric Corporation
WF	White Females
WINCO	Westinghouse Idaho Nuclear Corporation
WM	White Males
YOT	Year of Termination
ZPR	Zero Power Reactor
ZPPR	Zero Power Plutonium Reactor
630-A	High Temperature Marine Propulsion Reactor
710	Fast Spectrum Refractory

Acknowledgments

Funding for this study was provided through an agreement between the U.S. DOE and the U.S. DHHS. This technical report resulted from the cooperation and support of the DOE and its employees and contractors. Many employees at the Idaho National Engineering and Environmental Laboratory (INEEL) provided numerous sources of information that were required to complete this complex study. The INEEL professional staff has also answered many telephone calls and emails as source information was interpreted. This study particularly benefited from the assistance of the following individuals at the INEEL site: Ed Ahrens, Gordon Buttars, Paul Creighton MD, John Culley, George Espinoza, K.C. Gerard, Judy Hamilton, Marie Hill, Jim Jones, David Klepich, Rich Knighton, Bill Larsen, Stacey Madson, Tom Must, Henry Ng, Ken Puphal, Carl Robertson, Paul Ruhter, John Seward, Sharon Sorce and Joel Trent.

The staff at the Naval Reactors Facility (NRF), also located at the INEEL, was equally cooperative and supportive. We thank, in particular, Theron Bradley and Andy Richardson for providing access to data and records used in this study.

Current and former staff at DOE headquarters, including Marsha Lawn, Gerry Petersen, and Heather Stockwell, were instrumental for ensuring that NIOSH investigators were allowed access to all essential documents at the INEEL, as well as providing ongoing encouragement for completion of the study.

Former NIOSH scientists Paul Mills and M. Kathryn Brown carried out initial epidemiological work on the study. Current and former NIOSH employees John Cardarelli, Doug Daniels, Sam Glover, Tim Jiggens and Jim Neton assisted with several aspects of the radiation exposure assessment. Current and former NIOSH fellows and employees Robyn Anderson, Jenneh Burphy, Vicki Dryfhout-Ferguson and Sara Foster assisted with imputation of missing hire dates, validation of the migrant status variable, socioeconomic status classification, and determining adequacy of cohort follow-up, respectively.

We are also grateful to the many current and former government and contract staff working with NIOSH who have provided essential services for the collection, compilation and validation of records and data used in this study. These people include Steven Ahrenholz, David Back, Pi-Hsueh Chen, Betsy Dupree-Ellis, Larry Elliott, Mary Engle, Jean Geiman, Sam Glover, B.J. Haussler, Shannon Hiratzka, Kevin Hubbard, Kim Jenkins, Travis Kubale, Yanmei Li, Relada Miller, Frank Moran (deceased), Susan Nowlin, Dianne Reeder, Robert Rinsky, Mitch Singal, Henry Spitz, Melissa Tolson, Thurman Wenzl and James Yiin. Kim Collins, Susan Freking, Donna Pfirman and Erin Wickliff assisted in the preparation of this report.

Vital status and cause of death information was obtained from the National Death Index, with additional assistance provided by the vital records offices of each U.S. state. We also acknowledge the contribution of the U.S. Social Security Administration for providing demographic and vital status information for INEEL cohort members.

Finally, the quality of this report was substantially improved as a result of technical reviews that were provided by colleagues within NIOSH, including Misty Hein, Avima Ruder, Sharon Silver, and Tim Taulbee, as well as scientific peers outside of NIOSH.

1 Introduction and Background

1.1 Introduction

In December 1990 the Secretary of the U.S. Department of Energy (DOE) and the Secretary of the U.S. Department of Health and Human Services (DHHS) signed a Memorandum of Understanding that transferred authority for the conduct and management of all analytic epidemiologic studies of workers at DOE facilities to DHHS, specifically the National Institute for Occupational Safety and Health (NIOSH). The study of the Idaho National Engineering and Environmental Laboratory (INEEL) was conducted under that authority. This report describes the epidemiological research that NIOSH has undertaken among past and present employees at the INEEL.

There were several reasons for studying the past and present INEEL workforce. First, health risks among the workforce have never been examined through analytical epidemiology, and preliminary evidence from radiation dosimetry records indicated that sufficient exposures may have occurred to warrant concern for worker health. Second, since 1949 approximately 71,000 people have worked at the INEEL. A substantial number of these workers might have been affected by radiation and/or other hazardous substance exposures. Third, the Government Affairs Committee of the U.S. Senate and Governor Cecil Andrus of Idaho requested that the Centers for Disease Control and Prevention (CDC) evaluate the health effects that may have occurred in the area surrounding the INEEL and among the INEEL workforce.

In January 1993 the DHHS Advisory Committee for Energy-Related Epidemiologic Research met and reviewed the proposed research programs of NIOSH. At this meeting the Committee approved the INEEL epidemiologic study in principle and recommended that NIOSH proceed with protocol development.

The primary objective of this study was to assess potential associations between possible excess mortality in the INEEL workforce and exposures to ionizing radiation and/or other toxic elements at the worksite. Within this context, cancers at specific organ sites were of particular interest. To meet this objective, NIOSH conducted an all-cause cohort mortality epidemiologic study among INEEL employees.

Ionizing radiation exposure at the site appears to have resulted primarily from external radiation sources associated with fission products. Over the years of operation, however, INEEL contractors have conducted monitoring programs for internal deposition of radionuclides as well. A previous analysis of internal emitters at the INEEL failed to find extensive exposures to workers (Horan and Braun 1993). Therefore, this study primarily evaluated external ionizing radiation exposures; however, the study also examined the extent of internal radionuclide deposition and its contribution to ionizing radiation exposures for the INEEL cohort.

Although chemical exposures were not the primary focus of this study, potential confounding exposures merit evaluation. A complete list of classified documents of potential interest in assessing both radiological and chemical exposures was compiled and subsequently reviewed by NIOSH personnel with appropriate security clearances. Declassification of these documents was requested from DOE. If declassification was not granted, DOE was requested to provide copies of the documents with all classified information removed. Only documents that can be made available to the public were used for this study so that any results and conclusions may be critically evaluated by the scientific community and the public.

Permission from the U.S. Navy was received to include civilian Naval Reactors Facility (NRF) employees in the INEEL cohort analysis. The inclusion of civilian NRF workers is important because current and former NRF employees also have worked at other INEEL facilities, and NRF data were required in order to obtain complete work history information on these employees. In addition, estimates of exposure to radiation for NRF workers make up a substantial proportion of the total site dose at INEEL. U.S. Navy personnel who served only active duty training tours at the NRF were not included in the study.

1.2 INEEL Overview

Situated on approximately 890 square miles of land on the Snake River basin in southeastern Idaho (Figure 1-1), the INEEL is a government-owned, contractor-operated facility whose primary function has been to design, build and test nuclear reactors for the U.S. government. Originally called the National Reactor Testing Station when construction commenced in 1949, the facility was renamed the Idaho National Engineering Laboratory in 1974. In 1994 the words "and Environmental" were added to the title. Over the past decade many of the INEEL missions have changed with an increasing emphasis placed on environmental remediation and development of nuclear waste management technologies.

Since 1949 at least 71,504 people have worked at INEEL. At the time the study was being planned in 1992, the INEEL employed approximately 12,000 people who worked for seven different contractors: EG&G of Idaho (EG&G), Westinghouse Idaho Nuclear Corporation (WINCO), Westinghouse Electric Corporation (WEC), Babcock and Wilcox (B&W), MK-Ferguson (MK-F), Argonne National Laboratory-West (ANL-W), and Protection Technologies Idaho (PTI). The relationship of these contractors to the DOE Idaho Field Office is depicted in Figure 1-2. As can be seen in the figure, three DOE Field Offices share responsibility for administering the various on-site contractors. These contractors and numerous subcontractors are responsible for the nine primary and four secondary on-site operations areas. Although each area has a distinct mission, there are many inter-related activities that involve potential exposure to similar agents, for example external ionizing radiation. The operational timeline of major INEEL contractors is listed in Table 1-1.

Epidemiologic Study of Mortality and Radiation-Related Risk of Cancer Among INEEL Workers

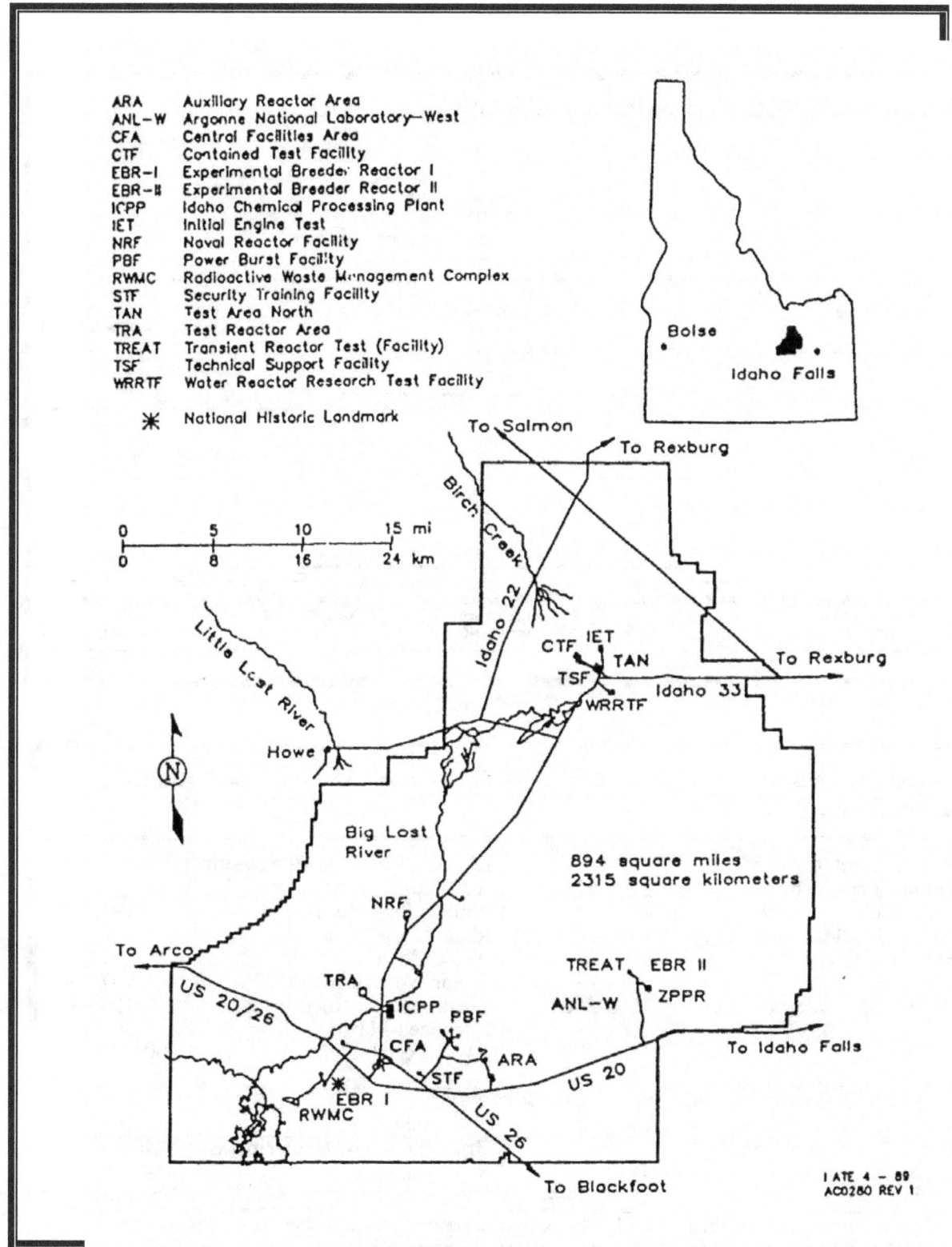

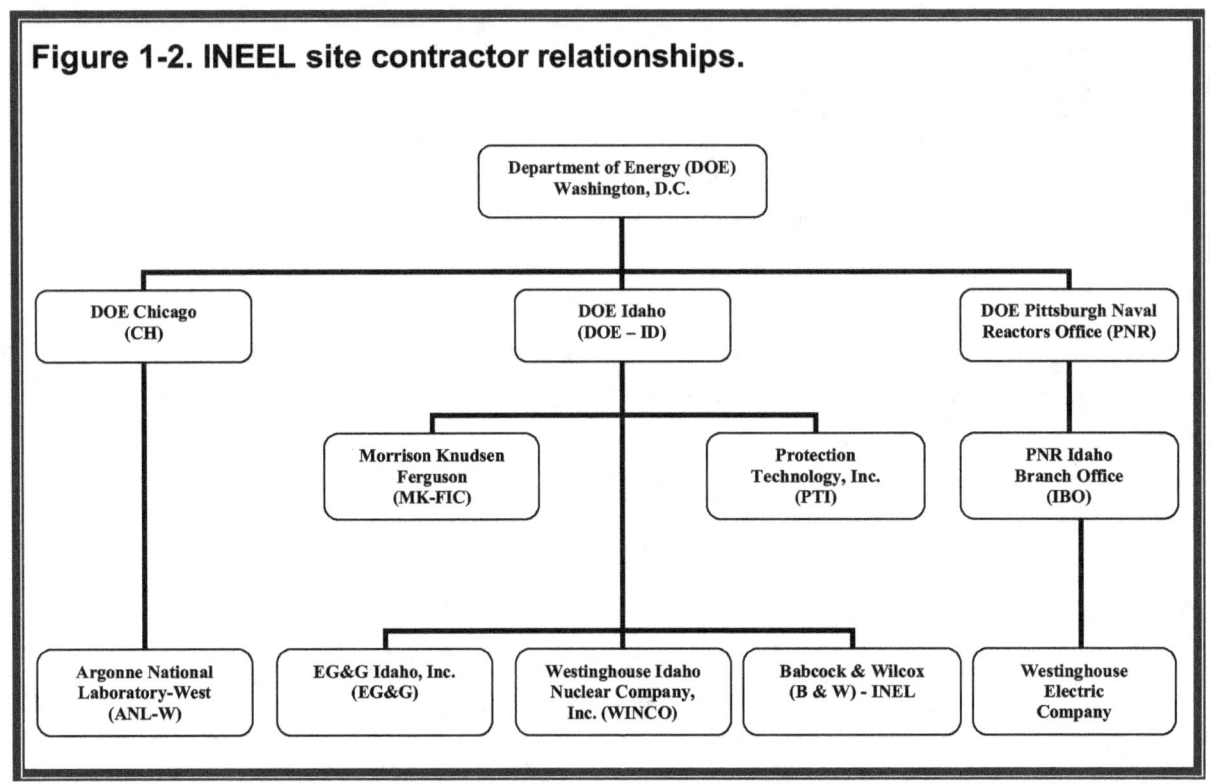

Figure 1-2. INEEL site contractor relationships.

Table 1-1. Major contractors at the INEEL since 1949 with periods of operations.

Prime Contractors*		Idaho Chemical Processing Plant	
Phillips Petroleum	1953-67	American Cyanamid	1950-53
Idaho Nuclear Co.	1967-72	Phillips Petroleum	1953-67
Aerojet Nuclear	1972-76	Idaho Nuclear Co.	1967-72
EG&G-Idaho	1976-94	Allied Chemical	1972-79
Lockheed-Martin	1994-99	Exxon Nuclear Idaho	1979-84
BWX Technologies	1999-2004	Westinghouse Nuclear Idaho	1984-94
		Lockheed-Martin	1994-99
		BWX Technologies	1999-2004
Argonne National Laboratory – West Complex†			
Argonne National Laboratory West		1949-2004	
Naval Reactors Facility		**Specific Manufacturing Capability**	
Westinghouse Electric	1953-99	Rockwell INEEL	1986-91
Bechtel Bettis	1999-2004	BWX Technologies	1991-2004

* Operated most of the facilities at the INEEL except Argonne National Laboratory Facilities, Naval Reactors Facility and the Idaho Chemical Processing Plant since 1953.

† Operated Experimental Breeder Reactor I at a site remote from current ANL-W complex.

The primary sources of ionizing radiation exposure have been from reprocessing spent nuclear fuels from military and civilian reactors and operation of 48 nuclear reactors that were present at the site prior to 1992. A total of twelve of these reactors remained operational in 1992 (Table 1-2). These reactors range in complexity from the Experimental Breeder Reactor I (EBR-I) at ANL-W facility, to U.S. Navy submarine reactors at the NRF to neutron radiographic reactors at ANL-W. An additional 38 reactors and critical devices have been removed from operational status (Table 1-3). Some of these reactors have been intentionally destroyed in testing procedures, and others have ceased operation as research programs have been completed. One (the SL-1 reactor) was destroyed in an accident in 1961 (described in more detail in the Appendix §A.1.9).

Table 1-2. Reactors and critical assembly facilities operating or operable as of 1992.

Name	Startup	Abbreviation	Operating Contractor
1. Advanced Reactivity Measurement Facility No. 1	1960	ARMF-1	EG&G
2. Advanced Test Reactor	1968	ATR	EG&G
3. Advanced Test Reactor Critical Facility	1959	ATRC	ANL
4. Argonne Fast Source Reactor	1959	AFSR	ANL
5. Coupled Fast Reactivity Measurement Facility	1968	CFRMF	EG&G
6. Experimental Breeder Reactor-II	1963	EBR-II	ANL
7. Large Ship Reactor Prototype "A"	1958	AIW-(A)	WEC
8. Large Ship Reactor Prototype "B"	1958	AIW-(B)	WEC
9. Natural Circulation Reactor	1965	S5G	WEC
10. Neutron Radiography Facility	1977	NRAD	ANL
11. Submarine Thermal Reactor	1958	S1W(STR)	WEC
12. Transient Reactor Test Facility	1959	TREAT	ANL

The main portions of the INEEL reactor facilities are located dozens of miles from population centers. This remote location has resulted in the need for extensive on-site services, including cafeterias, laundries, machine shops, welding shops, carpentry, and sheet metal fabrication, among many others. Correspondingly there are large numbers of workers in these varied occupations. In addition to the on-site operations, an estimated 4000 administrative and research personnel were employed in the city of Idaho Falls in 1992.

The INEEL is a complex DOE site with diverse and heterogeneous operations. The facility has been engaged in missions that have changed substantially through the years. Consequently, the population under study has been exposed to various agents and, therefore, comprises not one, but many subcohorts, each with a unique exposure history. By the very nature of the INEEL missions, workers have been exposed to ionizing radiation and other non-radiological agents in the workplace since construction began in 1949 and the first reactor fuels loaded in 1951. A summary description of the INEEL is presented in the Appendix §A.1.

Table 1-3. Dismantled, transferred, or standby status reactors and critical assembly facilities at the INEEL.

Name	Startup	Shutdown	Abbreviation	Operating Contractor
1. Advanced Reactivity Measurement Facility No. 2			ARMF-II, PPCo, INC	
2. Boiling Water Reactor Experiment No. 1	1953	1954	BORAX-I	
3. Boiling Water Reactor Experiment No. 2, 3, 4	1954	1958	BORAX-II, III, IV	ANL
4. Boiling Water Reactor Experiment No. 5	1962	1974	BORAX-V	ANL
5. Cavity Reactor Critical Experiment			CRCE	GE, INC
6. Critical Experiment Tank			CET	GE
7. Engineering Test Reactor	1957	1981	ETR	INC, ANC, EG&G
8. Engineering Test Reactor Critical			ETAC	INC, ANC, EG&G
9. Experimental Beryllium Oxide Reactor	Terminated		EBOR	GA
10. Experimental Breeder Reactor-I	1951	1964	EBR-1	ANL
11. Experimental Organic Cooled Reactor	Terminated		EOCR	PPCo.
12. Fast Spectrum Refractory			710	GE
13. Gas Cooled Reactor Experiment Metals Reactor	1960	1962	GCRE	AGC
14. Heat Transfer Reactor Experiment No. 1	1956	1957	HTRE-I	GE
15. Heat, Transfer Reactor Experiment No. 2	1957	1961	HTRE-II	GE
16. Heat Transfer Reactor Experiment No. 3	1958	1961	HTRE-III	GE
17. High Temperature Marine Propulsion Reactor			630-A	GE
18. Hot Critical Experiment			HOTCE	GE
19. Loss of Fluid Test Facility	1978		LOFT	EG&G
20. Materials Test Reactor	1952	1970	MTR	PPCo, INC
21. Mobile Low Power Reactor No. 1 (ARMY)	1961	1965	ML-1	AGC
22. Nuclear Effects Reactor	1967	1970	FRAN	INC
23. Organic Moderated Reactor Experiment	1957	1963	OMRE	AI
24. Power Burst Facility	1973	1989	PBF	EG&G

Table 1-3. Dismantled, transferred, or standby status reactors and critical assembly facilities at the INEEL.

Name	Startup	Shutdown	Abbreviation	Operating Contractor
25. Reactivity Measurement Facility			RMF	PPCo.
26. Shield Test Pool Facility			SUSIE	GE
27. SNAP 10A Transient 1	1963	1965	SNAPTRAN-1	AI/PPCo.
28. SNAP 10A Transient 2	1965	1966	SNAPTRAN-2	AI/PPCo.
29. SNAP 10A Transient 3	1964	1964	SNAPTRAN-3	PPCo.
30. Special Power Excursion Reactor Test No. 1	1955	1964	SPERT-I	PPCo.
31. Special Power Excursion Reactor Test No. 2	1960	1965	SPERT-II	PPCo.
32. Special Power Excursion Reactor Test No. 3	1958	1968	SPERT-III	PPCo.
33. Special Power Excursion Reactor Test No. 4	1962	1970	SPERT-IV	PPCo.
34. Spherical Cavity Reactor Critical Experiment			SRCE	ANC
35. Split Table Reactor			STR	GE, INC, ANC
36. Stationary Low Power Reactor No. 1	1958	1961	SL-1	CE
37. Zero Power Reactor No. 3			ZPR-III	ANL
38. Zero Power Plutonium Reactor	1969	1993	ZPPR	ANL

1.3 Research Strategy and Objectives

1.3.1 Specific Objectives

Specific objectives of the INEEL epidemiologic retrospective cohort mortality study were:

1. To determine whether cause-specific mortality patterns among INEEL workers (past and current) have differed from that of an appropriate, external population (total U.S. or a regional population).

2. To determine whether mortality patterns among INEEL workers who were monitored for external and internal ionizing radiation have differed from those who were not monitored.

3. To determine whether mortality patterns among INEEL workers with higher cumulative external ionizing radiation doses have differed from those with lower doses and if a dose-response relation exists between exposure and cause-specific mortality.

4. To assess the role of confounders and/or effect modifiers on any associations observed between external ionizing radiation exposures and outcomes. Such factors include smoking, exposure to other chemical and physical agents, and certain socioeconomic and demographic factors, which may contribute to the healthy worker effect observed in most nuclear worker cohorts.

1.3.2 Population Studied

The study cohort was defined as all INEEL employees with an assigned security number (i.e., all those with permanent passes or a security clearance) between January 1, 1949 and December 31, 1991. Workers were identified from several contractor sources: computerized, hard copy and microfilm/microfiche files of personnel, medical, security and dosimetry databases (see §2.2). Both males and females and all race/ethnic groups were included in the study; although, as in many DOE facilities, white males (WM) were expected to constitute the largest demographic group. Excluded persons consisted of military employees at the NRF, university employees on fellowship or temporary assignment to INEEL, DOE (or its contractors') employees permanently assigned to other sites, visitors, and persons who could not be followed up. Persons who could not be followed up consisted of those who lack information on both date of birth (DOB) and valid Social Security number (SSN) or those with very common names for whom NIOSH could obtain a DOB but no SSN (see §2.1). In addition, persons without date of hire (DOH) and date of termination (DOT) were excluded if these dates could not be imputed using site information (see §2.2.1).

1.3.3 Outcomes Studied

In this all-cause cohort mortality study, all fatal events occurring to members of the workforce were examined. However, given the *a priori* knowledge of the potential association between ionizing radiation and cancer, deaths from site-specific cancers were the disease outcomes of greatest interest. Those cancer sites categorized as having an established relation to external radiation sources according to the Committee on the Health Effects of Exposure to Low Levels of Ionizing Radiation, BEIR V (NRC 1990) were of particular interest and are listed in Table 1-4.

Table 1-4. Radiogenic cancers from BEIR V (NRC 1990).

Salivary glands	Urinary bladder
Esophagus	Kidney
Stomach	Brain
Colon	Central nervous system (CNS)
Rectum	Thyroid
Lung, Bronchus	Skin
Multiple myeloma	
Breast	
Leukemia (except chronic lymphocytic)	

In addition, elevations in rates of cancer of the liver, paranasal sinuses and bone appear to occur primarily after exposure to inhaled, injected, or ingested alpha and/or beta particles. These internal emitting particles can lodge in the respective organs and expose them to radiation over time. However, as noted previously, internal emitters appear not to have contributed substantially to total radiation exposures of workers at the INEEL. These cancers were of *a priori* interest among workers classified as likely having internal radiation exposure.

1.3.4 Workplace Exposures at INEEL

The large number of research, manufacturing, support, and nuclear fuel reprocessing operations at the INEEL has resulted in an expansive number of chemical and physical agents being present over the five decades of site operation. Exposures to these agents may have increased the risk of specific disease processes, including some cancers known to be associated with radiation exposures. Retrospective exposure assessment of the radiological and non-radiological agents at the INEEL was considered for examination of potential causal associations between exposures and health effects. Complete and accurate exposure assessment data are also necessary to appropriately assess potential confounding effects in the analysis of the radiological and chemical or physical agent exposures. However, currently available information was insufficient to determine the magnitude and extent of exposures to non-radiological agents.

1.3.4.1 Radiological Exposures

Radiological exposures to ionizing radiation above naturally occurring background levels have occurred at most facilities within the INEEL boundary. The exposures consist of both external and internal exposures originating from a large variety of radionuclides used or created during various processes and nuclear experiments at the INEEL. Exposure to external sources of ionizing radiation at the INEEL has been associated with research and development of nuclear reactors, fuel reprocessing, and waste disposal and management. Workers have received external exposure to gamma radiation, x-rays, neutrons and several different sources of beta radiation. Film badges were used initially as personal dosimeters until 1966, when thermoluminescent dosimeters (TLDs) were introduced. Sustained monitoring programs for internal emitters have also been established at the INEEL.

Because of the number of facilities and their changing missions throughout time, it is not feasible to list all potential sources of ionizing radiation, but they can be collapsed into general categories. These categories include, but are not limited to, fission products (noble gases, strontium and cesium), activation products (cobalt-60) and alpha, beta, gamma and neutron radiations. In §2.3.1 an in-depth discussion of availability of records and techniques to summarize these exposures is given.

External Exposures

Although the INEEL was established in 1949, the first recorded personal radiation exposure measurement occurred in September 1951 at the EBR-I. Dosimetry services utilized the traditional film badge dosimeters from the commencement of operations until 1966, when the INEEL became the first national laboratory to use TLDs. This change in technology increased the sensitivity of measurement and increased the ability to detect lower exposures.

The majority (>85%) of the available health physics data were whole body gamma and beta exposures, whereas neutron measurements were somewhat more limited. Improvements in neutron dosimetry occurred with the introduction of the TLD in 1966. In addition to the whole body exposure data, the INEEL has collected dosimetry data on extremity exposures since 1970.

Internal Exposures

Data on internal exposures exist from 1951 to the present time in two forms: whole body counting (WBC) results and bioassay analysis results. At time of initiation of this study, the relative importance of internal emitters among INEEL workers was unclear. Published reports and initial discussions with INEEL staff indicated that exposure to internal emitters at the INEEL has not been extensive. The INEEL currently has bioassay monitoring programs for approximately 1000 individuals. These programs differ substantially among contractors at the INEEL. Several hundred dose evaluations have been performed as a result of assay findings in excess of some regulatory or administrative limit or because of the occurrence of an incident. These limits have changed over time and have generally become more restrictive. As a result, many other doses below previous limits may be of interest in the epidemiologic study.

Current modeling techniques for the calculation of exposure were applied where possible to historical data where bioassay results or other supporting material were indicative of significant elevations. NIOSH collected available electronic data and supporting documentation on internal emitters located at the Radiation and Environmental Sciences Laboratory and other INEEL facilities for the evaluation of internal radionuclide exposure potential.

1.3.4.2 Non-Radiological Exposures

Industrial hygiene measurements in the INEEL work environment were very limited prior to the late 1970s. Most measurements have been made since 1985. Inspection of the industrial hygiene sampling records indicated that chemical exposures at the INEEL have included metals, common industrial and laboratory solvents, and some rather unique materials used in nuclear energy research activities as well as other research programs at the INEEL. An abbreviated list of compounds for which industrial hygiene data are known to be available appears in Table 1-5. This list is not inclusive of all potential exposures at the INEEL but represents more recent industrial hygiene concerns.

Available information was insufficient to estimate the chemical exposures of concern for this study. Among the industrial hygiene data reviewed, the quantity and quality varied substantially among the different contractors. However, all industrial hygiene monitoring and exposure records at all INEEL facilities were evaluated for this study.

Table 1-5. Agents with known industrial hygiene data at the INEEL.			
Chemical Agents			
Acids & Bases	**Gases**	**Solvents and Other Hydrocarbons**	
Acetic acid	Carbon monoxide	Acetone	Acrylamide
Hydrochloric acid	Chlorine	Aromatic hydrocarbons	Amyl acetate
Inorganic acids	Hydrogen fluoride	Benzene	Butanol
Phosphoric acid	Nitric oxide	Carbon disulfide	Butoxyethanol
Sodium hydroxide	Nitrogen dioxide	Carbon tetrachloride	Butylacetate
Sulfuric acid	Nitrogen oxides	Chloroform	Diisocyanates
	Ozone	Ethanol	Dipropylene glycol
		Halogenated hydrocarbons	Formaldehyde
			Hexamethylene diisocyanate
Metals & Metalloids	**Dusts & particulates**	Hexane	Limonene
Aluminum	Asbestos	Isoamyl acetate	Methyl methacrylate
Arsenic	Fly ash	Isopropanol	Polychlorinated biphenyls
Barium	Graphite	Methanol	Toluene diisocyanate
Beryllium	Nuisance dust	Methyl ether	Total hydrocarbons
Chromium	Portland cement	Methyl ethyl ketone	Trichlorotrifluoroethane
Copper	Silica	Methylene chloride	
Iron oxide	Wood dust	Styrene	
Lead		Tetrachloroethylene	
Magnesium		Toluene	
Mercury		Xylenes	
Mercury alkyl compounds			
Molybdenum	**Physical Agents**		
Phosphorus	Noise		
Welding fumes			
Zinc			

2 Methods

2.1 Cohort Definition and Identification

The INEEL cohort was defined as any civilian worker who was employed by the site or its subcontractors on or before December 31, 1991, who could be adequately identified for vital status follow-up purposes. Employment and other records were obtained as described in §2.2 for 101,998 persons who ever worked or were issued a security or dosimetry badge at the INEEL facility between its construction and the time that the data were collected (late 1993). The INEEL base cohort consisted of persons present at the INEEL site at some point between its construction (1949) and 1991, inclusive. A total of 33,793 individuals were excluded from the cohort, because they did not meet the initial definition of the cohort (Table 2-1). Excluded persons consisted of military employees at the NRF, university employees on fellowship or temporary assignment to INEEL, DOE (or its contractors') employees permanently assigned to other sites, visitors, and persons who could not be followed up. Persons who could not be followed up consisted of those who lacked information on both DOB and valid SSN and those with very common names for whom a DOB was available but no SSN. Of the 68,205 remaining workers, 4644 did not meet the cohort definition, as they began employment after December 31, 1991. Thus, the total number followed in the cohort was 63,561. In general, those excluded from the cohort were rather early, short-term workers (Figure 2-1, Table 2-1). The dosimetry statistics for the individuals removed from the cohort are shown in Table 2-1. These workers had very low average and collective exposure levels, compared to individuals included in the cohort. The overall collective dose of those not included in the cohort was less than 10%, and the average doses were approximately one-sixth, of those included in the cohort analysis.

It was discovered two years after creating the final cohort file and after all analyses were complete that 487 workers should have been removed from the cohort because they failed to meet the cohort inclusion criteria. These workers were predominantly "other DOE site" employees (N=475, or 13.8% of the total "other DOE site" employees), and the remainder were military employees (N=12, or 0.07% of the total military employees). Of the "other DOE site" employees inadvertently left in the cohort, 28% were Nuclear Regulatory Commission (NRC) Region IV employees, 15% were Knolls Atomic Power Laboratory (NY) employees, 14% were DOE employees of Argonne National Laboratory (IL), 13% were Office of Personnel Management employees, 11% were DOE-Washington employees, and 8% were Lawrence Livermore National Laboratory workers. Dosimetry information pertaining to the INEEL site exposures, as well as any off-site exposures recorded in the INEEL dosimetry system, was used in the epidemiologic analyses described in this report.

Active vital status ascertainment was used in tracking this cohort. Therefore, workers of unknown vital status were assigned a "date last observed" (DLO; see §2.5) beyond which they were considered lost to follow-up. The preferred (state) mortality rates used in the Life Table Analysis System (LTAS) begin in 1960. Therefore, workers whose death or DLO alive occurred before 1960 or were missing were ineligible for inclusion in the analyses that used general population rates for computation of SMRs. This number totals 1667, leaving 61,894 cohort members eligible for SMR analyses in which person-years begin in 1960 (i.e., all state rate comparisons). These individuals were included in analyses in which person-time begins in 1949 (e.g., §5.5.1). The overall collective dose among these largely excluded workers was very low (Table 4-1). Additional exclusions, for those lacking sufficient information to be included in the life table analysis, are described in Table 5-1.

Of the 63,561 total INEEL workers in the cohort, 3129 (4.9%) were missing a valid SSN (although just 1833 or 3.0% of 61,462 workers with follow-up after 1960 were missing SSN).

Table 2-1. Application of exclusion criteria to INEEL cohort records.						
Rationale for inclusion/exclusion		Total N	Radiation-monitored N (%)	Collective dose (Sv)	Avg dose (mSv) among monitored	Mean duration employed (years)
Total	Initial record count	101,998	61,651 (60.4%)	516.10	8.37	
Exclusions	Military only	17,492	17,360 (99.2%)	30.53	1.76	0.64
	Off-site DOE only	2977	2785 (93.6%)	5.54	1.99	3.94
	University only	530	32 (6.0%)	0.07	2.19	0.78
	Visitor only	27	13 (48.1%)	0.004	0.34	0.71
	Cannot follow-up	12,767	3589 (28.1%)	7.26	2.02	0.64
	Hired after 12/31/91	4644	1427 (30.7%)	2.86	2.01	1.63
Total eligible to be included	All	63,561	36,442 (57.3%)	469.83	12.89	--
	DLO before 1/1/1960	1667	711	1.53	2.15	--
	DLO missing	94	0	0	0	--
	Others	61,800	35,731	468.30	13.11	--

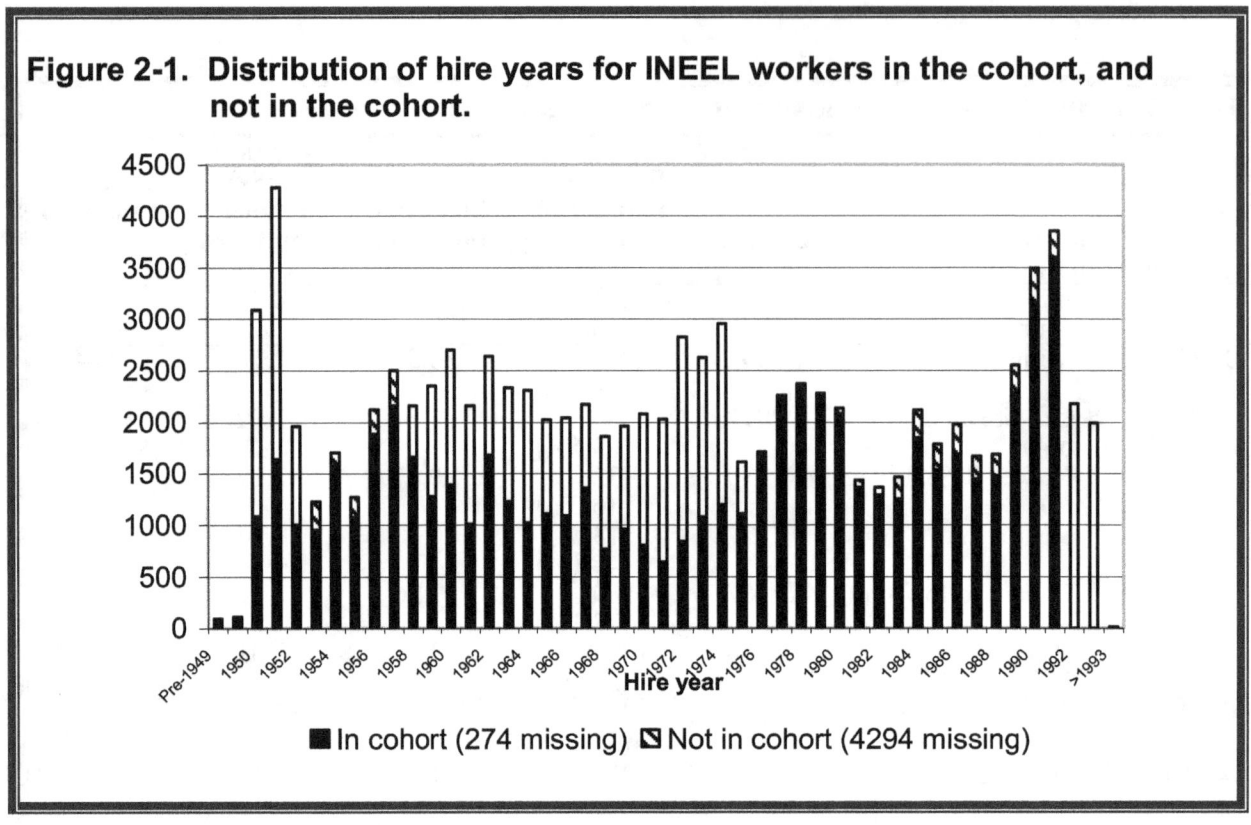

Figure 2-1. Distribution of hire years for INEEL workers in the cohort, and not in the cohort.

2.2 INEEL Record Capture and Coding

Several sources of records were obtained from the INEEL site for this study. Seven major sets of records and electronic files were used to identify workers, to obtain demographic and work history information and to estimate exposures for workers in the study. These data sources are shown in Table 2-2. The Roster file was created by NIOSH by keying data contained in the work history cards obtained on microfiche. Keyed fields included names, DOB, dates of first hire and last termination, employer name, and job titles (only the first job title was listed). The SSN and S-number (a sequentially issued identifier assigned by INEEL) were also keyed. All fields were double entered for quality control purposes, and any discrepancies between the entries were immediately identified to the second data entry clerk for correction. The remaining files identified in Table 2-2 were obtained from sources at the INEEL site.

Table 2-2. INEEL records sources used in cohort mortality study.

System Name	Record Format	Number of Records	Description	Types of data in record system (not all records contain all information)
Roster	Microfiche	50,985	Work history cards for all non-NRF site employees hired before 1978	Names, DOB, SSN, sex, S-number, DOH, DOT, job titles
SECIMS	Computer file	106,528 *	Security information monitoring system data	Names, DOB, SSN, sex, S-number, DOH, DOT, contractor, job titles
HRS	Computer file	15,882	Human resources system data for all non-NRF site employees hired in 1978 or later	Names, DOB, SSN, S-number, DOH & DOT, job titles
OMP	Computer file	19,275	Occupational medicine program data for all non-NRF site employees included in medical monitoring program, in the 1970s or later	Names, DOB, SSN, S-number, medical examination dates, job titles, employment status
NRF	Computer database	4900	Naval Reactors Facility data for all NRF employees	Names, DOB, SSN, S-number, DOH & DOT, job titles, badge-level dosimetry data from 1974-1998
MUD	Computer file	207,663 [1]	"Master Update Dump" of the INEEL site dosimetry data from 1949-1985, including NRF doses before 1974	Names, DOB, SSN, S-number, badge numbers, job titles, annual photon, beta and neutron doses, date of fission product, tritium, radionuclide or whole body count examination
RDS	Computer database	797,941 [1]	Radiation dosimetry system of the non-NRF doses received from 1986-1998	Names, DOB, SSN, S-number, badge numbers, job titles, badge-level photon, beta and neutron doses, monitoring results for fission product, tritium, radionuclide or whole body count examination

* Multiple records per individual.

2.2.1 Demographic and Employment Records

The primary sources of identifying and demographic information, including names, SSN, DOB, sex, race and ethnicity, for each worker were (in decreasing order of importance) the Roster, SECIMS, HRS, NRF and OMP files. The dosimetry files contained some demographic information as well; however data such as DOB were frequently found to have been incorrect or filled with nonsensical data (e.g., many dates of birth were reported to have been 11-11-11).

DOH was required to begin person-year calculations (see §2.6.2) and DOT was needed to estimate mortality risk by duration of employment. DOHs and DOTs were obtained from each of the files in Table 2-2. A hierarchy of dates in these files was used to determine both first DOH and last DOT. This algorithm is described in the Appendix §A.2. In general, workers were assigned DOH using the earliest first DOH from any of the files, and DOT using the last employment date available in any file. There were a number of workers who were missing DOH after completing this process. For these individuals, the S-number was used to impute DOH, through a process described in detail in the Appendix §A.3.

The life table program requires a DOT for all workers in the study, for analyses based on employment duration. Decedents missing a DOT were assigned the date of death as a DOT. Termination dates were unavailable for workers who were still employed in about 1993, when the study databases were obtained. However, complete radiation monitoring data were obtained for site workers through 1998. For these individuals a presumptive DOT of December 31, 1998 was used if the worker was being monitored for radiation exposure as of the end of 1996, so that duration of exposure could be calculated. For workers whose last radiation monitoring date occurred before that time, the last radiation monitoring date was assigned as a DOT. For non-radiation-monitored individuals missing DOT, a DOT was left as missing. Such individuals were excluded in life table analyses that used duration of employment, but have no effect on Poisson regression analyses, which were conducted only for radiation-monitored individuals. The distribution of DOH and DOT sources is shown in Table 2-3.

Table 2-3. Distribution of hire and termination dates by source file.

Source	Number of cohort members with Source as:	
	Hire date	Termination date
Roster	39,325	23,781
SECIMS	14,556	13,252
NRF	2474	1216
HRS	3202	571
MUD	1412	6444
RDS	1898	12,068
OMP	0	1788
Imputed	420	2632
Death date	--	497
Missing	274	1312

Job titles and contractor affiliations were collected from each of the INEEL source files listed in Table 2-2. More than 20,000 unique job titles and approximately 28,000 contractor codes were identified from these sources. The collapsing and usage of these data are discussed in §2.3.2 and §2.4.2 below.

2.2.2 Radiological Record Sources

NIOSH received three data systems containing health physics records for workers at INEEL. The Master Update Dump (MUD) contained dosimetry information for workers at INEEL prior to 1986 by year. Since 1986 the Radiation Dosimetry System (RDS) from the Operational Dosimetry Unit (ODU) in the Radiological and Environment Sciences Laboratory has stored detailed external and internal radiation monitoring data by badge exchange period. The third set of files contained dosimetry and demographic information for civilian contract workers at the NRF. This system was established in 1974. Files within the MUD, NRF and RDS systems were linked by various key variables. Information in the files can be divided into 4 groups:

1. **Demographic information** – Personal information, for example names, SSN, Security number (S-number), DOB and sex.

2. **Dosimetry information** – Files containing dosimetry data, for example badge number, area code, penetrating dose, non-penetrating dose, neutron dose, extremity dose, film/TLD read date, bioassay results and WBC.

3. **Descriptive information** – Ancillary data, for example area code, contractor code and organ code.

4. **Supporting files** – Other data, for example area code histories, contractor code histories, reporting information and corrections to files.

Considerable effort was required to locate and eliminate substantial duplication of data among the three systems. MUD and NRF contained overlapping identical data for several workers between 1951 and 1973. The dosimetry data sets also contained numerous records outside specified data ranges.

The original RDS database delivered to NIOSH in 1994 included monitoring data for the years 1986-1993. An updated RDS was obtained in 2001, extending the years of coverage through 1998. The updated database had more dose measurements than the original during the same years of coverage. Differences were resolved prior to the application of an algorithm to calculate annual and cumulative external dose estimates. Also, an NRF update was obtained at the same time. The contents were reconciled and added to the previous NRF database.

2.2.2.1 Off-site Dose Adjustments

Off-site dose was labeled OFFSITE in MUD and AREA CODE 999 in RDS. These doses were ostensibly received at some other facility while a person was employed at INEEL. Area code 999 refers to non-INEEL exposures in almost all cases. However, doses from visitor badges were also assigned to area code 999. There were nine cases in which a person's summed MUD dose was recorded as the RDS area code 999 dose value for a given period of years. Of the nine, three were corrected by NIOSH in the updated RDS database, and the remainder were corrected by the INEEL site.

A total of 313 persons had dose assigned to area code 999 only. Records for a sample of 15 people from this group with the highest doses were sent to the site for evaluation. Four persons were found to have worked at the Grand Junction site (not included in this study), which used INEEL as their dosimeter processor. The ODU supplied the code that identified those people, and they were removed from the study files. The remaining 11 people had area code 999 doses and no other information about them at the site. They may have been at the site working for a subcontractor before 1986. Dosimetry badges would not have been required in these cases. They accounted for 5% of the total (about 0.86 Sv) dose assigned to area code 999.

In numerous cases individuals with several doses assigned to area code 999 had duplicate dosimetry information. For example, a person was assigned doses 17.64 mSv and 12.94 mSv, but the 12.94 mSv was actually part of the 17.64 mSv. The ODU was certain that this problem was not isolated, but its extent could not be estimated. A random selection of 20 persons with more than one area code 999 assigned dose was sent to the ODU for evaluation. When compared to physical dosimetry records it was determined that not more than 0.5% of the records contained this type of error so further adjustments were not made.

2.2.2.2 The Updated RDS Database

When compared with the original RDS database sent to NIOSH, the updated RDS had greater numbers of measurements for many workers. Many of these workers had additional measurements for periods prior to 1986, when the RDS system was first adopted. To evaluate changes in the dosimetry system data in the updated version of RDS, spreadsheets were generated for the years 1981, 1986 and 1992 that contained all dosimetry results for each worker from MUD and both the initial and updated RDS. Numerous extremity and area code 999 doses were found to have been added to the updated database. No additional penetrating, non-penetrating or neutron whole body doses were identified. When the dose algorithm was applied separately to the initial and updated RDS (also including MUD data) dose estimates matched and the updated database was deemed reliable.

2.2.2.3 The NRF Database

A systematic review of the NRF dosimetry data began by combining the two NRF databases, checking for duplicates and then dividing the dose data into periods of one month to ascertain the wear periods (i.e., the apparent period of time over which the badge was worn):

- 90.7% of the wear periods were at 1 month or 3 months.
- 8.0% of the wear periods were between 3 month and 12 months.
- 0.1% of the wear periods were between −2 months and −359 months.
- 1.2% of the wear periods were between +13 months and +599 months.

The negative month wear periods resulted from erroneous end dates in a sequence as shown in Table 2-4 (the corrected date follows the erroneous date in the table).

The 1.2% positive wear periods that were greater than 13 months were labeled in the NRF database as previous doses (i.e., doses received prior to starting employment at INEEL). However, nearly all (99.8%) of these doses were found to match the

Table 2-4. Examples of errors in end date sequences in NRF file. The end year of 1991 is incorrect.	
BEG DATE	**ENDDATE**
1994-08-01	1994-08-31
1994-09-01	1991-09-30
1994-09-01	1994-09-30

person's total dose in MUD or a combination of MUD and RDS and were doses received either at the NRF before 1974 (when the NRF database was developed) or during other employment at the INEEL. Approximately 0.2% did not match to the MUD or RDS databases and were reconciled on an individual basis.

Dosimetry data for NRF workers who terminated before 1960 were recorded in MUD. These doses were identified as 'previous dose' in the new NRF data system and were removed to avoid duplication.

2.2.2.4 Internal Dose Data

The internal dose assessment for the INEEL cohort was based on evaluation and manipulation of three main SAS files: bioassay_areacode, exp_hist, wbsampprm_wbnuclprm. These databases encompassed the MUD and RDS periods. The data consisted of sample results, sample date, name, data flags noting errors, isotope and area code. No information concerning minimum detectable activity was available. Comments were found on less than 0.1% of the data cells. Two other files exist that duplicated much of the material in the three main files. The exp_hist file contained information on bioassay and WBC, but only by frequency of sampling for the MUD period. The bioassay_areacode and wbsampprm_wbnuclprm files contained result information for the periods covered by RDS (>or=1986) and MUD (<1986) for those persons whose work years overlapped the two periods.

2.2.3 Record Linkage

The seven primary files listed in Table 2-2 were linked together using a variety of techniques. The S-number was thought to have been a unique, global identifier across all files and was used initially to link individuals within the files. However, several problems were discovered: (1) not every record was identified by S-number; (2) some S-numbers were re-used; and (3) some individuals were issued more than one S-number. As a result, other identifiers, such as SSN and first name, last name and DOB, were additionally used to link records among the files. DOH and DOT were also used to facilitate linkages.

Following all linkage efforts, a unique identifier (NIOSH_ID) was assigned to every record associated with a unique individual in the cohort and was used to track all information on cohort members.

2.3 Exposure Assessment

2.3.1 Radiological Exposures

2.3.1.1 External Dosimetry

History of INEEL external dosimetry program

The personnel monitoring program began in August 1951 (Puphal 1996). The type of dosimeter, limit of detection (LOD), and wear period varied from 1951 to the present (Table 2-5) as did the accuracy (Figure 2-2). During the early 1950s several types of detection devices were provided by the Personnel Metering Branch (PMB). The function of the PMB was to supply external dosimetry monitoring equipment and keep records of individuals' exposures. Personnel monitoring devices used during this period were the film badge, pocket chambers (PC), and self-reading dosimeters. Finger rings and wrist dosimeters were used but were limited to specific jobs with potential for exposure to the extremities.

Table 2-5. Summary of dosimeter characteristics over time (Perry 2002). B-G represents Beta and gamma monitoring; N represents neutron monitoring.

Time Span	Radiat. Type	Type of Device or System	Areas Affected	Approx. monitoring period (days)	Worker Classification	Neutron LOD (mSv)	Beta LOD (mSv)	Gamma LOD (mSv)
1951-03/1958	B-G	552 Du Pont Film	All	7	Rad & Non Rad		0.3	0.3
03/1958 - 12/1966	B-G	558 Du Pont Film	All	7 & 30	Rad & Non Rad		0.3	0.1
1951 - 1974	N	Kodak Type A	All	30	Radiation	0.1-0.14		
1975 - Present	N	"Albedo"	All	30 & 90	Radiation	0.15		
12/1966 - ~1975	B-G	TL – Disk	All	90 & 180 & 360	Non Radiation		0.15	0.15
03/1958 - 09/1973	B-G	558 Du Pont Film & Kodak Type 3	CPP	30	Radiation		0.3	0.1
03/1958 - 02/1974	B-G	558 Du Pont Film & Kodak Type 3	ANC	30	Radiation		0.3	0.1
03/1958 - 02/1974	B-G	558 Du Pont Film & Kodak Type 3	Argonne	30	Radiation		0.3	0.1
09/1973 - 02/1974	B-G	TL – Disk	CPP	30	Radiation		0.15	0.15
02/1974 - 12/1974	B-G	ATLAS	CPP	30	Radiation		0.3	30
02/1974 - 02/1975	B-G	ATLAS	ANC	30	Radiation		0.3	30
02/1974 - 04/1975	B-G	ATLAS	Argonne	30	Radiation		0.3	30
~12/1974 - 12/1985	B-G	TL – Chips	All	30 & 90 & 360	Rad & Non Rad		0.15	0.15
01/1986 - 07/1986	B-G	Panasonic	All	30 & 90	Rad & Non Rad		0.15	0.15
07/1986 - ~09/1989	B-G	Panasonic	All	30 & 90	Rad & Non Rad		0.3	0.1
~09/1989 - 11/1993	B-G	Panasonic	All	30 & 90	Rad & Non Rad		0.3	0.15
11/1993 - Present	B-G	Panasonic	All	30 & 90	Rad & Non Rad		0.3	0.1

The single filter element film badge was used at the site until 1958. This badge was unable to differentiate between low-energy non-penetrating photons and beta radiation. The type 552 film packet (containing 502 type sensitive film and type 510 insensitive films) was used with a metal holder that contained front and rear 1" x 1 ¼" x 1 mm thick cadmium filters over the film. A ½" x ⅝" rectangular "window" with no cadmium shield was located at the bottom of the film for beta and "soft" gamma radiation detection. The transition region between low and high energy is somewhat arbitrary but appears to have been approximately 200 keV where the photoelectric effect in silver decreases (Cember 1983). At this energy, the ratio between rad and roentgen is essentially unity for muscle and bone. Low-energy photons can produce as much as a factor of 25 over- response in the film because of the photoelectric absorption with the silver halide of the emulsion compared to beta and high energy photon radiation. Site documents (F.V. Cipperley, Personnel Dosimetry SOP, 1958) indicate that all low-energy photon radiation would have been included in the beta exposure, and all high-energy penetrating photons would have been considered as gamma. Beta exposures were likely to have been overestimated because of the low photon energy over-response.

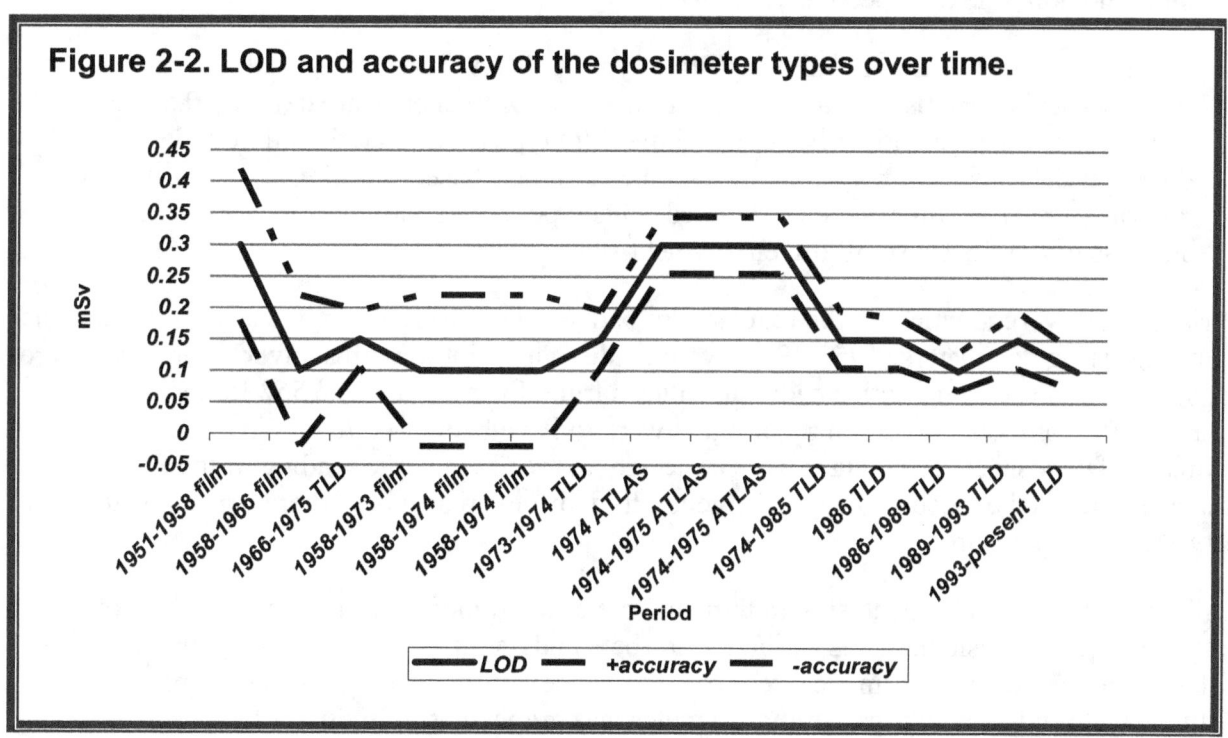

The minimum detectable dose during this period was 0.30 mSv (30 milli-roentgen) and routine badges were analyzed weekly. Conversion from Roentgen to rem and then to Sievert (Sv) in SI units required information on radiation type and energy to obtain deep dose equivalent at 10 mm tissue depth (NCRP 1995). For calculation of dose estimates in this study, the conversion was assumed to have been 1:1 for badges prior to 1986. Even with this assumption, the absorbed dose value would not differ by more than 10% from the deep dose equivalent at photon energies greater than 250 keV and by only 3% at photon energies greater than 600 keV.

The film badge holders used before 1958 were made of stainless steel with dimensions 1-⅞" long, 1-⅜" wide and ¼" thick. The upper portion of the badge contained a 1 mm thick cadmium plate for measuring penetrating gamma dose and a lower portion containing the open window for beta measurement. If the dose was only counted as due to gamma rays the neutron contribution may have been underestimated. Three types of film were used in the badge (Puphal 1996):

1. DuPont beta-gamma type 552 film badge was used in most areas. Each packet contained a sensitive and insensitive type of emulsion. The sensitive emulsion was used for low dose measurement, and its range was between 0.30 mSv and 50 mSv. The insensitive emulsion was used for the higher doses and had a useful range between 1.50 mSv and 300 mSv. The cadmium filter element has a high cross section for thermal neutrons; therefore, extra film darkening would have occurred.

2. DuPont type 558 film badge was used at two reactor areas (not specified in Puphal 1996). The sensitive and insensitive emulsions had a range between 0.10 mSv and 30 mSv and 15 mSv and 8000 mSv, respectively.

3. Kodak Personal Neutron Monitoring Film, type A (NTA) badge was used for neutron monitoring. The emulsion was sensitive to neutrons with energy greater than 0.5 MeV and was worn in a badge with either of the above film types. The stated minimum detectable dose was 0.14 mSv but may actually have been higher, based on information from other nuclear sites of the time, such as Hanford. This type of emulsion has problems with fading and loss of dose because of low-energy neutrons.

Wrist badges were identical to the regular film badge except were worn like a wrist watch. The first finger ring dosimeters (1951-1958) were made of aluminum with a silver filter. They were replaced by a plastic ring with a cadmium filter during this period. Type 552 film was used in all rings. PC were distributed on request and were to have been used to obtain a real-time estimate of exposure. Film badges were collected and read when the reading of the PC indicated elevated exposure. They were generally worn in areas where there was a greater probability of exposure.

Calibrations were performed using radium and metallic uranium for gamma and beta exposure, respectively. The dosimeter was calibrated for beta and gamma exposures using only two dose values. For each batch of film received from the vendor, film density (darkening) measured with an analog visible light transmittance meter was graphed as a function of exposure (Bennett 1957). All calibrations of beta and gamma sensitive and insensitive film were plotted on the same graph. No adjustment for backscatter was made in these early days of the INEEL monitoring program. Gamma exposure was obtained by using the density under the cadmium shielded portion of the badge. The beta reading was obtained from the density difference of the shielded and the open window portions of the badge. An unexposed film processed with the personnel films was used as a blank reading and subtracted from the exposed film density.

The film badge, as modified in 1958, had four filter positions: open window, 0.0393" cadmium, 0.005" silver and 0.0191" aluminum. DuPont type 558 film was used for its higher sensitivity, which lowered the minimum detectable dose to 0.10 mSv. The use of a four-filter badge gave greater energy discrimination for photon radiation and reduced the potential for over-response of the film badge in the low-energy photon region. The neutron film remained the same; that is, NTA film by Kodak.

During this period badges were collected and read on a bi-weekly and monthly schedule. Calibration film was still exposed to radium but at 12 different dose levels ranging from 0.50 mSv to 8.0 Sv. Energy response of the badge was tested using National Bureau of Standards low-energy x-rays and cobalt-60. Testing of the film badge showed that for photon energies greater than or equal to 40 keV there was a contribution to the density under the cadmium filter that varied with photon energy. The magnitude of this contribution was not sufficiently reflected in the radium calibration exposure conditions.

In 1966 the change to TLDs began. They were worn by persons who were expected to receive less than 5 mSv in one year (Cusimano and Cipperley 1968). The remaining monitored workers continued to use film dosimeters. Dosimeter exchange periods were extended from 3 to 6 months for two-thirds of those using TLDs in mid-1968. In 1969 the Automatic Thermoluminescence Analyzer System (ATLAS) was developed. The ATLAS badge was a modified version of the film badge with the same filter configuration except the holder was made of plastic. The dosimeter elements themselves were lithium fluoride Teflon discs made in a homogeneous mixture 30% lithium fluoride phosphor and 70% Teflon by weight. The Teflon element was placed in contact with a heating bar that induced thermoluminescence. The light was then transformed into an electrical signal by a photomultiplier tube. The lower LOD of the system was determined to have been 0.10 mSv for radium gamma rays and 0.10 mSv for uranium beta rays (Cipperley and Gammill 1959). Initial problems with identification and temperature control caused delay in the use of the system. However, by 1974 all persons were monitored for photon and beta radiation with TLDs.

In the early 1980s the DOE (USDOE 1986) began developing a program for external dosimetry laboratory accreditation called DOELAP (DOE Standard for the Performance Testing of Personnel Dosimetry Systems). Up to this time all doses were reported as gamma and beta or neutron with calibrations being performed in air. After DOELAP doses were reported as deep and shallow. The deep dose, Hp(10), was reported at 1 cm depth in tissue and shallow dose, Hp(0.07), was reported at 0.007 cm depth in tissue. Calibrations were then performed on an anthropomorphic phantom.

The current beta-gamma dosimeter used for personnel dose monitoring was implemented in 1986. The Panasonic dosimeter is a multi-filtered type with 3 elements of $Li_2B_4O_7$ and one element of $CaSO_4$. This dosimeter provides beta-gamma low and high energy discrimination at shallow and deep tissue depth. The dosimeter provides gamma dose monitoring from 17 keV to Cs-137+ energy levels and beta monitoring from Tl-204 to Y-90 energy levels. The system was installed in January 1986 with a beta and gamma minimum reporting level of 0.15 mSv (carry-over from the 2-chip system). This limit was changed in July 1986 to 0.10 mSv gamma and 0.30 mSv beta. In approximately September 1989, as part of the DOELAP accreditation, the studies required by DOE/EH-0027 (USDOE 1986) established a Lower Limit of Detection (LLD) of 0.15 mSv for gamma and 0.30 mSv for beta. This average was a practical interpretation of the following derived values (INEEL/EXT/01-00636).

From 1951-1975 slow neutron chambers were used for the detection of slow and thermal neutrons, and fast neutron exposures were estimated using Kodak NTA film emulsion. The fast neutron dose was determined by counting the number of tracks in the NTA emulsion under high magnification with an oil immersion microscope. Exposures greater than 0.20 Sv would saturate the film and make track counting ineffective. Also, large gamma exposures tended to fog the NTA film making track counting more difficult. In 1975 the neutron monitoring program that used Kodak Type A film from its inception was moved to a new 'Hankins' albedo-style TLD, which used three TLD-700 and three TLD-600 phosphors enclosed in a polyethylene case inside a cadmium box. This dosimeter has provided neutron monitoring since 1975 and is currently still in use. The LOD for the albedo dosimeter is 0.15 mSv (Report # 960112, *Technical Basis of the INEL Personal Neutron Dosimeter*, Idaho State University, January 1996). The dosimeter error was documented as ± 100% at 0.15 mSv, ± 50% at 1.0 mSv and ± 30% at 10 mSv (May 28, 1980 memorandum from D. Jones to John Barry).

Missed Dose

Consideration is now given to dose missed by a dosimeter because of the limit of its detection capabilities (Atwood *et al.* 1990). The magnitude of the missed dose depends primarily on two factors. The first is the smallest dose that is positively recorded on a film with an appropriate level of certainty. This limit depends on film type, that is the composition of the film emulsion, and type, energy and geometry of radiation incident on the film. The second factor is related to the wear period or badge exchange frequency. At INEEL, film dosimetry was the prime recorder of radiation exposure until 1966 when TLDs were introduced. By 1974 TLDs completely replaced film. From 1951 until 1958 the badge exchange frequency was weekly and the LOD of the dosimeter was 0.30 mSv (see §2.3.1). Because of the high detection limit and short badge exchange frequency this period was considered to have the greatest amount of missed dose. During this period the policy of reporting a dose as zero when below detection level was also enforced. After 1958 exchange frequencies were increased to monthly and quarterly, and limits of detection dropped by factors of 2 and 3. Also, with the advent of TLD use the accuracy increased and dose values below detection limits were recorded. The cross-over time point for reporting of less than LOD values was 1975 (Figure 2-3).

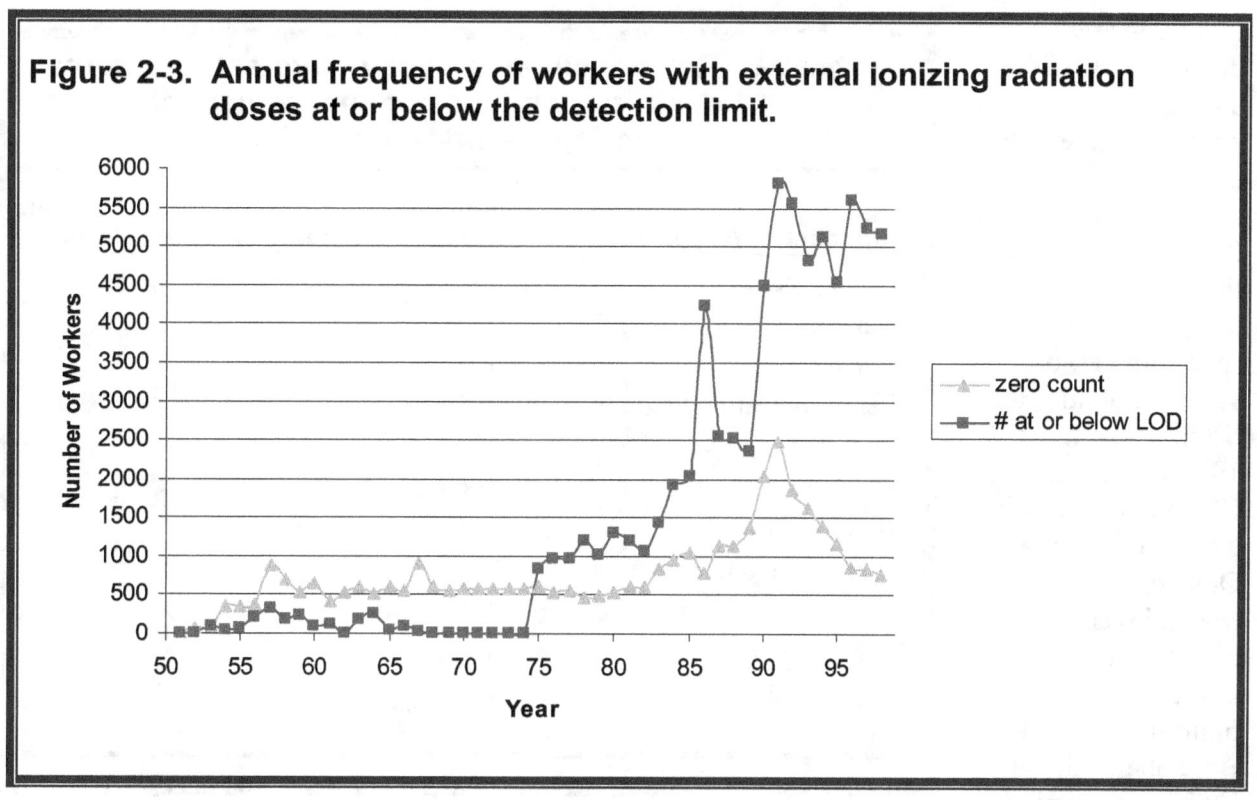

Figure 2-3. Annual frequency of workers with external ionizing radiation doses at or below the detection limit.

Assignment of External Exposures to INEEL Study Cohort Members

The dose values in the electronic database (MUD) before 1986 consisted of annual sums for each worker (see §2.2.2). After this time doses were recorded by the period of badge exchange (RDS). In this study dose assignment to cohort members was made on an annual basis because of the limiting nature of the MUD data. Within that year a person may have worked in multiple facilities. Area code was used as surrogate for work location to differentiate off-site exposure from those that occurred onsite (see §3.1). The final annual doses were not corrected to the deep dose equivalent at 1 cm depth as explained above. For the epidemiological analyses, doses were accumulated over the workers' entire work history. No adjustments to individual doses were made from possibly undetected badge dose.

2.3.1.2 Internal Dosimetry

Bioassay methods have been used at INEEL since 1951 (Puphal 1994). Urinalysis was the main technique for assessing uptake of radionuclides. Sampling frequency varied between quarterly and yearly with the expected potential for uptake. The primary radionuclides initially of concern were uranium and its isotopes, and fission products. Later the number increased to include plutonium and other radionuclides (Table 2-6). Counting was done for gross alpha and beta activity, but it was later determined that iodine and ruthenium could not have been detected by these methods. Therefore, gross gamma counting of the samples was added to assess iodine exposure. In the early 1960s fecal analysis was added to the methods of determining internal exposure. At the same time new methods for detecting plutonium in urine were developed. In-vivo or WBC began as a full program in 1962. Sodium iodide was used mainly for the detection of fission and activation products. Later the use was extended to measuring U235 and Pu239. The detection limits for WBC, using a gamma spectrometer, of various nuclides are listed below as of 1971 in the Idaho Chemical Processing Plant (ICPP) Internal Dose Assessment Manual were: uranium (total) 30 micrograms; plutonium-239, 481 Bq; cobalt-60, 7.4 E07 Bq; plutonium-238, 296 Bq; cesium-137, 7.4 E07 Bq; strontium-90, 1110 Bq (from skull bremsstrahlung).

Table 2-6. Radionuclides/groups of interest for internal dosimetry and the commencement of monitoring methods at INEEL.

Radionuclide/Group	Monitoring Began	Urinalysis	Fecal	Whole Body
Fission products	1951	X		
Uranium	1951	X		
Actinides	1953	X		
Fission/Activation Products	1962			X
Actinides	1965		X	
Strontium	1967		X	
Tritium	1968	X		
Uranium, Plutonium				X
Iodine 125	1968		X	
Strontium Skull	1968			X
Fission products	1971		X	
Uranium Isotopic	1971	X		

Assignment of Internal Exposures to INEEL Study Cohort Members

As discussed in §2.2.2 above, records for internal dosimetry at INEEL were too voluminous to feasibly copy for a cohort of this size. Additionally, internal dose estimates have been calculated based on results for analytical methods that have changed substantially over the years, and the greatest risk of exposure was likely to occur when relatively insensitive methods were used. Therefore, for the purposes of this study, the population of workers was divided into groups based on monitoring status for internal dosimetry: not monitored, monitored but no intake detected, and monitored with intake detected.

An algorithm was developed for the study to determine whether a worker had the potential for a positive bioassay and hence a potential internal dose. Prior to 1986 the bioassay data available in MUD consisted only of annual sampling frequencies with radionuclide type unspecified. Quantitative internal monitoring data had not been computerized by the site. No attempt was made to quantify dose, as the data required (e.g., initial versus follow-up samples, dates, solubility, chronic or acute exposure) were not readily available. The following paragraphs give a brief overview of the methods involved in determining a person's internal dose status.

Positive internal dose status was inferred for those who were monitored but with unknown results using the internal dose calculation software, Integrated Modules for Bioassay Analysis (IMBA 2001), and a graphical method for LOD approximation. These techniques were necessary in the absence of an actual LOD value, as radionuclides like plutonium may contribute substantial dose to lung tissue even when dose is reported to have been at or below the LOD, because of the insensitivity of counting equipment for many early monitoring years. Since the purpose of this procedure was to identify persons with a potential internal dose, it was decided that any error should overestimate exposure potential. Actual dose was not estimated because of the paucity of electronic data.

Internal dosimetry records were available in MUD, RDS and NRF. The MUD data set only indicated the number of *in vivo* or *in vitro* results by calendar year. Tests were listed for apparent routine whole-body and thyroid. Additionally, the numbers of special tests for whole-body, thyroid and urine samples were also listed. Workers employed during this period of time were considered monitored for internal deposition if any number of samples were listed unless they were single events during initial and terminal years of employment only. Only those with multiple *in vivo* or *in vitro* tests within the same year were classified as exposed. All others employed only during the time covered by the MUD data set were considered not monitored for internal doses, along with those who had only single initial and terminal year.

In RDS substantially more data were available on the internal dosimetry results, including dates of individual samples, quantities found, and radionuclide, among others. However, the LOD for the assay result, which was needed to identify those workers who could be categorized as exposed-positive, was not available. As an alternative, the amount of radioactive material necessary to produce 1 mSv to the most exposed organ over a 50-year period was estimated. For this calculation, a chronic exposure to a radionuclide was assumed and was set equal to the period of external dose monitoring. This base concentration was then compared to the actual data to determine a virtual LOD. Any value above this base was classified as a positive intake. The external monitoring period was used, since it is certain that the person was working at INEEL facilities with radioactive materials during this time period.

All ICRP 66 defaults in the software were used in the calculations as well as an assumption of inhalation as the entry pathway. All materials were presumed insoluble. The frequency of each test result value was graphically displayed to select the LOD. NIOSH-IMBA (IMBA 2001, ICRP 1990, ICRP1994, ICRP 1995) was then used to calculate the sample result that would cause a cumulative effective dose equivalent of 1.0 mSv to the maximally exposed organ. This value was selected as the *de minimus* dose for those who would be placed in the positive intake category. One exception to this strategy was made for plutonium-239 in which case the lowest non-zero reported result was 8 nCi. Since the dose associated with this level of intake cannot have been considered *de minimus*, all individuals with non-zero positive bioassay results for plutonium-239 were categorized as exposed. Detection limits beyond those WBC numbers stated above were not available for this study.

2.3.2 Non-radiological Exposures (Subcohort Identification)

Walk-through evaluations were completed in most of the operating buildings at INEEL in the early 1990s. Workers who had been employed for various periods since 1949, as identified through site contacts, described occasions when higher past exposures might have occurred. Industrial hygiene records at the site and seven boxes identified in storage at the Federal Records Center in Seattle, Washington were reviewed by NIOSH investigators. Discussions also occurred with contractor health and safety staff and with DOE in Idaho.

2.3.2.1 Contractor Code Collapse and Identification

The Master File Cards received from the prime INEEL contractor, EG&G-Idaho, were the primary source of work history information for many workers employed prior to 1983 when information on these cards was used by the INEEL to build SECIMS. According to EG&G, only workers employed in 1983 and after had complete work history information entered into SECIMS. NIOSH used the information on the cards to build a more complete work history file called Roster.

Early attempts to key information into Roster revealed substantial inconsistencies in the source information. Many records showed changes in employment status without associated dates. Other dates were inconsistent with the overall work history. Most workers had only an initial job title. Location and department level assignments were frequently absent.

One parameter, contractor, was relatively complete for the vast majority of workers. These data were entered into Roster in the same order as they appeared on the MFC. Multiple abbreviations were used for each contractor, some of which could not be interpreted. Once coding was completed, all contractor name abbreviations from all record sources were exported to a separate file. Over 28,000 unique contractor name character strings were identified and subsequently collapsed into nearly 3000 business entities with employees who had worked at the INEEL. Each of these entities was assigned a unique identifier, called a "contractor code."

These business entities (contractor codes) were sorted by the number of employees in the combined work history records. An employer category was assigned to each employer with more than 10 employees at the INEEL. Many businesses had recognizable titles although others required further investigation to identify the type of work most likely for their employees. Business entities with sufficient available information were assigned to a category listed in Table 2-7. These categories, and in some cases contractor codes, were then used to help define many of the subcohorts used in this study (see §2.3.2.3).

Table 2-7. Categories assigned to INEEL employers and the total number of individuals with work history records that identify an employer in the category (workers may appear in more than one category).

Category	Employer Category Description	# Workers
1	Janitorial, Laundry	663
2	Office, Computer Services	918
3	Engineering, Technical	2063
4	General Building Services, Plumbing, Flooring	1467
5	General Mechanical Services, Elevators, Instruments	1061
6	Grounds	240
7	Professionals, Medical, Journalism, Insurance	939
8	Auto, Aviation, Pest Control	6307
9	Transportation	903
10	Foods	521
11	Ranch, Land Management	121
12	Laboratories and Environmental Services	757
13	Utilities	1387
14	Electrical Services	232
15	Radiological Services	2423
16	Painting	458
30	Construction – Excavation, Drilling, Paving, Masonry	2800
31	Construction – Insulation, Surface Coatings	732
32	Construction – Not Otherwise Specified	20,652
33	Construction – Electrical	2219
34	Construction – Metals, Welding	1616
35	Construction – Mechanical	2156
36	Construction – Roofers	429
40	Suppliers – Chemicals	304
41	Suppliers – Office	45
42	Suppliers – Miscellaneous	139
50	DOE – Government Employees	11,230
51	DOE – Off-site Contractor Employees	571
52	INEEL Primary Contractors Employees	30,027
53	NRF Contractor and Government Employees	8133
60	Educational Institutions and Organizations	1820
61	Unions	11
63	Federal Government Employees (non-DOE)	1318
64	State of Idaho	4083
65	Local Government	31
66	Other States	23
67	Foreign Nationals	95
68	Security	1251
70	Military	496
	Unassigned	1663
	TOTAL	112,304

2.3.2.2 Job Title Collapse and Identification

Job titles were used both to identify exposure-based subcohorts of interest (along with contractor categories described in §2.3.2.1 above) and to classify workers by socioeconomic status (SES). Job titles were available within each of the seven source files used in creation of the cohort roster. These files varied substantially, however, in the completeness and comprehensibility of job title information (Table 2-8). The most complete sources of job titles were the Roster, OMP, HRS and NRF files. The SECIMS and RDS files were very incomplete sources of job title information.

Table 2-8. Availability of job title fields by source file for INEEL cohort.

File name	Type of field	Number of unique values	Approximate number of workers in file	Approx. number of workers with job title	Percent completeness of file
Roster	Job titles (alpha)	11,553	63,000	57,400	91.1%
SECIMS	Job codes (2-letter codes)	40	72,173	6275	8.7%
	Job descriptions (alpha)	40			
OMP	Job codes (alphanumeric)	95	19,273	19,008	98.6%
HRS	Job codes (2 or 3 digit numeric code) Job titles (alpha) Work discipline (alpha)	1775	15,037	13,423	89.3%
NRF	Job titles (alpha)	4619	4900	4900	100%
	Job code	4619			
RDS	Craft codes (numeric)	53	60,700	1016	1.7%
MUD	Craft codes (numeric)	131	50,000	21,242	42.5%

Several of the source files (e.g., Roster, HRS and NRF) had extremely large numbers of unique job titles, over 18,000 in total. To facilitate the development and analysis of subcohorts and SES measures, and to ensure consistency among job titles in the various files, each unique job title was collapsed into one of approximately 500 job titles (Table 2-9). These job titles were collapsed initially by an industrial hygienist familiar with the site's activities. These assignments were then checked by an epidemiologist and sociologist working on the SES classification for the study. Any conflicting assignments were resolved by the team, using information, where available, that was specific to the INEEL site.

The 500 collapsed job titles were each assigned a 1980 Census Bureau Occupational code (see Appendix §A.4, as shown in Table 2-9). In general, supervisors were assigned the same Census occupational code as the related employee, when no more specific classification could be found. These classifications commonly occurred in the professional and technical occupations; for example, technician and engineering supervisors were assigned the same Census codes as technician and engineer, respectively. For occupations with different classifications among supervisors and supervisees, for example, the skilled trades, the more specific Census codes were used for supervisors or foremen.

2.3.2.3 Identification of Exposure-based Subcohorts

The contractor codes and categories, collapsed job titles and assigned 1980 Census codes were all used to identify subcohorts of interest. These codes are identified in Table 2-10. Because no change dates were available for the complete cohort, no analyses by duration of employment in these categories were possible.

Employer types ("prime" employers, subcontractors, and multiple employer types) were identified using contractor codes and categories. The "prime" employers consisted of DOE-Idaho, NRF, ANL-W, or any of the prime contractors operating the INEEL facility over its history (Table 1-1).

Chemical workers were identified as those ever employed at the ICPP as indicated by contractor category, or those ever having a collapsed job title associated with possible exposure to chemical agents (e.g., chemists, mixing operators, Table 2-10).

Construction and service workers were identified both by contractor category and by 1980 Census Occupation code. This group consisted of workers who met at least one of the two following qualifications: (1) were ever employed in one of the contractor categories associated with construction or service work (e.g., maintenance, repair or mechanical support); (2) ever worked in a job related to cleaning, building service, mechanic, repair work, or construction (Table 2-10).

Painters were identified as anyone employed in a contractor category or collapsed job title associated with painting work. Asbestos workers were identified as workers ever employed by a contractor performing insulation work, or whose collapsed job title indicated possible exposure to asbestos on the job (including asbestos workers, insulation workers, boiler operators and pipefitters; Table 2-10).

Security guards and other protective service workers consisted of those ever employed in Census categories associated with these occupations. It was hypothesized that accidental deaths might have been elevated among this subcohort. In addition, the stress due to maintaining responsibility for securing a large, heterogeneous nuclear facility might cause an increase in cardiac or other stress-related causes of death.

Reactor workers were identified solely by collapsed job title; the group consisted of anyone ever employed as a reactor engineer, operator or technician, or a power plant operator. Reactor maintenance workers were not included, as reactor technicians could not be adequately identified by job title or contractor category. Lastly, drivers consisted of anyone identified as a truck or bus driver using the 1980 Census occupation codes associated with motor vehicle occupations.

Category assignments were not mutually exclusive; that is, workers may have been assigned to more than one subcohort (e.g., all asbestos workers were also included in the category of construction and service workers). However, within a subcohort, workers were assigned to just one category based on ever or never meeting the subcohort definition.

Table 2-9. Collapsed job titles and 1980 Bureau of Census occupational codes (indicated by BOC) used for each collapsed job title.

Collapsed Job Title	BOC	Collapsed Job Title	BOC	Collapsed Job Title	BOC	Collapsed Job Title	BOC
Laboratory Director	004	Engineer Welding	045	Engineer Research	059	Film Producer	194
Controller	007	Metallurgist	045	Engineer Specialist	059	Radiology Tech	206
Treasurer	007	Engineer Chem	048	Engineer Student	059	Xray Technician	206
Manager Personnel	008	Engineer Chem Super	048	Engineer Super	059	Health Technician	208
Supervisor Procurement	009	Engineer, Fuels	048	Engineer, Installation	059	Medical Technician	208
Accountant Super	019	Engineer Nuclear	049	Engineer, Sanitary	059	Technologist	208
Administration	019	Engineer Nuclear Super	049	Engineer, Scheduling	059	Electronics Technician	213
Administration Super	019	Engineer Reactor	049	Engineer, Service	059	Operator Electrical	213
Construction Manager	019	Engineer Architect	053	Junior Engineer	059	Radio Technician	213
Contractor	019	Engineer Civil	053	Cartographer	063	Metrology Assistant	214
Facilities Management	019	Engineer Civil Super	053	Engineer Survey	063	Technician Manufact	214
Laboratory Manager	019	Engineer Constr	053	Surveyor	063	College Student	216
Management Technical	019	Engineer Structural	053	Surveyor Super	063	Engineer Mfg	216
Manager Constr	019	Engineer Agriculture	054	Engineer Computer	064	Engineer Plant	216
Manager Hvac	019	Engineer Elec	055	Engineer Systems	064	Engineer Safety	216
Manager Reactors	019	Engineer Elec Super	055	Engineer Systems Super	064	Engineer Technician	216
Manager Research	019	Engineer Electronics	055	Systems Analyst	064	Engineer, Assistant	216
Production Super	019	Engineer Industrial	056	Mathematician	068	Health Physics Tech	216
Research Admin	019	Engineer Quality	056	Health Physicist	069	Health Phys Tech Super	216
Superintendent	019	Engineer Safety Super	056	Health Physicist Super	069	Industrial Hygienist	216
Supervisor	019	Engineer, Field	056	Physicist	069	Nuclear Technician	216
Supervisor Plant	019	Metrologist	056	Chemist	073	Draftsman	217
Supervisor Research	019	Engineer Hvac	057	Chemist Nuclear	073	Draftsman Super	217
Accountant	023	Engineer Mech	057	Meteorologist	074	Cartographer Techn	218
Auditor	023	Engineer Mech Super	057	Geologist	075	Agriculture Techn	223
Cost Analyst	025	Engineer Test	057	Scientist	076	Biological Technician	223
Financial Officer	025	Engineer Marine	058	Scientist Super	076	Analyst	224
Records Analyst	026	Consultant	059	Biologist	078	Technician Chem	224
Staff Analyst	026	Engineer	059	Conservationist	079	Technician Manager	224
Instructor Corporate	027	Engineer Acoustic	059	Physician	084	Data Analyst	225
Labor Representative	027	Engineer Asbestos	059	Psychiatrist	084	Geologist Technician	225
Personnel/Labor Rel	027	Engineer Assembly	059	Nurse	095	Laboratory Technician	225
Buyer	033	Engineer Design	059	Dietitian	097	Research Technician	225
Buyer Super	033	Engineer Envir	059	University Professor	154	Pilot	226
Inspector Const	035	Engineer Hot Cell	059	Vocational Counselors	163	Computer Programmer	229
Management	037	Engineer Inspection	059	Librarian	164	Computer Technician	235
Military Commissioned Officer	037	Engineer Instrument	059	Archaeologist	169	Inspector	235
Architect	043	Engineer Maintenance	059	Attorney	178	Operator Technician	235
Facilities Designer	043	Engineer Operating	059	Writer Technical	184	Reactor Tech	235
Engineer Aerospace	044	Engineer Process	059	Artist	188	Reactor Tech Super	235
Engineer Materials	045			Photographer	189	Sales Manager	243

Table 2-9, *Continued*. Collapsed job titles and 1980 Bureau of Census occupational codes (indicated by BOC) used for each collapsed job title.

Collapsed Job Title	Boc	Collapsed Job Title	Boc	Collapsed Job Title	Boc
Insurance Agent	253	Mechanic Apprentice	549	Repairman, Installation	549	Painter	579
Real Estate Agent	254	Construction Mechanic	553	Brickmason Foreman	553	Painter Apprentice	579
Advertising Agent	256	Erector	554	Tile Setter Super	553	Plasterer	584
Salesman	257	Mechanic Air Compr	555	Carpenter Foreman	554	Pipefitter	585
Engineer, Sales	258	Mechanic, Maintenance	555	Electrician Foreman	555	Piping Detailer	585
Sales Clerk	275	Greaser	555	Electrician Super	555	Plumber	585
Cashier	276	Oiler	556	Lineman Foreman	556	Steamfitter	587
Clerical Super	303	Mechanic Radio	557	Lather Foreman	556	Pipefitter Apprentice	587
Office Manager	303	Repairman, Pump	557	Painter Foreman	557	Plumber Apprentice	588
Computer Oper Super	304	Repairman Computer	558	Pipefitter Foreman	557	Cement Finisher	588
Computer Operator	308	Telephone Lineman	558	Plumber Foreman	558	Cement Finisher Appren	588
Operator Peripheral	309	Installer Telephone	558	Plumber Super	558	Concrete Finisher	589
Library Assistant	329	Repairman Telephone	558	Asbestos Foreman	558	Glazer	593
Blueprint Clerk	335	Telephone Installer	558	Cement Finisher Foreman	558	Applicator	593
Bookkeeper	337	Furnace Repairman	558	Concrete Foreman	558	Asbestos Mechanic	593
Payroll/Clerical	338	Hvac	558	Construction Foreman	558	Asbestos Worker	593
Farm Management Advisor	475	Installation Refrig	558	Construction Super	558	Insulation Worker	593
Farm Manager	475	Mechanic, Refrig	563	Driller Foreman	564	Insulator	594
Farm Foreman	477	Repairman, Refrig	565	Driller Super	565	Roller Operator	595
Gardener	486	Locksmith	567	Fence Foreman	569	Roofer	597
Groundskeeper	486	Repairman, Office Machine	573	Flooring Foreman	575	Ironworker	597
Animal Caretaker	487	Elevator Construction	575	Foreman	576	Steel Construction	597
Camp Tender	494	Elt Inst.	576	Glazer Foreman	576	Steelworker	598
Forester	495	Mechanic Elev	577	Ironworker Foreman		Struct Metal Crftsmn	599
Logger	496	Construction Millwright		Ironworker Super		Driller	599
Animal Control	499	Millwright		Paving Foreman		Construction	599
Erector Foreman	503	Blacksmith		Roofer Foreman		Fence Construction	615
Installation Supervisor	503	Mechanic Acoustic		Steelworker Foreman		Installer Constructn	615
Installer Foreman	503	Mechanic, Installation		Brickmason		Jackhammer Operator	617
Instrumentman Super	503	Pump Installer		Brickmason Apprentice		Explosives	633
Maintenance Foreman	503	Repairman, Tool		Floor Layer		Powderman	633
Maintenance Super	503	Rigger		Tile Setter		Miner	633
Mechanic Foreman	503	Steeple Jack		Carpenter		Boilermaker Foreman	633
Mechanic Super	503	Installer		Carpenter Apprentice		Boilermaker Super	633
Millwright Foreman	503	Instrument Repair		Drywall Installer		Boilerman Foreman	633
Oiler Foreman	503	Instrument Technician		Electrician		Craftsman Foreman	633
Pump Foreman	503	Instrumentman		Lineman		Inspector Super	633
Repairman Foreman	503	Machine Mechanic		Electrician Appr		Machinist Foreman	633
Repairman Super	503	Maintenance		Lineman Appr, Television		Machinist Super	633
Rigger Foreman	503	Mechanic		Lineman Apprentice		Operator Foreman	633
Mechanic Automotive	505	Repairman		Tower Construction		Operator Supervisor	633

Table 2-9, *Continued.* Collapsed job titles and 1980 Bureau of Census occupational codes (indicated by BOC) used for each collapsed job title.

Collapsed Job Title	Boc	Collapsed Job Title	Boc	Collapsed Job Title	Boc
Production Foreman	633	Operator Nuclear	699	Decorator	789	Bulldozer Operator	855
Reactor Operator Super	633	Reactor Operator	699	Sign Painter	789	Grader Operator	855
Sheetmetal Foreman	633	Utilityman	699	Lab Assistant	795	Heavy Equip Operator	859
Sheetmetal Super	633	Lathe Operator	704	Checker Manufacturing	796	Warehouseman	859
Shipfitter Foreman	633	Drilling Operator	708	Inspector Welding	796	Yardman	859
Utilityman Foreman	633	Grinder	709	Radiographer	797	Laborer Bus Mgr	863
Machinist	637	Metal Plater	723	Radiographer Asst	797	Laborer Foreman	863
Boilermaker	643	Saw Operator	727	Radiographer Super	797	Labor Insulation Super	863
Shipfitter	646	Printer	734	Weigher	798	Laborer Super	863
Sheetmetal Worker	653	Typesetter	736	Driver Truck Foreman	803	Service Manager	863
Sheetmetal Apprentice	654	Laundry	748	Driver Truck Super	803	Warehouseman Super	863
Wood Finisher	658	Laundry Mgr	748	Driver Truck	804	Mechanic Helper	864
Sailmaker	674	Mixing Operator	756	Delivery	805	Carpenter Helper	865
Glass Blower	675	Operator Machine, Mixing	756	Teamster	805	Hody	865
Craftsman	684	Operator Chemical	757	Milkman	806	Surveyor Helper	866
Craftsman Apprentice	684	Compressor Operator	758	Driver Bus	808	Construction Laborer	869
Military	684	Operator Compressor	758	Chauffeur	809	Laborer	869
Operator, Rad Waste	684	Finisher	759	Driver	814	Laborer Flooring	869
Baker	687	Crusher Operator	768	Locomotive Engineer	824	Laborer Insulation	869
Inspector Hot Cell	689	Cutting Machine Operative	769	Railroad Foreman	826	Machinist Helper	873
Inspector Mechanical	689	Cutting Operative	769	Heavy Equip Super	843	Stock Handler	877
Production Tester	689	Film Processor	774	Equipment Operator	844	Chip Spreader	878
Boilerman	695	Operator Air Tool	777	Derrickman	848	Service Worker	878
Operator Power Plant	695	Operator Machine	777	Hoistman	848	Fuel Handling Techm	885
Boiler Operator	696	Tool Operator	779	Winch Operator	848	Gas Station Attendant	885
Pump Operator	696	Welder	783	Craneman	849	Cleaner	887
Stationary Fireman	696	Welder Foreman	783	Backhoe Operator	853	Driox Service	887
Operative, Diesel	699	Welder Super	783	Dragline Operator	853	Porter	887
Operator	699	Panel Assembler	785	Excavating Operator	853	Packer	888
Operator Hot Cell	699	Marble Craftsman	787	Blade Operator	855		

Table 2-10. Identification of exposure-based subcohorts of interest using contractor categories, collapsed job titles and 1980 Bureau of Census occupational codes (BOC_1980)

Name	Code	Subcohort Description	Identification method
Employer type	1	Unknown contractor	Contractor code=(R5, 73, 0 or missing)
	2	Prime contractor, NRF, ANL-W, or DOE-ID employee	Contractor category=52, 53, 3 or 1
	3	Subcontractor	All other contractor categories and codes
	4	Multiple employers	Multiple contractor categories (within groups 2-3)
Chemical workers	1	ICPP or chem worker	Ever had contractor code=(N72, N429, 146, 169, 892) OR Collapsed job title=(CHEMIST; CHEMIST NUCLEAR; DRIOX SERVICE; ENGINEER CHEM; ENGINEER CHEM SUPER; METAL PLATER; MIXING OPERATOR; METALLURGIST; OPERATOR CHEMICAL; OPERATOR MACHINE, MIXING; OPERATOR, RAD WASTE; TECHNICIAN CHEM)
	0	Non-chem worker	All other workers
Construction or service worker	1	Construction or service worker	Ever had contractor category=(3x + 4, 5, 9, 14, 16) OR BOC_1980=(453-455, 503-599, 684)
	0	Non-construction worker	All other workers
Painters	1	Painter	Collapsed job title=(PAINTER; PAINTER APPRENTICE; PAINTER FOREMAN; SIGN PAINTER) OR Contractor category=16
	0	Non-painter	All other workers
Security guard	1	Security guard or police	Ever had BOC_1980=(418-427)
	0	Non-security worker	All other workers
Asbestos workers	1	Asbestos worker	Ever had collapsed job title=(APPLICATOR; ASBESTOS FOREMAN; ASBESTOS MECHANIC; ASBESTOS WORKER; ENGINEER ASBESTOS; INSULATION WORKER; INSULATOR; BOILER OPERATOR; BOILERMAKER; BOILERMAKER FOREMAN; BOILERMAKER SUPER; BOILERMAN; BOILERMAN FOREMAN; PIPEFITTER; PIPEFITTER APPRENTICE; PIPEFITTER FOREMAN) OR Contractor category=31
	0	Non-asbestos worker	All other workers
Reactor workers	1	Reactor or nuclear operator	Ever had collapsed job title=(ENGINEER REACTOR; MANAGEMENT, REACTOR; MANAGER REACTORS; REACTOR ENGINEER; OPERATOR NUCLEAR; OPERATOR POWER PLANT; REACTOR OPERATOR; REACTOR OPERATOR SUPER; REACTOR TECH; REACTOR TECH SUPER)
	0	Non-reactor worker	All other workers
Drivers	1	Truck or bus driver	Ever had BOC_1980=(803-809)
	0	Non-driver	All other workers

2.4 Confounder Assessment

2.4.1 Demographic Variables

For each member of the cohort, DOB was determined using site records, supplemented by information obtained from the NUMIDENT file system of the U.S. Social Security Administration (SSA). The NUMIDENT files contained information presented in a worker's initial application for a Social Security card, as updated to reflect new information as a worker pays into Social Security, retires, makes a claim, or provides information to the Internal Revenue Service (http://www.ssa.gov/policy/about/epidemiology.html). Information on sex and race/ethnicity was also obtained from site records, where available, and from the SSA's NUMIDENT file system.

In cases where disagreements occurred in DOB, sex or race/ethnicity among records, the information contained in the majority of unique record sources was selected as the most probable value. If no majority of reliable records was identified, the information obtained from the SSA was used. For data analyses, those of unknown race were assumed to be white.

DLO was identified either as the date of death or as the end of vital status follow-up (December 31, 1999) for those of known vital status. For those of unknown vital status, date of last medical examination, date of receipt of retirement information, DOT, or date of last radiation monitoring, whichever occurred latest, was used as the DLO.

Membership in the Church of Jesus Christ of Latter-Saints (LDS) was thought *a priori* to be an important potential confounder in the cohort, as the prevalence of LDS membership is high in the INEEL region, and LDS membership has been found to be associated with many healthier lifestyle behaviors and low mortality rates, such as reduced smoking-related cancer and heart disease mortality (Lyon *et al.* 1976, 1978). Using information obtained from obituaries of former INEEL employees in the Idaho Falls region, a separate analysis determined that workers local to the INEEL region (as identified by state of issue of the SSN) were about four times as likely to have been LDS members as those not local to the region (Burphy *et al.* 2004). Therefore, the state of issue of the SSN (identified by the first three digits of 516-520; or 528-529; or 646-647, which are issued to Idaho, Utah, Wyoming and Montana) was used as a surrogate of local status to the INEEL region. All other SSN issue numbers were used to identify migrants to the INEEL region.

2.4.2 Socioeconomic Status

Many studies of nuclear workers have evaluated the importance of social class or SES (here, used interchangeably) as determinants of health status that may also be independently associated with exposure (Cardis and Kato 1993, Cardis *et al.* 1995, Gilbert *et al.* 1992, Frome *et al.* 1997, Wing *et al.* 2004). Nuclear worker studies have typically classified workers by status as of first (Gilbert *et al.* 1992), longest-held (Wing *et al.* 2004), or last (Beral *et al.* 1985) job title. In the present study, the primary objective in evaluating SES was to consider it as a surrogate of potential behavioral factors such as smoking. Most INEEL employees had just a single SES assigned (i.e., they had just one job title, or the same assigned SES class across multiple job titles). For workers with more than one job title crossing SES categories, first job title was employed, as it was thought to better-reflect behaviors like smoking, which may be established in early adulthood (Poulton *et al.* 2002).

SES was assigned to workers in the INEEL cohort, based on a classification of the job at first hire, in a manner similar to that used in other studies of DOE workers (Gilbert *et al.* 1992). Each collapsed job title was assigned a 1980 U.S. Bureau of Census occupational code, as described in §2.3.2.2. These Census occupational codes were assigned to one of the following 6 SES groups: professional, intermediate, skilled non-manual, skilled manual, partly skilled and unskilled (Table 2-11). Approximately 21% of cohort members had no job title available from which to assign an SES category.

Table 2-11. 1980 Bureau of Census occupational codes assigned to SES groups for INEEL cohort analysis.

SES Category Description	Professional	Intermediate	Skilled non-manual	Skilled manual	Partly skilled	Unskilled
1980 Census Occ. Codes	003-005	006	253-389	207	405-406	407
	007-008	009-013		403-404	415	437-443
	014-015	016-019		413-414	424-436	448-455
	023	024-037		416-423	444-447	459
	043-057	058		473-476	456-458	464-467
	059	063		497	463	469
	064-089	095		503-679	468	477-485
	096	097-106		684-699	486-488	494-496
	113-154	155-164		824	683	498-499
	165-169	173-177		843	703-823	864-889
	178-179	183-206		849-855	825-829	
		208-243			834	
		489			844-848	
		833			856-863	

For this study, job titles were hypothesized to reflect educational levels, which are strongly associated with lifestyle-related health behaviors, access to health care, and other factors predictive of health outcomes (Cella *et al.* 1991, Lantz *et al.* 1998, Marchand *et al.* 1997, Steenland *et al.* 2002). Educational level was not available for most INEEL workers in this study, with the exception of 4900 NRF employees and approximately 14,000 workers with data in the HRS file. Therefore, the association between assigned SES class and educational level was evaluated for NRF engineers, technicians and several other collapsed job categories within the INEEL cohort, as well as for study cohort members with data in the HRS file. The HRS file is a relatively recent database and is likely to be more reflective of associations with SES for more recent workers than for those employed in the early years of the INEEL's operation.

The NRF engineers as a class were observed to be well-educated (Figure 2-4). Only manufacturing, safety and plant engineers had median educational levels below a bachelor's degree. Therefore, these three engineering types were assigned to the Intermediate category.

All other engineers were assigned to the Professional category.

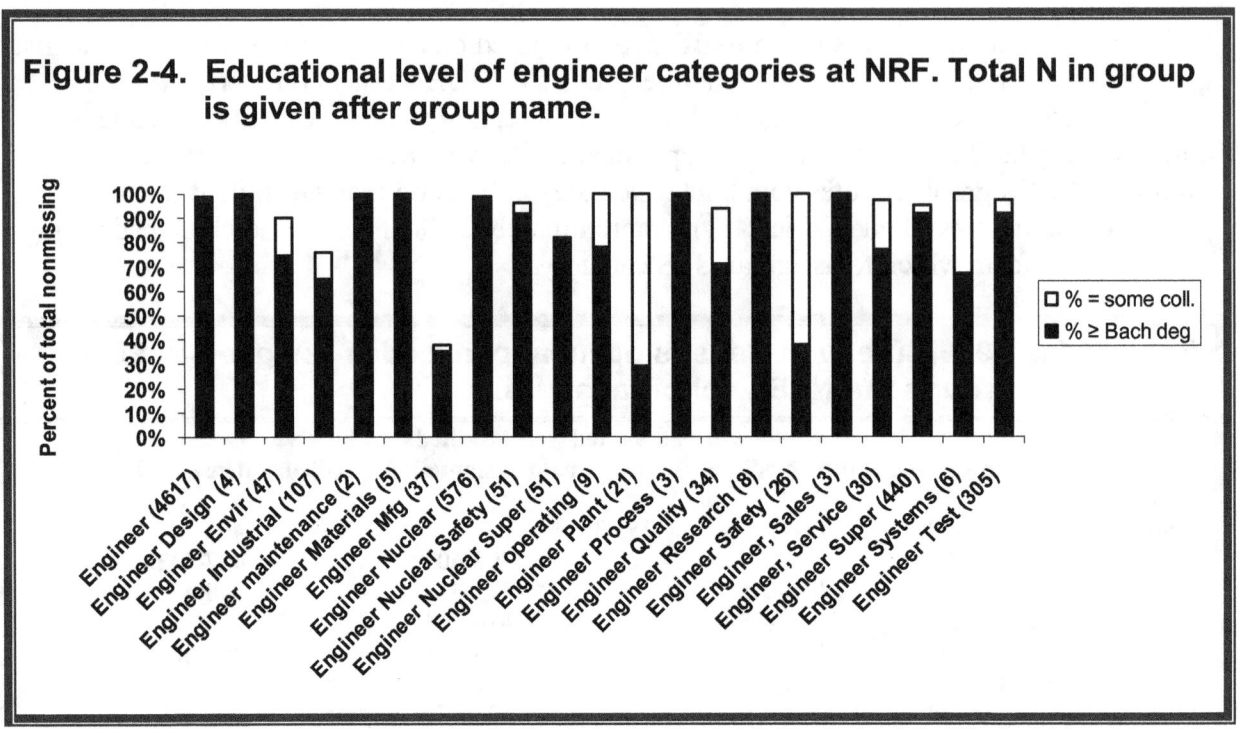

Figure 2-4. Educational level of engineer categories at NRF. Total N in group is given after group name.

In contrast, technicians as a group were less well-educated (the target for this group, classified as SES Intermediate, was "some college"). Of the technician group, only medical technicians, military commissioned officers, and technologists had a median education of less than "some college" (Figure 2-5). Therefore, all technicians except medical were assigned SES of Intermediate.

SES was consistently associated with highest educational level, for the 14,077 workers in the HRS file (Figure 2-6). Nearly 87% of Professional workers had a bachelor's degree or higher, compared to 42% of Intermediate workers, about 10% of Skilled workers, and 6-7% of Partly skilled and Unskilled workers. Skilled workers were much more likely to have had special training or some college, as compared to partly skilled and unskilled workers (Figure 2-6).

The evaluation of SES replicability across job titles by individual was assessed for several collapsed job titles, by evaluating a random sample of workers with at least one mention of this job title (Table 2-12). The most problematic job titles were Analyst, Engineer Maintenance, Inspector and Surveyor (when the words "rodman" and/or "chainman" were included in the original job title). The resolution of these assignments is shown in Table 2-12. In addition, while reviewing these random samples, it was noted that in some instances there were two different job title SES assignments for the same first day of employment, because the job titles differed in two or more source files. These conflicting records were reconciled on an individual basis.

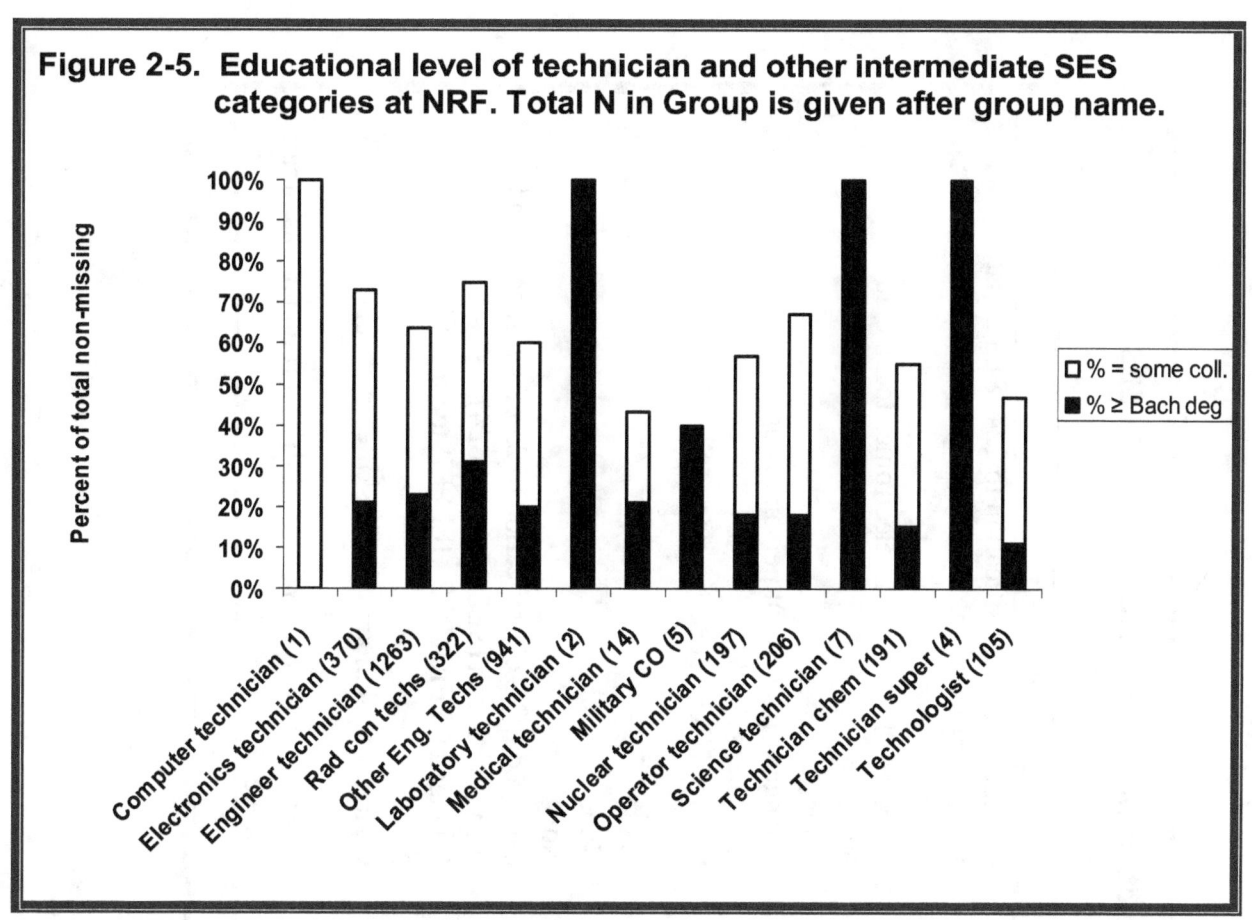

Figure 2-5. Educational level of technician and other intermediate SES categories at NRF. Total N in Group is given after group name.

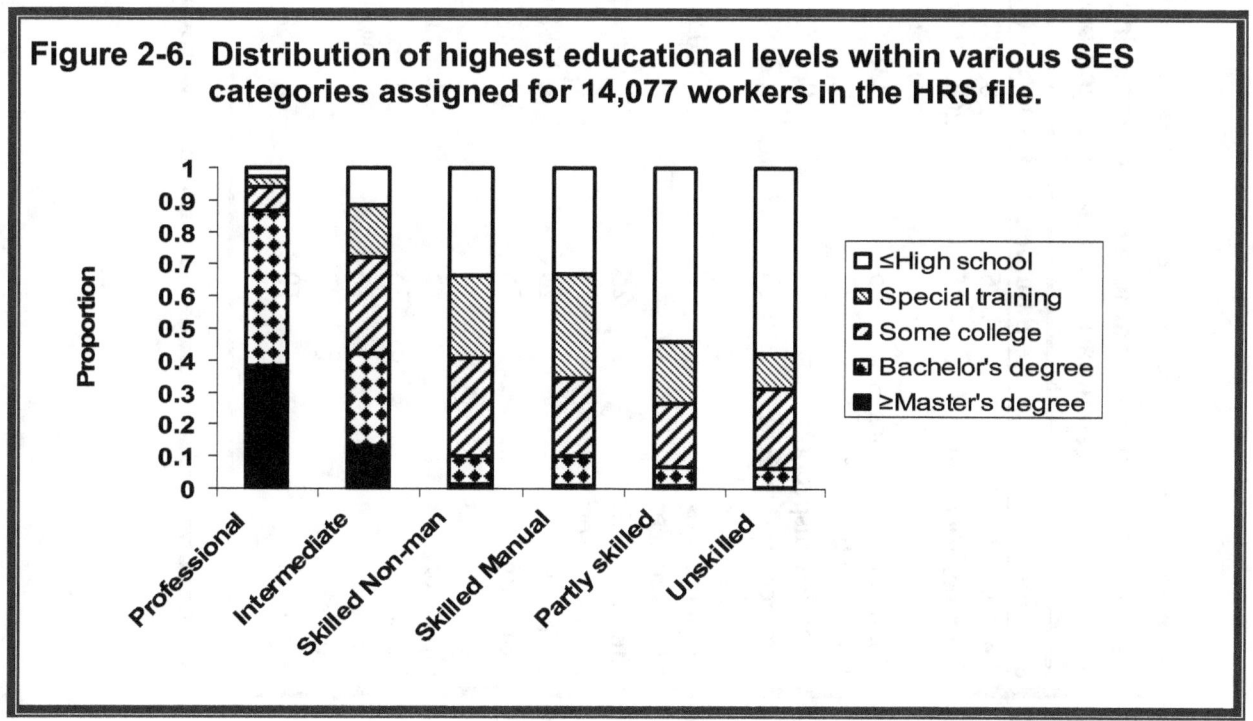

Figure 2-6. Distribution of highest educational levels within various SES categories assigned for 14,077 workers in the HRS file.

Table 2-12. Random sample of problematic job titles with respect to conflicts in SES determinations, and changes used to solve the problem.

Job title	# Sampled	Incorrect (#, %)	Change implemented
Analyst	25	20, 80%	Changed BOC classification to 224 (SES 2 instead of 1)
Engineer Electronics	25	5, 20% (2 were electricians)	When conflicting job title is "electrician," defer to it
Engineer Electronics	25	2, 8%	Changed "Elec tech" to electronics technician
Engineer Industrial	25	3, 12%	None
Engineer Inspection	3	0, 0%	None
Engineer Instrument	13	4, 31%	Changed "Instrument person & engineer" in MUD/RDS to "Engineer instrument"
Engineer Maintenance	3	3, 100%	Changed BOC classification to SES 1
Engineer Safety	25	5, 20%	None
Engineer Test	25	0, 0%	None
Foreman	25	4, 16%	None
Inspector	25	18, 72%	Changed BOC classification to 235 for Inspector and Inspector Mech (SES 2 instead of 4)
Operator	25	2, 8%	Changed OPERATOR HELPER to OPERATOR
Scientist	25	0, 0%	None
Surveyor: Rod/chainman	16	8, 50%	Changed to skilled or partly skilled manual (SES=4 or 5)
Surveyor	10	3, 30%	Changed to SURVEYOR HELPER if rodman, chainman or aide

2.5 Vital Status and Cause of Death Ascertainment

The follow-up period employed in this study was January 1, 1949 to December 31, 1999. The primary data sources used for ascertaining deaths were the SSA Death Master File (DMF), the National Death Index (NDI), and Pension Benefits Incorporated, a commercial service. In addition, several sources were searched to determine whether individuals who were not found to be deceased were actually alive at the end of follow-up, including the SSA's "presumed living" search, the Internal Revenue Service, the Health Care Finance Administration (HCFA), and information on retirees who were monitored in the Occupational Health program at the INEEL.

Cause of death (COD) was ascertained from several sources. For deaths occurring in 1979 or later, COD was obtained directly from NDI-Plus. For deaths occurring before 1979 (i.e., prior to the availability of NDI), death certificates were obtained from the state in which death occurred. The state of death was identified in the SSA DMF in about half of the cases; when not identified, a search algorithm was employed to find the most likely state of death. Several states have developed computerized or microfiche indexes of deaths. These states, which include many common retirement states for the cohort, such as California, Washington, Oregon, Texas, Nevada, Arizona and Florida, were searched first to identify decedents. If the date of death was outside the range of years covered by the state death index, or if there was not death index coverage in a state, then the following hierarchy was employed in searching for the death certificate: (1) Idaho; (2) Utah; (3) State of issue of SSN; (4) Other likely states of out-migration (based on death information obtained from NDI), including Colorado, Oklahoma, Illinois and Pennsylvania.

Death certificates were coded by a qualified nosologist, using revision 9 of the International Classification of Diseases (ICD-9). Deaths occurring in 1999 were available from NDI in the tenth revision of the ICD but were recoded into ICD-9 using a translation table modified from Anderson *et al.* 2001 (see Appendix §A.5).

2.6 Statistical Analysis

2.6.1 Radiological Data

Log probability regression of annual dose was used to estimate the fraction of collective dose attributable to missed dose. A regression line was drawn for a portion of the natural log of annual dose versus standard normal deviation curve in the region between 1 mSv and 5 mSv.

Between these points the curve remains essentially linear. The equation is in the standard form of $y=mx+b$, where y is the natural log of the annual dose and x is the Z-score. The slope and intercept were determined using the least squares fit method and the data extrapolated to zero annual dose. Below this region the curve trends to zero because of the LOD of the dosimeter. Above this region the curve tends to flatten because of administrative controls at the site that limited the dose a worker could receive at the high end. All analyses were conducted using SAS version 8.02 (SAS Institute Inc. 1999).

2.6.2 Epidemiologic Data

Descriptive analyses of the cohort were conducted using SAS software (ver 8.02). Associations between covariates in the analysis were evaluated using the correlation and general linear modeling procedures in SAS.

The primary form of epidemiologic analysis for this cohort consisted of standardized mortality ratio (SMR) and standardized rate ratio (SRR) analyses, supplemented with Poisson regression for diseases suspected to be associated with external ionizing radiation. The latter tool allowed greater flexibility to specify the form of the risk function for external ionizing radiation, and the multivariate modeling environment allowed incorporation of important covariates (such as sex, race, duration of employment, SES classification, likelihood of internal exposures) and facilitated comparisons of results with other radioepidemiologic studies. Poisson regression is also less dependent on complete ascertainment, since "expected" values were computed using an internal comparison group.

2.6.2.1 Analytic Variables

The exposure variables and covariates varied in their completeness across the cohort. Sex, DOH and DOB were the most complete covariates and were missing for 0.03%, 0.4% and 1% of the cohort, respectively. The exposure variables and covariates considered in the analysis are given below.

Exposure
- Monitoring status (required)
- Annual external radiation (not required)
- Internal radiation category (not required)
- Contractor code (not required)

Covariates
- Sex (required, but unknown were assumed male)
- Race (required, but unknown were assumed white)
- Attained age (required)
- Year of birth (required)
- DOH (required)
- Duration of employment (not required)

2.6.2.2 SMR and SRR Analysis

The SMR and SRR analyses were conducted using ver. 1.0d of the NIOSH LTAS for the PC (Cassinelli *et al.* 2002). The LTAS program calculated expected numbers of deaths in the cohort by applying age-, race-, sex- and calendar year-specific mortality rates from a standard population to the corresponding person-years at risk in the cohort. These expected numbers were then compared with the number of deaths actually observed in the cohort (or exposed portion of the cohort). For SMR analysis, the standard population was generally the U.S. or a regional population, and indirect standardization was used. SRRs were calculated whenever comparisons were made between subcohorts or groups of workers in the INEEL cohort. For SRR analysis, the standard population was the entire cohort under analysis, and direct standardization was used (Steenland *et al.* 1990). The use of direct standardization in comparing two groups within the cohort was preferred (over simply calculating ratios of SMRs) for several reasons: first, the results are expected to be more statistically valid because between-group differences in distributions of stratification variables (e.g., age and calendar year) will not distort the accuracy of the rate ratios. Second, the variability in both the numerator (exposed category) and denominator (unexposed category) is explicitly taken into account in the directly standardized SRRs. Third, the directly standardized rates may be used in a weighted regression analysis to evaluate any potential linear trend across the exposure categories.

In the INEEL cohort analysis, SMRs were calculated for overall mortality as well as for cause-, sex-, calendar time-, and exposure-specific mortality. Two-sided 95% CIs for the SMRs were obtained assuming a Poisson distribution. CIs (either 95% or 99%) that exclude 1.00 may be of particular interest to some who interpret the report so they are highlighted in the result tables for convenience. All life table analyses were based on underlying cause alone.

Analyses were standardized by age, sex, race and calendar year. Analyses stratified by duration of employment were also conducted. For SMR analysis, the comparison population was the general U.S., as well as a regional population consisting of the residents of Idaho, Montana and Wyoming, from 1960-1999 (the only years for which these rates are available).

The main analysis focused on comparisons of mortality risk (using SRRs, as described above) among various groups within the INEEL cohort, for example, individuals who were issued a dosimetry badge compared to those who were not. Other subgroups of interest were workers monitored for exposure to one or more of fission products, tritium and/or radionuclides (all classified as Internal dose status), the various contractor groupings (e.g., construction, supply or operations contractors), and workers employed at various facilities (e.g., the NRF or the ICPP). Identification of various subcohorts and stratification groups of interest is shown in Tables 2-10 and 2-13.

Because of the expected importance of regional population differences in cancer rates (particularly for cancers related to smoking and other behavioral risk factors), analyses were conducted using both U.S. and regional rates. The SMR analysis assumes that the denominator (i.e., comparison population) variability is zero; this assumption may not be met by using Idaho, a state with a very small population (a large minority of which was included in this cohort), and most rates within the state are based on small numbers. Therefore, a set of combined states near the INEEL facility (Idaho, Montana and Wyoming) was used as a regional comparison population. Utah was excluded from this group because of its unusually low lung cancer rates (Nelson *et al.* 1994) perhaps reflective of lifestyle differences related to LDS membership among a large percentage of residents of that state. Idaho and other western U.S. states do also have a substantial minority of residents who are LDS members, as reflected in this cohort by the large percentage of workers whose SSNs were issued in Idaho, Montana and Wyoming who were LDS members.

Table 2-13. Additional subcohorts of interest in the INEEL cohort analysis.

Name	Code	Subcohort Description	Identification Method
Internal dose status	1	Not monitored	See §2.3.1.2
	2	Monitored with no dose	
	3	Monitored with pos dose	
Local vs. migrant population	0	From ID, WY, UT, MT	Local workers: first 3 digits of SSN=(516-520; or 528-529; or 646-647)
	1	Migrant from other state	Migrant workers: first 3 digits of SSN=(001-515; or 521-527; or 530-645; or 648-728)
SES categories	1	Professional	SES class 1 assigned to job title at first hire
	2	Intermediate	SES class 2 assigned to job title at first hire
	3	Skilled non-manual	SES class 3 assigned to job title at first hire
	4	Skilled manual	SES class 4 assigned to job title at first hire
	5	Partly skilled	SES class 5 assigned to job title at first hire
	6	Unskilled	SES class 6 assigned to job title at first hire

Special rate files for the NIOSH-LTAS analysis were also created to evaluate certain diseases of particular interest in some analyses, including: (1) all brain tumors combined; (2) leukemia other than chronic lymphocytic; (3) cancers of peritoneum and pleura combined. The categories for these combined rate files, by ICD revision, are given in Table 2-14. As mentioned above, all INEEL death certificates were coded into ICD revision 9, but the rate files for the comparison population were available in the revision in effect at the time of death. For all special rate files except combined peritoneum and pleura, the comparison population was the states of Idaho, Montana and Wyoming. For peritoneum and pleura, the comparison population consisted of these states plus Utah, so that the population sample size would be increased for these very rare cancers, and mesotheliomas identified in ICD-10 were included, as well.

Table 2-14. ICD codes used in the creation of special rate files for all brain tumors, all leukemia other than Chronic Lymphocytic Leukemia (CLL) and pleura and peritoneum cancer combined.

	ICD codes		
ICD revision	Brain tumors	Non-CLL leukemia	Pleura & Peritoneum Cancer
7th	193.0, 223, 237	204.1-204.4	158, 162.2
8th	191, 225.0, 238.1	204.0, 204.9, 205.0-205.9, 206.0-206.9, 207.0-207.9	158, 163.0
9th	191, 225.0, 237.5, 239.6	204.0, 204.2-204.9, 205.0-205.9, 206.0-206.9, 207.0-207.8, 208.0-208.9	158, 163
10th	C71, D33.0, D33.1, D33.2, D43.0, D43.1, D43.2	C91.0, C91.2-C91.9, C92.0-C92.9, C93.0-C93.9, C94.0-C94.7, C95.0-C95.9	C38.4, C45.0, C45.1, C48

Certain causes of death, such as lung cancer, oral cancer, emphysema and chronic bronchitis, alcoholism and alcoholic cirrhosis of liver, and ischemic heart disease, caused predominantly by tobacco and alcohol use, poor diet and/or lack of exercise (McGinnis and Foege 1992) were evaluated as markers of potential lifestyle-related differences among subcohorts or other groups of workers.

Workers were identified by badged monitoring status (for external radiation exposure) as some studies have found badged workers to be healthier than unbadged workers (e.g., Wilkinson *et al.* 2000, Silver *et al.* 2004). The dates of first monitoring and first positive exposure were required for each worker, to apportion person-time correctly by badging status. Therefore, a set of these dates was established using the badge issue and pull dates as available in MUD, RDS and NRF dosimetry data files. Proper analysis of these data in the NIOSH life table program required the assignment of a trivial dose (2E-6 mSv) to each worker on the dates of first employment (for all workers) and first monitoring (all monitored workers). Groups of workers were then compared using direct standardization to compare all badged workers (categories 2-7 in Table 2-15) to unbadged workers (category 1 in Table 2-15) and to compare badged workers with a positive dose (categories 3-7) to unbadged workers.

In general, subcohort analyses were conducted by assigning all person-time to the specific subcohort of interest as of the first DOH of the workers. However, because monitoring dates were specifically known, all analyses comparing badged status were conducted by beginning badged person-time accrual as of the date of first monitoring and attributing prior unbadged person-time to the unbadged category. Similarly, for the analyses comparing badged workers with positive dose to unbadged workers, person-time prior to receiving a positive dose was assigned to the badged-zero-dose group and to the badged-positive dose group thereafter. The analyses were repeated by assigning all person-time for persons who were ever badged to the badged group (and all person-time to those with positive dose to the badged positive group), and very similar results were seen (Appendix Tables A-4 and A-5). Because most badged workers were radiation-monitored when hired at the INEEL, only 1809 additional person-years (0.4% of the total) were considered unbadged when allocating by person-time. Thus, SMRs and SRRs were virtually identical using the two methods.

The SRR analyses employed trend tests for cumulative exposure to high-energy photon radiation, using the category cut-points identified in Table 2-15. A Z-score was computed as the ratio of the mean to its standard error; this ratio was compared to a standard normal distribution for evaluation of statistical significance for a linear trend in dose-response (using a two-tailed alpha of 0.05).

A SES comparison also was done for internal dose categories, defined as workers never monitored; workers monitored but with no evidence of exposure; workers monitored with some indication of exposure to transuranics, tritium, or fission products (see §2.3.1.2). For this analysis, the dates at which workers received their change in internal monitoring status were taken into account (i.e., workers' person-time was considered unmonitored for internal exposures during the period before their first monitoring event); however, once a worker achieved a higher monitoring category, person-time and deaths were assigned to that category thereafter.

Table 2-15. Cumulative dose category cut points for standardized rate ratio (SRR) analysis of INEEL cohort

Category Number	Dose Category	Interpretation
1	0 to 2E-6 mSv	Non-monitored group
2	3E-6 to 0.0099 mSv	Monitored; never received a dose (the minimum reported dose in a year was 0.01 mSv)
3	0.01 to 0.99 mSv	Monitored; received a cumulative dose at specified value
4	1.00 to 9.99 mSv	Same as #3
5	10.00 to 49.99 mSv	Same as #3
6	50.00 to 99.99 mSv	Same as #3
7	$\geq$ 100 mSv	Same as #3

Exposure lag periods were evaluated [5, 10 and 20 years for solid cancers, lymphoma and chronic lymphocytic leukemia (CLL), and 2, 5 and 10 years for non-CLL leukemias]. Only external dose categories 2-7 were used in the SRR analysis to test for trends in cancer with dose. Categories 2 and 3 were combined to form the baseline group (for increased sample size) for the SRR trend tests.

Analytic Strategy for SMR and SRR Analysis

The primary purpose of the life table analysis was to estimate mortality rate ratios, including 95% CIs, for different groups of workers. These results were interpreted according to the approach put forth by Rothman and Greenland (Modern Epidemiology, 1998, Chapter 12) that, according to modern epidemiologic practice, does not limit the interpretation to significance/hypothesis testing. Rothman and Greenland write that "epidemiologic applications need more than a decision as to whether chance alone could have produced an association. More important is estimation of the magnitude of the association, including an assessment of the precision (or its inverse, the variability) of the estimation method" (pg.183). They further state that "results that are not significant may be compatible with substantial effects. Lack of significance alone provides no evidence against such effects" (pg. 192).

Following this approach when evaluating life table results, the authors considered the magnitude of the observed SMR or SRR in relation to the width of the CI. Special consideration was given to those causes of death of *a priori* interest when results were discussed. Some of these point estimates are described as elevated or reduced when the 95% CI includes one. Point estimates and their corresponding CIs are provided in the data tables.

2.6.2.3 Multivariate Regression Modeling Analysis

For workers who were monitored for external radiation exposure, AMFIT software (Epicure, release 2.10; Preston *et al.* 1993) was used to conduct Poisson regression analysis for all cancers, individual cancers, and for groupings of cancer that were of *a priori* interest for high-energy photon radiation, the primary exposure of concern. These cancers included categories rated as "frequently or occasionally associated with radiation with authoritative or valid risk estimates" by Boice *et al.* (1996), including leukemia (except CLL), multiple myeloma (each analyzed separately), and cancers of thyroid, female breast, lung, stomach, colon, esophagus, bladder and ovary (grouped together). CLL was also evaluated separately. Recent studies have also suggested that cardiovascular diseases are affected by exposure to ionizing radiation (Hancock *et al.* 1993, Hauptmann *et al.* 2003, Paszat *et al.* 1998, Shimizu *et al.* 1999); therefore, an assessment of dose-response association for all cardiovascular disease (adjusted for age, sex, race and SES) was conducted. In addition, cancers and other causes of death that showed positive association with external radiation in the SRR analysis were evaluated using Poisson regression. Emphysema and certain smoking-related cancers, including lung, oral and nasal cavities, pharynx, larynx, esophagus, stomach, pancreas, liver, urinary tract and cervix (IARC 2004) were evaluated to assess the possibility of confounding by smoking within the cohort.

The potential for exposure to internal radiation was assessed by estimating risk for workers in three categories of internal dose: never-monitored (or monitored only once as part of a hire or termination examination), monitored but no measured dose (or only a small likelihood of dose), and monitored with potential for internal dose, as described in §2.3.1.2. Internal dose monitoring status was considered a potential confounder. Assessment of other important confounders was conducted as well. The interaction of certain potential confounders (e.g., SES and migrant status) as predictors of cancer risk was evaluated using a likelihood-based hypothesis test consisting of a comparison of the deviance between nested models. A p-value is reported for the chi-square statistic, with the degrees of freedom equal to the number of interaction terms added to the model.

The measure of interest was an estimation of the excess relative risk (ERR) per unit dose, adjusted for various confounders. For the Poisson analysis in AMFIT, the observed number of events in a cross-tabulation of the risk stratifiers was treated as a Poisson variable, with expected numbers computed from the product of the person-time in the cell and the fitted rate from the regression model parameters. The person-year weighted average dose within the dose category, cross-classified on all the stratification and modeled covariates, was used as the independent variable in the Poisson regression analysis (Preston *et al.* 1993).

To facilitate comparisons with other radiation epidemiology studies, such as the Japanese Life Span Study cohort (NRC 1990, Shimuzu *et al.* 1999) and other studies of nuclear workers (e.g., Cardis *et al.* 1995; Yiin *et al.* 2004) the preferred form of the regression model is the linear ERR model, which takes the following general form:

$\lambda(Z,d) = e^{\beta Z} \cdot [1+f(d)]$, where
    d is dose in mSv
    $\lambda(Z,d)$ is the risk of death from disease due to dose
    β is a vector of parameter estimates associated with model covariates (e.g., SES, migrant status, internal monitoring status, defined by vector Z
    $f(d)$ is a linear function ($\alpha_1 \cdot d$) for solid cancers and (in some instances) is a quadratic function ($\alpha_2 \cdot d + \alpha_3 \cdot d^2$) for leukemia

To avoid having to quantitatively specify the form of the age- and calendar-year-specific baseline cancer rates, all models stratified on age group (14 strata), calendar time (8 strata), duration of employment (<10 years and >10 years), and sex (Preston *et al.* 1993, Cardis *et al.* 1995).

Certain models also used categorical analysis of the dose groups, to directly estimate relative risk (RR) with each dose group. These models take the form: $\lambda(Z,d) = e^{\beta Z} \gamma D$, where dose is included in the log-linear term, D is a vector of different dose categories for which RRs were evaluated, and γ is the corresponding vector of parameter estimates

Models were fit using maximum likelihood methods. A common finding when large numbers of strata are fitted is that the iterative model fitting techniques in AMFIT do not converge on a set of parameter estimates. If this condition existed with the INEEL cohort data, then a log-linear modeling approach was considered. Parameter estimates for the dose-response linear ERR coefficients were estimated, along with their CIs, using profile likelihood methods (Moolgavkar and Venzon 1988, Preston *et al.* 1993). The appropriate tests for statistical significance of dose-response were a comparison of the improvement in fit with the addition of the dose-response coefficient; that is, the difference in deviance in nested models (with and without the test variable) compared to a chi-square distribution with degrees of freedom equal to the number of parameters added to the nested model.

Building of Baseline Model: General Approaches

The INEEL cohort is a component of the International Agency for Research on Cancer (IARC) fifteen-country combined international nuclear worker study (Cardis and Kato 1993). This analysis may be compared to the IARC study analyses since all results were stratified on calendar year, age group, sex and duration of employment (less than 10 years, and greater than or equal to 10 years). In addition, SES (in four classes, plus unknown as a separate class), migrant status to the region, the interaction of these two variables, and internal dose category were considered as potential confounders. Thus, an attempt was made to determine which factors should be considered in the model.

A baseline model was specified, based on knowledge derived from previous epidemiologic studies of DOE workers and other radiation-exposed cohorts. This model included attained age, sex and calendar time, as numerous studies of nuclear workers have shown confounding and/or effect modification by these variables (e.g., Cardis *et al.* 1995, Wilkinson *et al.* 1987). Other covariates were evaluated for confounding of the association between external radiation exposure and disease, including race, SES, migration to the INEEL region (as determined by state of issue of the SSN), and internal exposure status. Both on-site and off-site doses were included in most dose-response analyses. Alternative models with similar fits to data were evaluated and are presented.

Sensitivity Analyses

Factors evaluated in the creation of the baseline model are reported as part of a sensitivity analysis. These analyses included restriction by duration of employment to longer-term workers, evaluation of SES and migrant status, analysis by disease subcategory, and for leukemia alone, exploration of optimal lag periods and choice of categories for the dose-response analysis.

3 Radiation Exposure Assessment Results

3.1 External Dosimetry

The photon penetrating dose distribution among workers varied by year. The highest mean exposure occurred in 1965, but the highest exposure to an individual worker occurred in 1961 (Table 3-1). About 97% of the workers in each year did not exceed 35 mSv. A discussion of exposure sources by decade follows based on the Annual Reports of the Health Services Laboratory, for various years, National Reactor Testing Station, Idaho Operations, U.S. Atomic Energy Commission. However, the worker doses cited are based on the cohort under study.

1950-1959 The first work with radioactive materials began in 1951 with the EBR-I and the MTR. One or two new reactors per year were brought on line through 1956. As a result, the fission product inventory and the materials processed at the ICPP increased, leading to an increase in the collective site photon and neutron dose. In July 1954 at the BORAX-I facility, INEEL performed its first destructive test of a reactor and began the decommissioning process. The first criticality accident occurred on November 29, 1955 at the EBR-I facility and resulted in a core meltdown. On July 23, 1956 an operator received 216 mSv to the whole body at the MTR facility while working inside the reactor dome during maintenance. In 1957 and 1958 10 new research reactors came on line, and the Radioactive Lanthanum (RALA) facility began operation in 1956 at the ICPP. Over the next three years 36 runs were made with radioactive material. Annual collective dose during this period among INEEL cohort members peaked in 1958 at 19.89 person-Sv.

1960-1969 In 1960 three new reactors were started up and an all time peak of 23 reactors were operating at INEEL. Collective dose reached 19.39 person-Sv. In 1961 the SL-1 accident with three fatalities occurred, which increased the collective dose by 7 person-Sv. High penetrating and non-penetrating doses were received to accident responders and cleanup workers. Approximately 1000 workers were involved in the recovery and cleanup effort. In 1965 because of increased work load the collective dose increased to an all time high of 32.08 person-Sv. There were no accidents or unusual occurrences during this year. After the 1960s the person-Sv remained below 20 even with increasing personnel.

1970-1979 In the early 1970s the ICPP and Waste Calcining Facility had reached 20- and 10-year anniversaries, respectively, and maintenance activities increased as pumps and valves began to age. The collective dose was directly related to the maintenance shutdown time at these facilities. Collective dose dropped in the mid-1970s due continuous plant operations and minimal maintenance. Because of adherence to the principle of 'As Low As Reasonably Achievable' (ALARA), the collective dose continued to decrease.

1980-1998 Over this period the collective and average dose decreased even when the number of workers increased (Figure 3-1). This decrease was largely due to the decrease in the number of nuclear projects, shutdown of the Engineering Test Reactor, shutdown of the Waste Calcining Facility and a shift in the number of workers to Idaho Falls from the site.

Table 3-1. Penetrating photon dose (mSv) percentiles by year for monitored workers

Year	0%	25%	50%	75%	90%	97.5%	99.5%	100%	MEAN	STD DEV	NUMBER
1951	0	0	0	0	0	0.00	4.00	4	0.022	0.292	188
1952	0	0	0	0	0.3	2.22	7.27	12.75	0.213	1.002	913
1953	0	0	0.6	2.5	5.6	10.90	20.97	75.7	2.002	4.034	1408
1954	0	0	0	2.1	9	24.98	40.77	57.6	2.838	6.567	2449
1955	0	0	1	5.9	16.565	32.37	51.62	84.7	5.157	9.314	2946
1956	0	0	0	2	10.15	25.55	42.67	220.6	3.142	8.423	3209
1957	0	0	0	0.85	5.4	15.13	30.45	51.2	1.706	4.517	4695
1958	0	0	0.4	3.25	12.6	28.90	43.78	105.1	3.766	8.015	5079
1959	0	0	0.4	2.6	9.6	23.47	41.27	218.5	3.099	7.402	5344
1960	0	0	0.5	2.55	10.75	26.47	37.34	50.1	3.268	6.755	5827
1961	0	0	0.4	4.2	17.04	35.66	49.72	272.6	5.063	12.806	5192
1962	0	0	0.3	1.75	11.15	32.25	50.48	98.85	3.571	8.942	5339
1963	0	0	0.25	1.8	10.795	29.54	40.22	51	3.180	7.292	5520
1964	0	0	0.1	2.15	12.765	31.11	39.80	48.15	3.538	7.886	5446
1965	0	0	0	4.3	22.045	43.90	60.18	98.15	5.790	12.107	5520
1966	0	0	0.2	3.75	15.595	34.67	44.66	60.45	4.383	8.926	5180
1967	0	0	0	1.7	10.725	30.84	43.77	48.05	3.194	7.702	6304
1968	0	0	0	2.15	11.4	31.04	41.94	52.95	3.364	7.781	4922
1969	0	0	0	2.4	11.95	27.30	39.82	44.5	3.279	7.151	4758
1970	0	0	0	1.75	10.28	26.24	42.07	46.8	2.952	7.020	5051
1971	0	0	0	1.7	8.2	18.99	32.52	47.1	2.357	5.427	4764
1972	0	0	0	1.61	8.55	23.75	38.35	46.65	2.606	6.340	4762
1973	0	0	0	1.15	6.525	21.26	39.09	52	2.185	5.878	4494
1974	0	0	0	1	5.15	16.96	29.93	40.65	1.734	4.574	4878
1975	0	0	0.06	0.9	4.05	14.12	27.89	39.45	1.531	4.071	5025
1976	0	0	0	1.1	5.06	16.46	26.43	41.45	1.712	4.294	5489
1977	0	0	0	0.94	4.85	19.47	33.85	107.7	1.869	5.260	5677
1978	0	0	0.09	0.87	3.94	15.35	28.72	43.86	1.563	4.295	6551
1979	0	0	0	0.74	3.69	14.60	27.32	41.8	1.419	4.064	6863
1980	0	0	0	0.47	2.77	10.62	21.12	169.3	1.073	3.579	7380
1981	0	0	0	0.46	2.52	8.24	16.74	32.89	0.876	2.358	6722
1982	0	0	0	0.38	1.95	6.38	15.00	29.04	0.715	2.040	6556
1983	0	0	0	0.35	1.87	5.16	9.76	15.77	0.582	1.473	6610
1984	0	0	0	0.34	1.79	5.77	11.98	22.85	0.619	1.766	7476
1985	0	0	0	0.39	2.19	8.72	12.40	24.15	0.820	2.260	7917
1986	0	0	0	0.24	1.92	7.73	17.49	93.38	0.770	2.623	8568
1987	0	0	0	0.25	1.55	7.33	14.17	31.58	0.659	2.098	8575
1988	0	0	0	0.22	1.45	5.37	10.86	30.86	0.545	1.641	8667
1989	0	0	0	0.16	1.01	5.59	13.15	78.11	0.516	2.169	8848
1990	0	0	0	0.12	0.954	5.66	12.66	27.28	0.490	1.852	10165
1991	0	0	0	0	0.37	2.49	6.94	45.77	0.232	1.094	10742
1992	0	0	0	0	0.37	1.69	3.76	12.76	0.157	0.563	9571
1993	0	0	0	0	0.65	3.42	10.60	15.35	0.311	1.233	9048
1994	0	0	0	0.1	0.786	3.35	7.29	13.94	0.322	1.121	8473
1995	0	0	0	0.11	1.031	4.49	11.53	18.44	0.428	1.533	7818
1996	0	0	0	0.16	1.08	4.21	7.90	13.68	0.393	1.215	6459
1997	0	0	0	0.1	0.81	2.59	5.74	11.08	0.261	0.820	6280
1998	0	0	0	0.02	0.62	2.08	4.62	8.44	0.205	0.650	5875

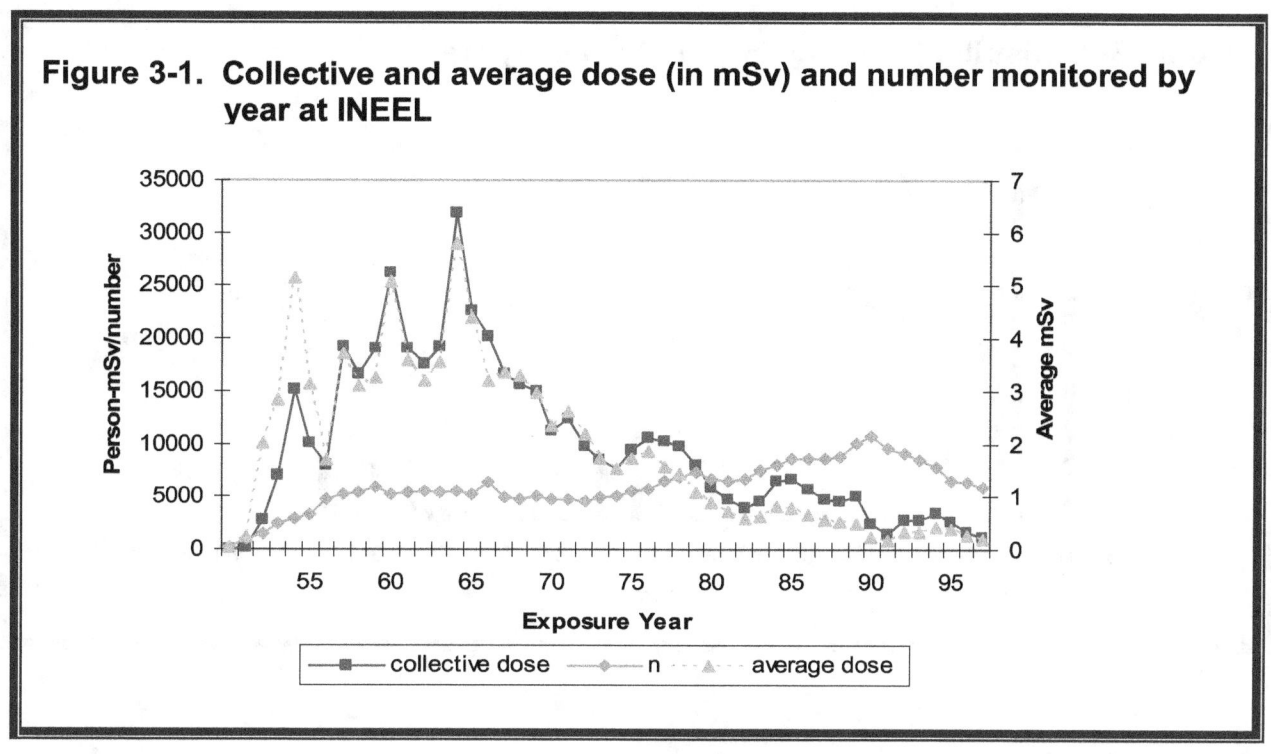

Figure 3-1. Collective and average dose (in mSv) and number monitored by year at INEEL

Personnel at the site worked in facilities designated by the dosimetry program as area codes. Workers are listed in the dosimetry database by area code and year. Area codes for facilities with the greatest number of dosimetry badges are indicated in Table 3-2. Before 1986 only the last area code worked was listed in the database; that is, if a person had worked several area codes, only the last appears in MUD. After 1986 the RDS recorded all area codes in which a person worked (Figure 3-2). The percentage of dosimeters used by an area code is an indicator of the number of workers assigned to that facility or area. The Chemical Processing Plant remained a potential source of radiation exposure all through its operation.

Table 3-2.	Facilities that used greater than 5% of the dosimeters by Area Code
Area code	Area description
42	Test Reactor Area
53	Chemical Processing Plant Monthly
55	Chemical Processing Plant Quarterly
263	Argonne National Laboratory West
772	Test Area North
774	Secure Manufacturing Monthly

Area code 999 is the designation for dose received by workers at sites other than INEEL. The identity of the site at which the dose was received is unknown. The total off-site dose was 58.80 Sv, which corresponds to approximately 10% of the on-site collective dose. Fewer than 5% of workers in any year had an off-site dose in their dosimetry record (Figure 3-3). These off-site doses are incorporated into some of the epidemiological analyses (see §6.3-§6.6).

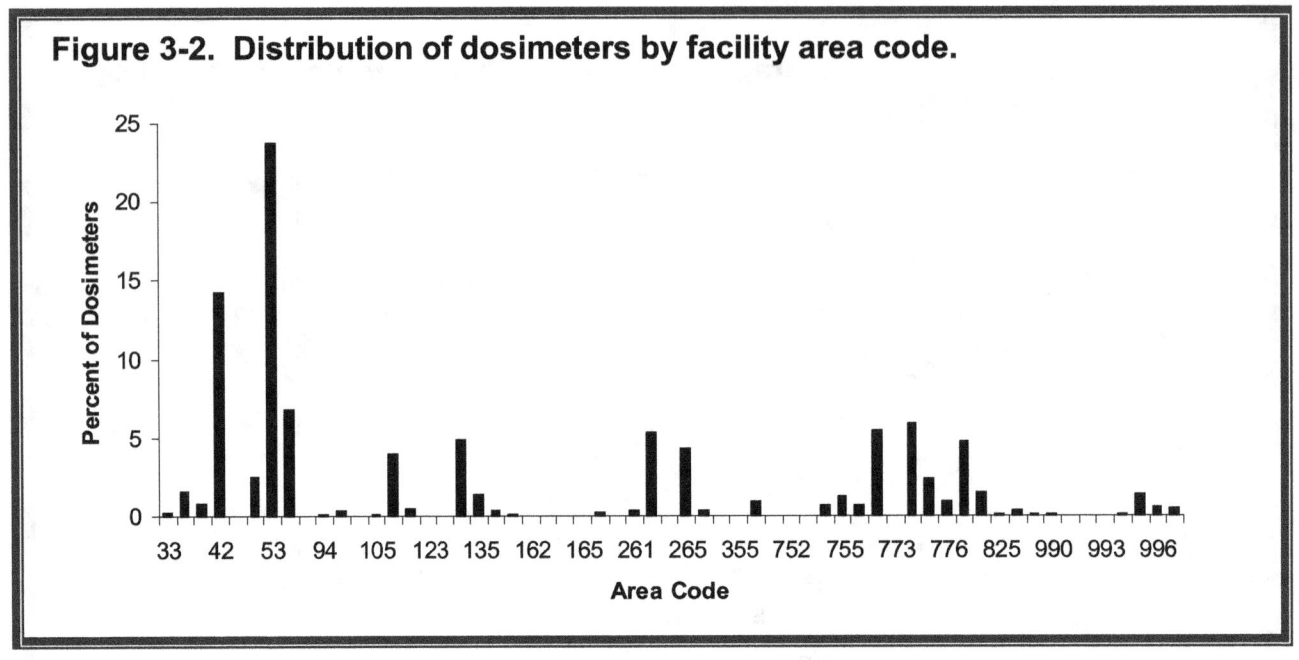

Figure 3-2. Distribution of dosimeters by facility area code.

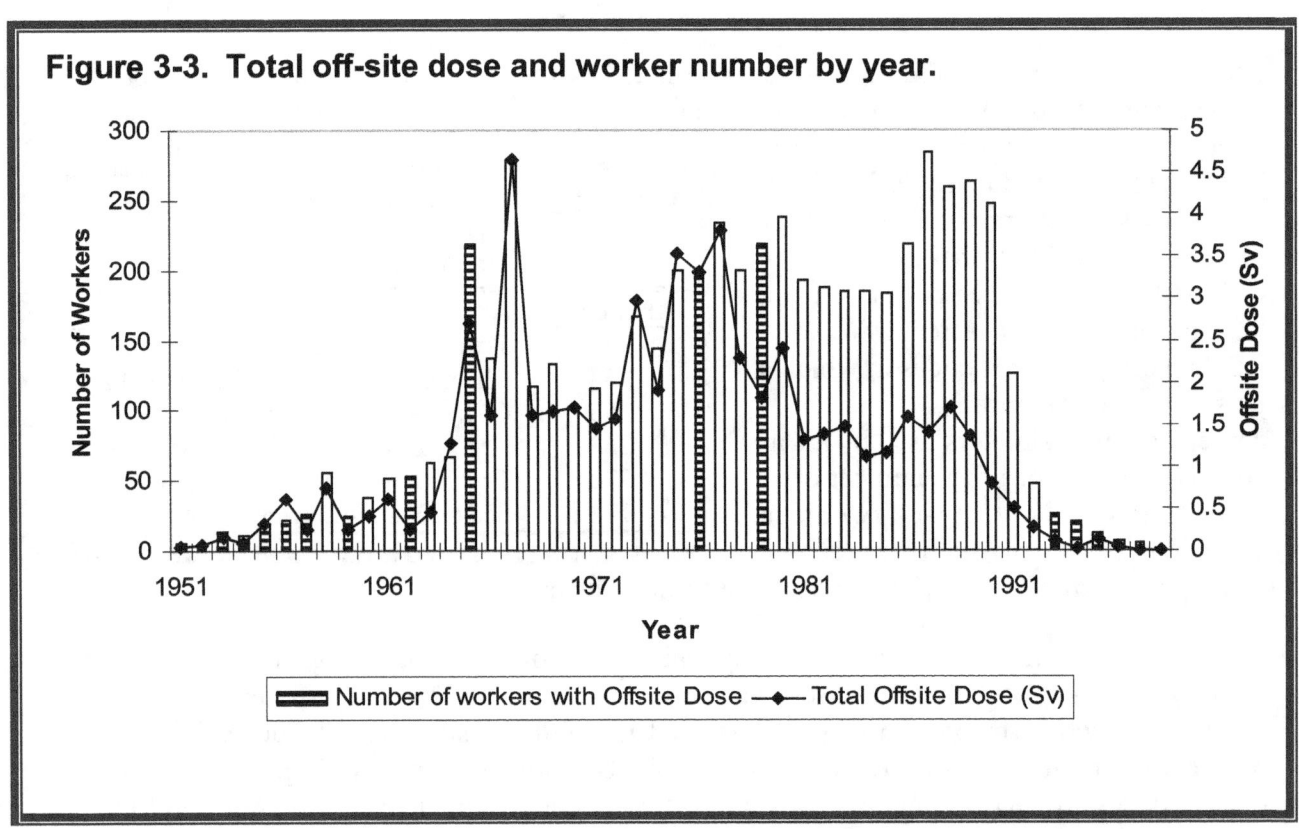

Figure 3-3. Total off-site dose and worker number by year.

The on-site penetrating photon dose ranged from 0 to greater than 0.50 Sv. The neutron dose ranged from 0 to 0.50 Sv (Figure 3-4, Figure 3-5).

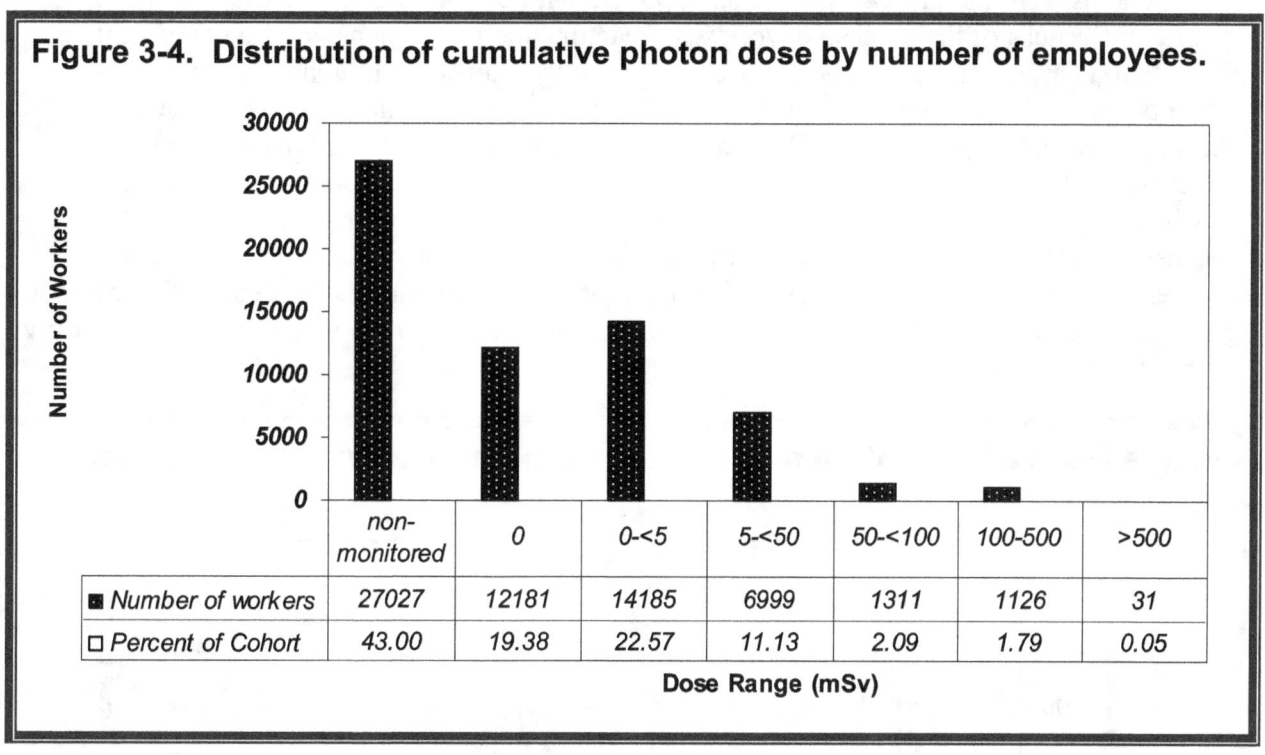

Figure 3-4. Distribution of cumulative photon dose by number of employees.

Dose Range (mSv)	non-monitored	0	0-<5	5-<50	50-<100	100-500	>500
Number of workers	27027	12181	14185	6999	1311	1126	31
Percent of Cohort	43.00	19.38	22.57	11.13	2.09	1.79	0.05

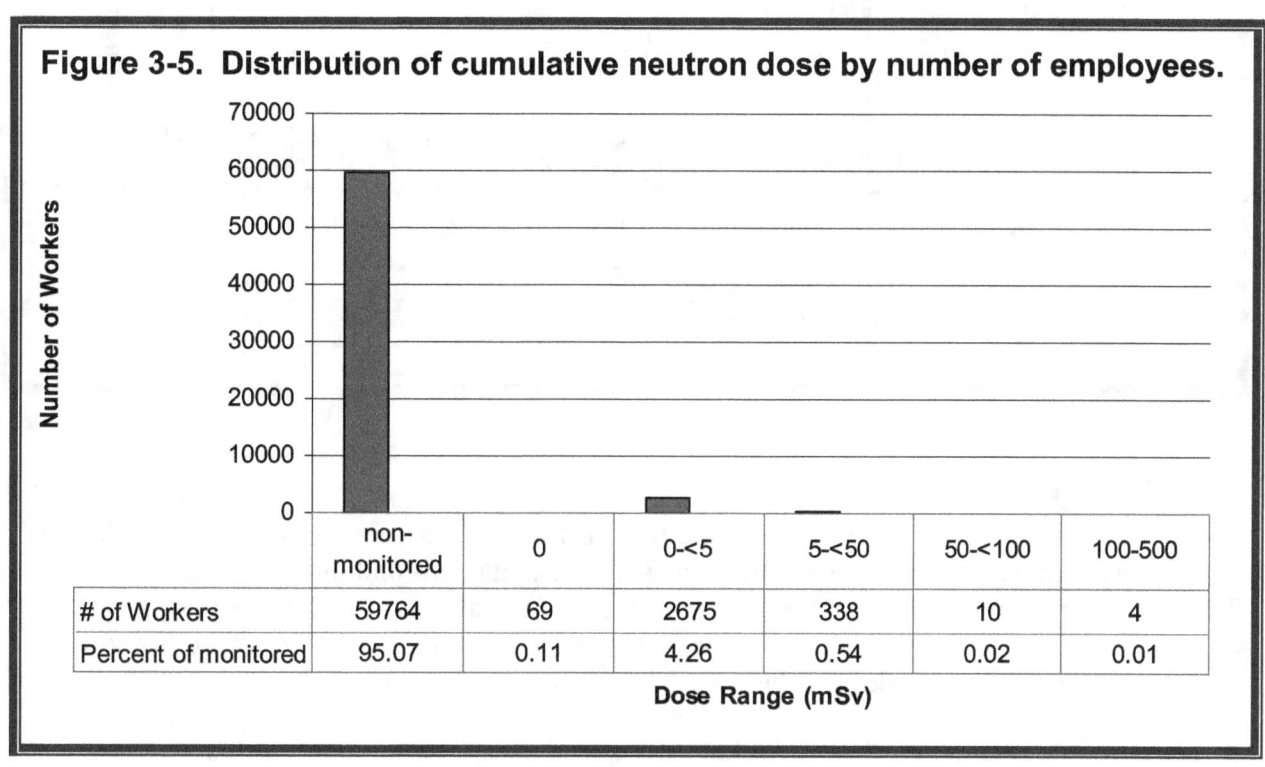

Figure 3-5. Distribution of cumulative neutron dose by number of employees.

Dose Range (mSv)	non-monitored	0	0-<5	5-<50	50-<100	100-500
# of Workers	59764	69	2675	338	10	4
Percent of monitored	95.07	0.11	4.26	0.54	0.02	0.01

As a result of a policy of recording a dose as zero if it fell below the detection limit of the dosimeter, more than 12,000 workers had zero dose recorded for all years of work. Just over 3000 persons were monitored for neutron exposure with only 69 having zero neutron dose recorded. Because of the extensive work with reactors and maintenance of the equipment, it is likely that a larger number of workers were exposed to neutrons. Though the number of workers exposed to neutrons may have been larger, the average annual neutron dose a maximized at 0.33 mSv in 1956. The change from NTA film to the Hankins albedo neutron dosimeter in 1975 does not reveal an increased missed dose due to insensitive detection capability (Figure 3-6). There is an increase in average neutron dose between 1969 and 1972 because of increased maintenance activities on facilities that reached their 15th and 20th anniversaries of operation. Less than 2% of the employees had cumulative doses of 50 mSv or greater for neutron and photon radiation combined and just over 74% received less than 5 mSv for photon radiation.

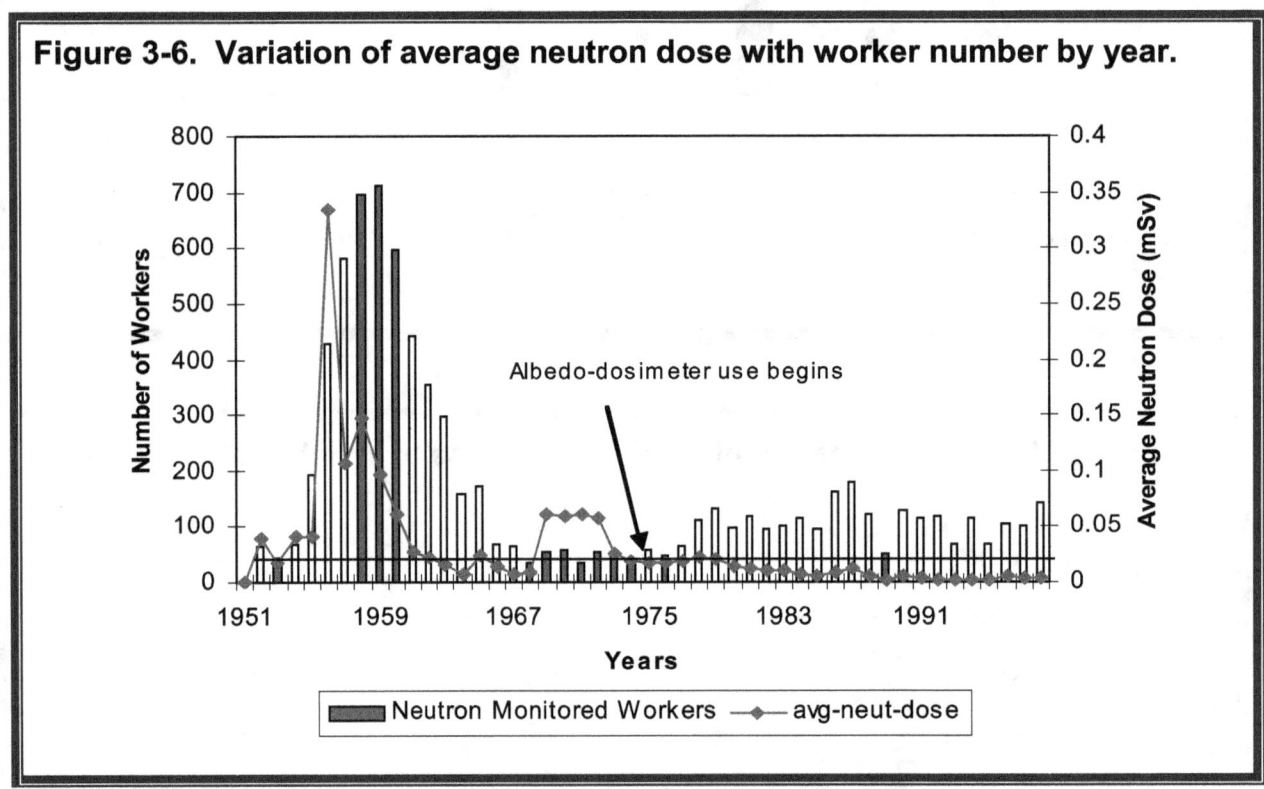

3.2 Internal Dose

Three methods of sampling were used at the INEEL facility to determine whether a worker had inhaled radioactive material. They were urinalysis, fecal analysis and lung counting (WBC). The first two methods required a worker to leave a biological specimen for analysis although the third method was non-invasive and involved placing a detector over the worker's chest. Each method produced a value for the activity present in the body.

A model was developed for this study that categorized each internal monitoring result as positive or negative for potential internal dose. A set of dose curves was generated for each radionuclide at several baseline activity values using IMBA software for each sampling method since detection limits for some radionuclides and sampling methods were not available. For example, a theoretical Pu-238 dose was computed for the maximally exposed organ using its associated theoretical activity value for urine, fecal and lung monitoring methods (Figure 3-7). Using the LUNG COUNT set of curves, it is seen that a dose of 1 mSv is delivered to the lung over 50 years by an activity amount of about 0.1 Bq. For Pu-238, the detection limit activity for lung counts among workers is known to be approximately 300 Bq. Unless the worker had urine or fecal samples that indicated less activity in the body, the worker was assumed to have had a potential for a positive dose from Pu-238. Less than 1% of the workers who were monitored for internal dose had Pu-238 lung counts and no other monitoring data. This method was applied for all workers who had sample data and had radionuclides that were in the library of IMBA.

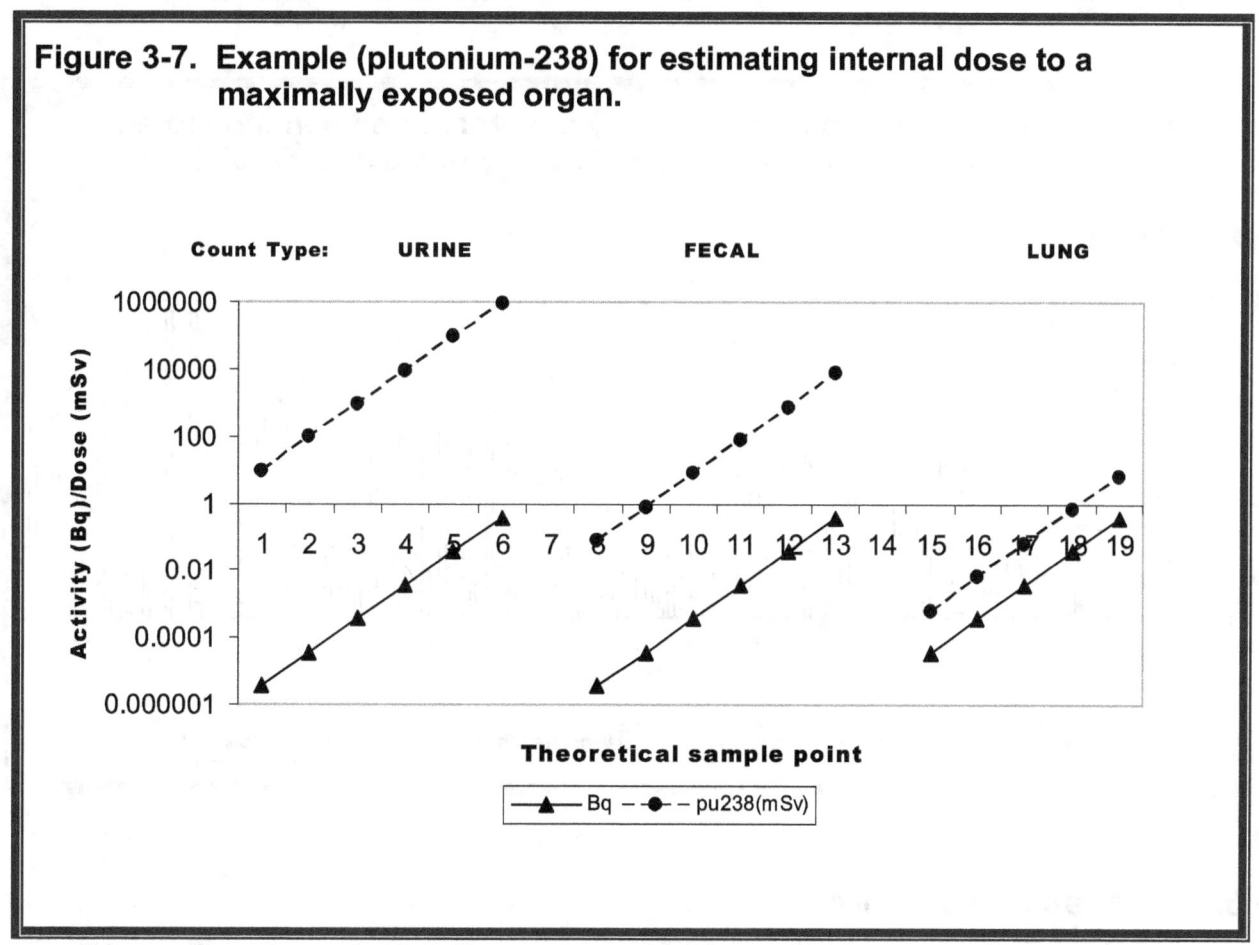

Figure 3-7. Example (plutonium-238) for estimating internal dose to a maximally exposed organ.

Based on this method three internal dose status categories were created from 91,856 observations for 63,561 employees.

1. **Monitored with positive internal dose** – 9426 employees monitored with potentially positive internal dose. The dates of first monitoring and first positive dose were estimated for each worker.

2. **Monitored without positive internal dose** – 12,580 employees monitored with potentially no internal dose. The date of first monitoring was estimated for each worker.

3. **Not monitored for internal dose** – 41,555 employees were not monitored.

Approximately 75% of workers up through the early 1960s who were monitored for external radiation were assigned to the positive internal dose category (Figure 3-8). However, inferences about significance of dose are limited because of a lack of readily available information on radionuclides and the analytical method results. The number of positive bioassay assignments each year appears correlated with the number of workers monitored for external radiation over this same time period. The assignment of workers to the positive internal dose category is highest in the early 1960s during the years of the SL-1 accident and a very large plant work load and follows a near-monotonic decrease after that time.

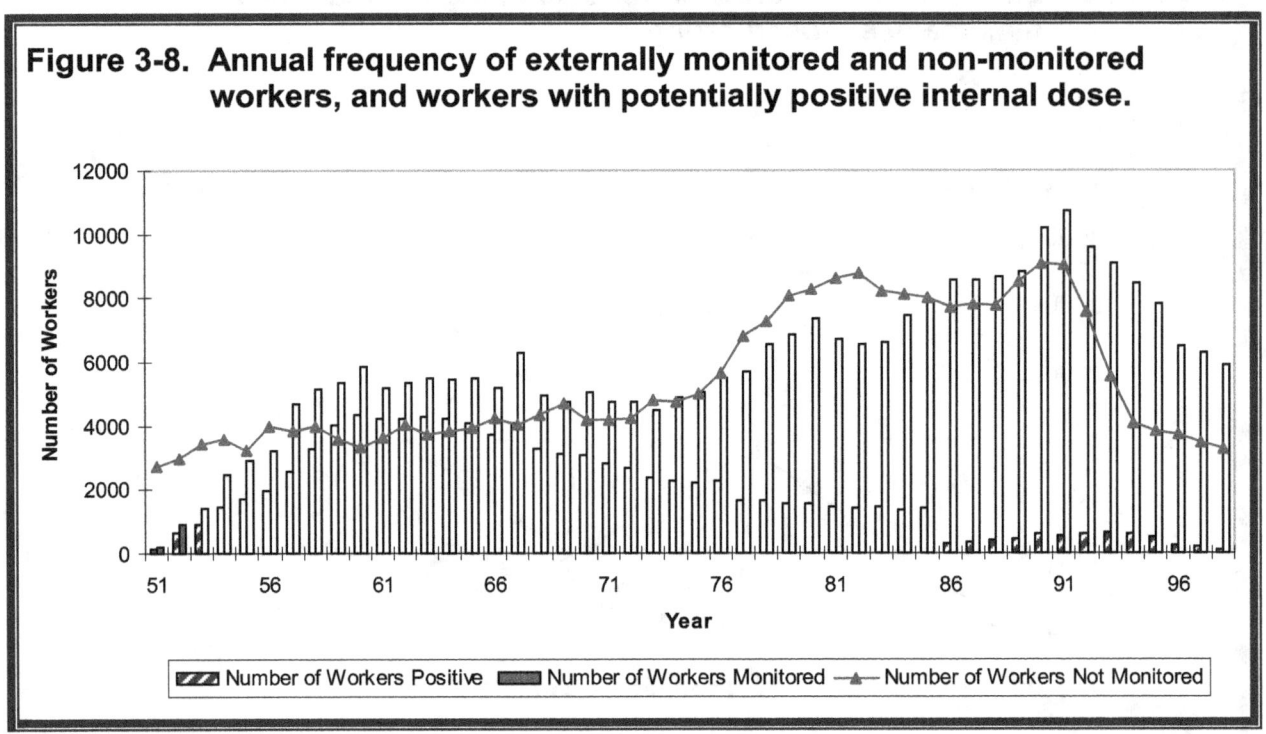

Figure 3-8. Annual frequency of externally monitored and non-monitored workers, and workers with potentially positive internal dose.

3.3 Missed Dose Estimate

The collective missed dose was extrapolated from the log-linear regression of annual doses between 1 mSv and 5 mSv (ordinate) and Z score (abscissa). The correlation coefficient was greater than 0.99 for this range. There are 20,886 values below the lower bound, 7019 values between upper and lower bounds and 3348 values above the upper bound. The estimated missed dose is found by subtracting the sum of recorded doses below 1 mSv from the sum of estimated doses below 1 mSv (Figure 3-9).

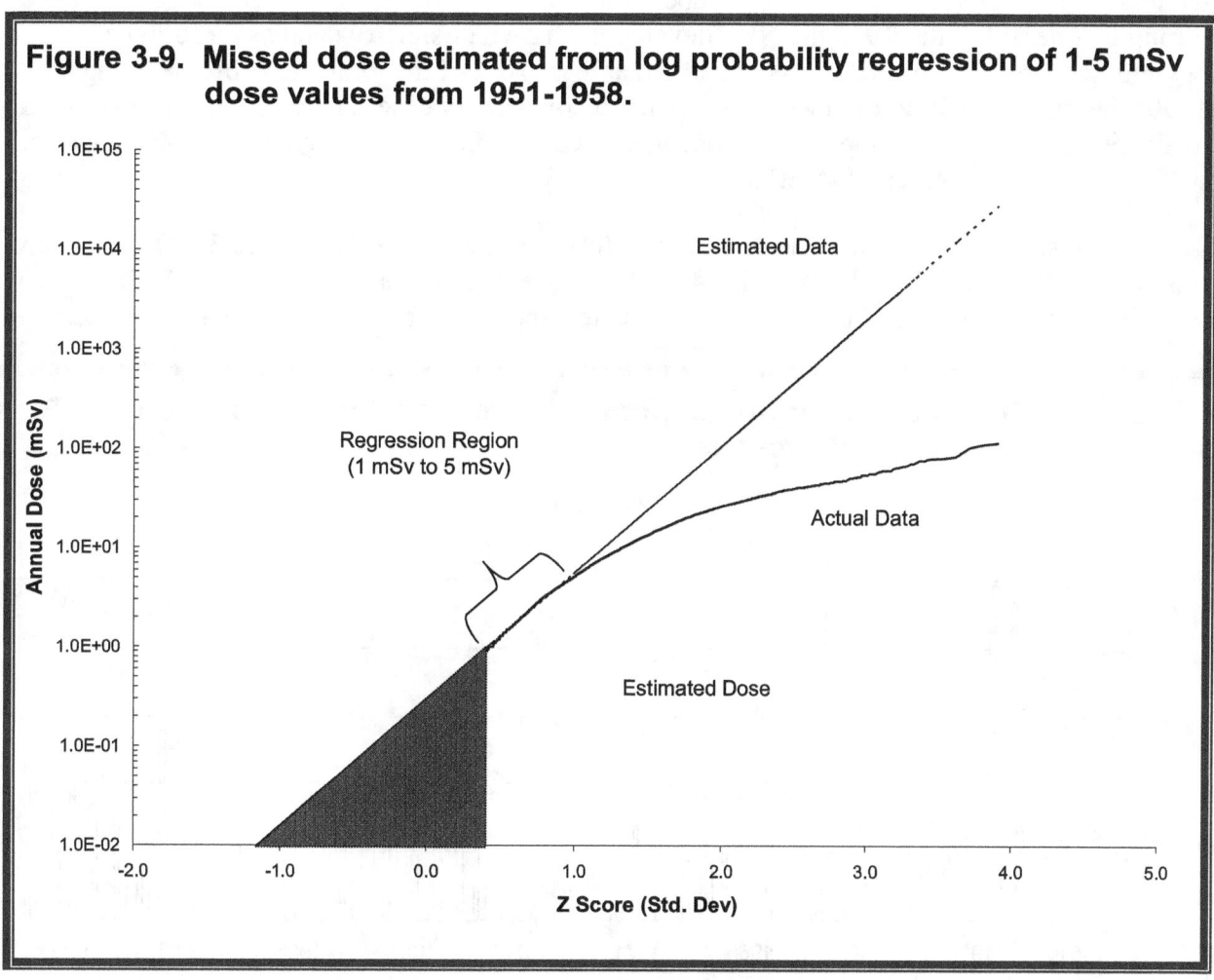

Figure 3-9. Missed dose estimated from log probability regression of 1-5 mSv dose values from 1951-1958.

The total collective dose for the period 1951-1958 was 62,376 mSv. The estimated missed collective dose for this period ranges between 2% and 10% of the total measured dose. In later years, dosimeter accuracy was increased and exchange frequency decreased so that the estimated missed dose would be a smaller percentage of the collective dose for that period. The uncertainty in the missed dose estimate is due to the inability to determine the number of values below the detection limit from the annual dose assignments in the source records.

3.4 Doses for NRF and Construction/Service Workers

3.4.1 NRF Doses

The NRF is made up of four major installations: the Submarine Prototype (S1W), the Large Ship Reactor (A1W), the Natural Circulation Submarine Prototype (S5G), and the Expended Core Facility (ECF). The prototype facilities are used for training naval personnel in nuclear plant operations. Fission products and neutrons were the main source of exposure.

NRF dose information was recorded in the MUD dosimetry database prior to 1975. Between 1959 and 1975 there is uncertainty about who worked at NRF based on the MUD dosimetry database. Therefore the dose/worker number data (Figure 3-10) may have been underestimated during that period. After 1975 the NRF developed its own dosimetry database. From the beginning of INEEL, the NRF has constituted at least 10% of the total work force. During the 1990s the ratio of NRF to total workers remained constant, but the relative contribution of NRF to site cumulative penetrating photon dose increased, due in part to the reduction of radioactive work at the other facilities of INEEL.

There were several years where neutron dose information was lacking (Figure 3-11). Data were missing for 1968 and 1970-1974. In 1964 neutron dose was comparatively high, which is likely an artifact of the dosimetry database problem stated above for the penetrating gamma dose.

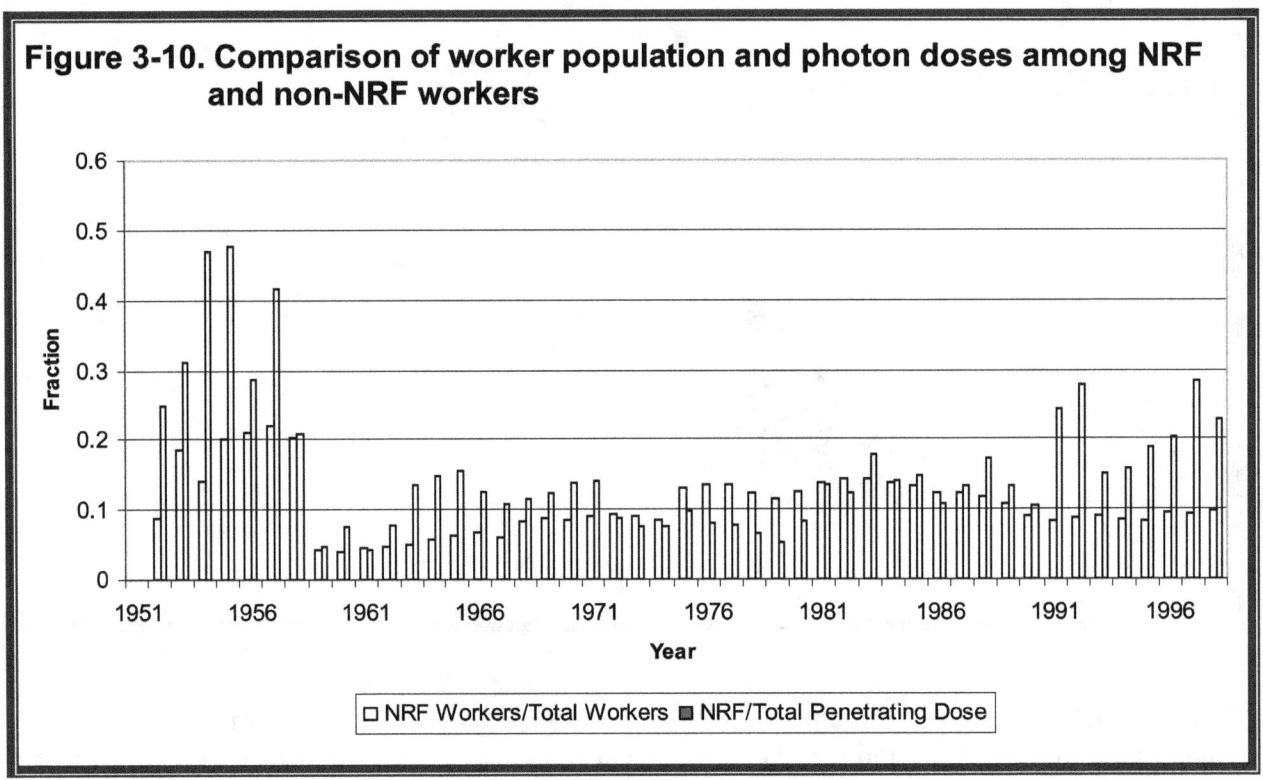

Figure 3-10. Comparison of worker population and photon doses among NRF and non-NRF workers

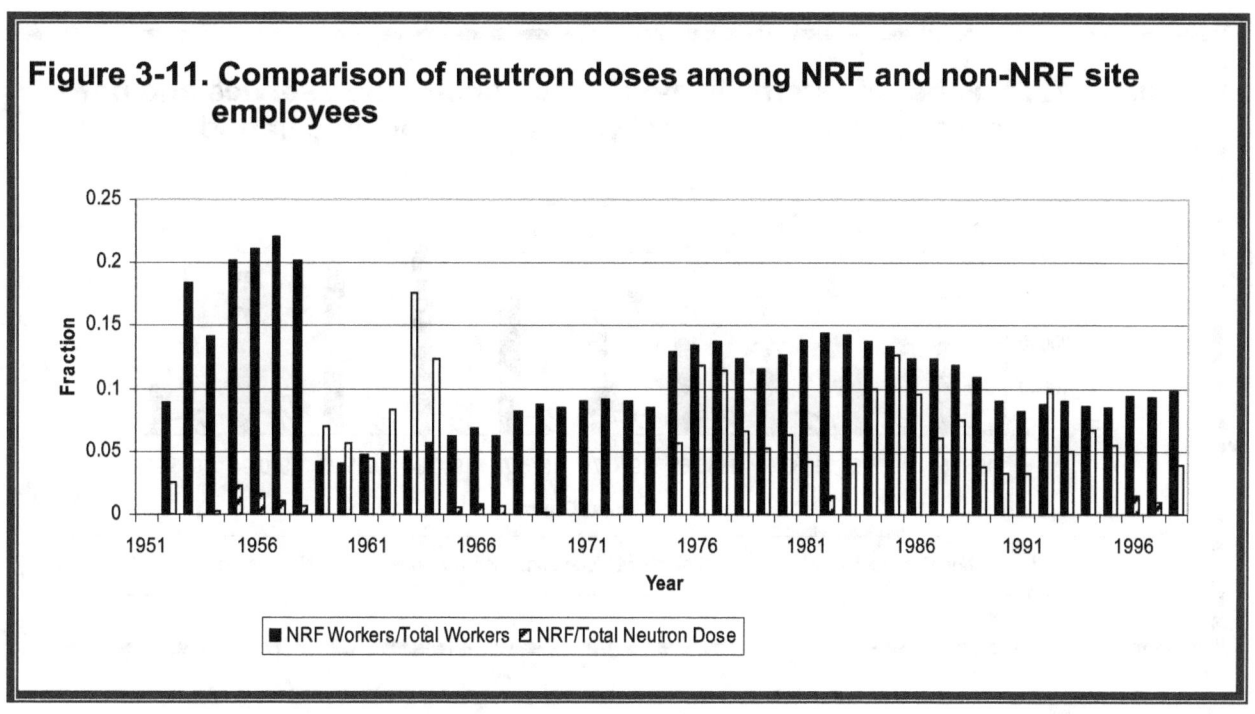

Figure 3-11. Comparison of neutron doses among NRF and non-NRF site employees

3.4.2 Construction and Service Worker Doses

Internal and external doses received by construction and service workers were compared to all other badged workers. From 1950 until 1965 the number of externally monitored construction and service workers exceeded that of workers not in construction or service jobs (Figure 3-12). During these years the INEEL site expanded rapidly, and jobs involving radioactive materials increased as well. Until about 1965 construction and service workers had relatively higher percentages of internal dose than non-construction/non-service workers (Figure 3-13). This finding is most likely explained by the early construction of the site facilities in the early 1950s, a leveling of construction during the operation peak of the mid-1950s to mid-1960s, and the removal and construction of facilities after 1965. In addition, maintenance service workers (who were included in construction worker doses) were in continuous demand across the facility. The ratio of doses (both average and cumulative) of construction and service workers to those of non-construction workers changed through these periods, also accounting for part of the increased external dose (Figs. 3-14, 3-15).

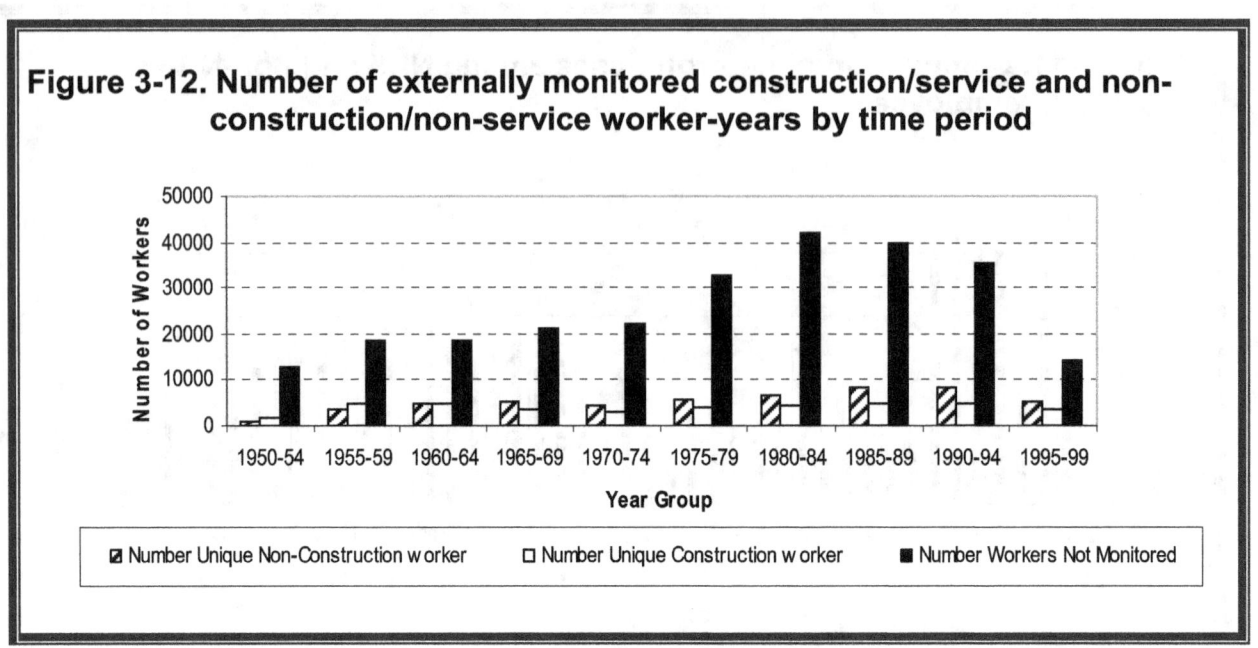

Figure 3-12. Number of externally monitored construction/service and non-construction/non-service worker-years by time period

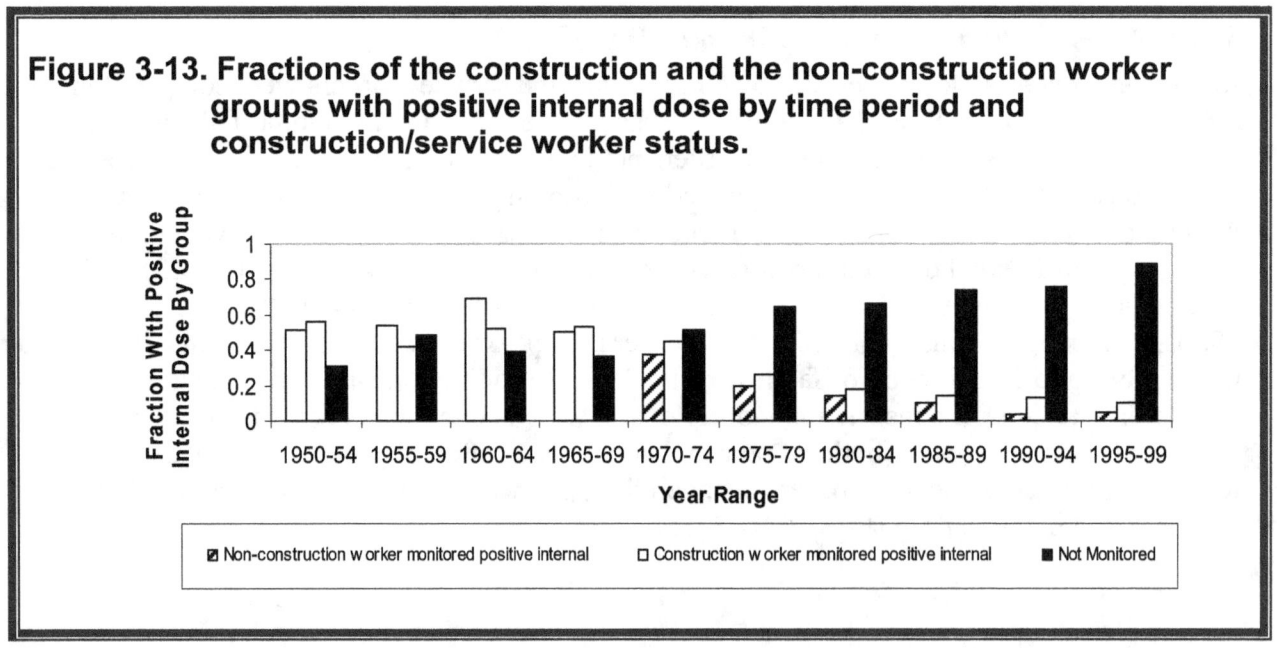

Figure 3-13. Fractions of the construction and the non-construction worker groups with positive internal dose by time period and construction/service worker status.

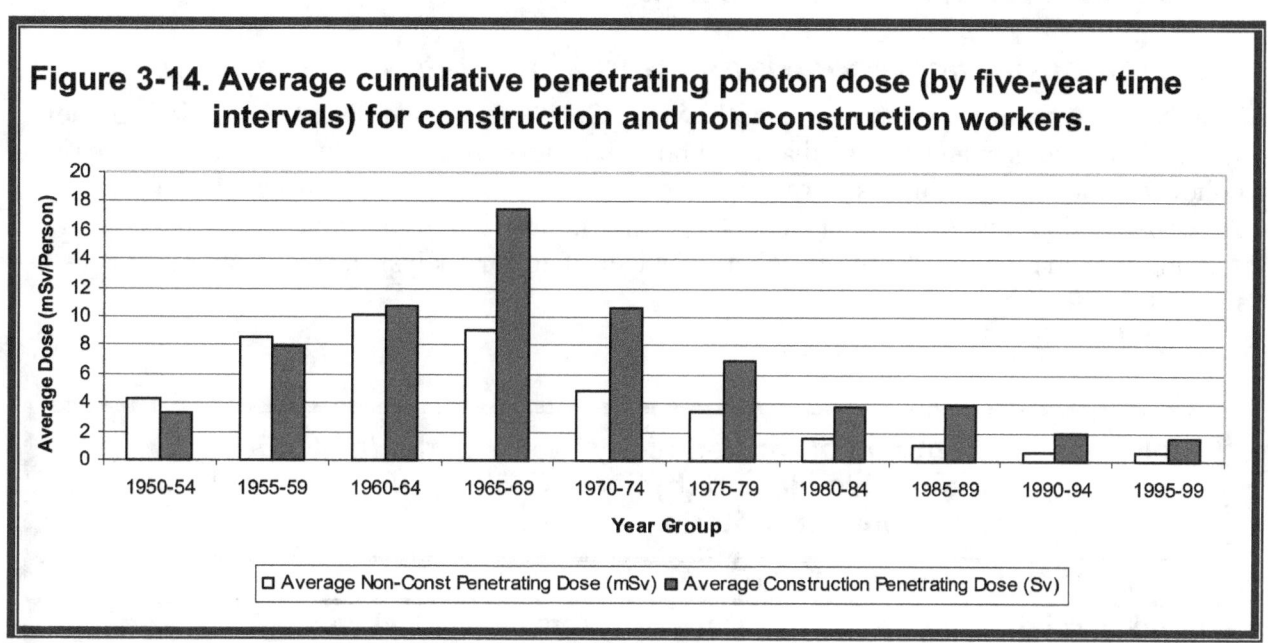

Figure 3-14. Average cumulative penetrating photon dose (by five-year time intervals) for construction and non-construction workers.

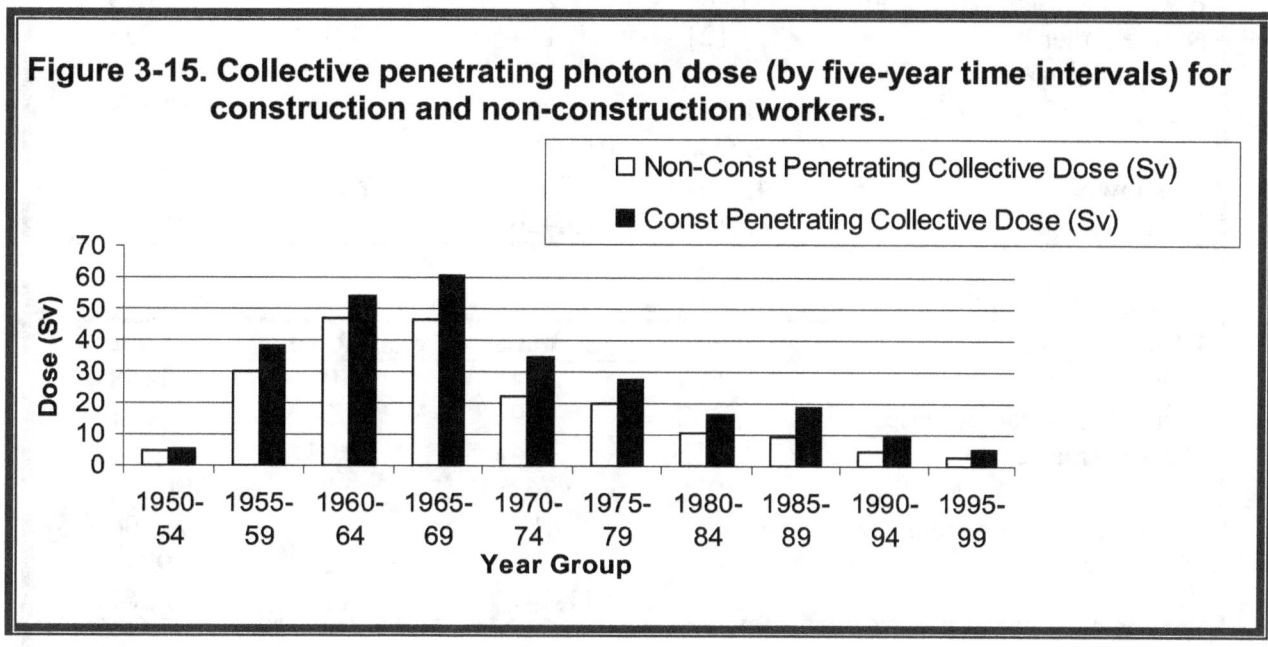

Figure 3-15. Collective penetrating photon dose (by five-year time intervals) for construction and non-construction workers.

4 Cohort Descriptive Results

4.1 Demographic, Work History and Vital Status Information

The INEEL workforce was predominantly white (96.1%) and male (80.6%) from information for workers with sex and race available. In both the entire INEEL workforce and the analytic cohort, the non-white workers were predominantly Asian, followed by Black or African-American and Native American (Table 4-1). After imputation of race and sex for those with missing information, 96.7% of the cohort was identified for analysis as white, and 82.2% as male (Table 4-2).

Table 4-1. Distribution of race/ethnicity by sex in the (a) INEEL demographic file and (b) final INEEL cohort after application of cohort eligibility criteria.

a) Total INEEL demographic file				
Ethnic Group	Male	Female	Unknown	Total
Asian	894	333	0	1227
Black or African-American	460	229	0	689
Native American	121	114	0	235
Hispanic (race unspecified)	15	15		30
White (% of those of known ethnicity)	42,511 (96.6%)	10,069 (93.6%)	0	52,580 (96.0%)
Unknown	41,341	2666	3230	47,237
Total	85,342	13,426	3230	101,998

b) INEEL cohort only				
Ethnic Group	Male	Female	Unknown	Total
Asian	874	324	0	1198
Black or African-American	452	219	0	671
Native American	113	104	0	217
White, incl. Hispanic (% of those with known ethnicity)	41,936 (96.7%)	9766 (93.8%)	0	51,702 (96.1%)
Unknown	8827	934	12	9773
Total	52,202	11,347	12	63,561

The entire cohort was relatively young, with a median age of 54 at the DLO (end of follow-up, date of death or DLO alive for those of unknown vital status). WM had substantially earlier birth years than white females (WF) and non-whites (Table 4-2). The average length of follow-up (21.1 years for the entire cohort) also differed somewhat by sex and race. WM were followed on average one year longer than females and non-white males (NWM). The median years of first employment were also quite different among WM and others: WM were hired six years earlier on average than WF and non-whites. Of the full cohort, 57% were monitored for external radiation exposure. The highest percentage of external monitoring occurred among WM (59%); the lowest monitoring frequency occurred among non-white females (NWF) (42%). Similarly, WM tended to have been monitored earlier than any other group (Table 4-2). The median duration of employment was similar for males and females, at three years for whites and two years for non-whites; however, a larger proportion of WM compared to other groups worked for longer than ten years. The median year of first radiation monitoring was earlier than the median year of first employment for whites and NWM (Table 4-2) because a much greater number of workers hired more recently were not monitored for radiation exposure (as observed in Figure 3-10).

In the full cohort, 10,906 (17.2%) workers were identified as deceased, and 48,821 were determined to have been alive as of the end of follow-up (December 31, 1999). Approximately 6% of the cohort had unknown vital status at the end of follow-up (Table 4-2). As expected from their earlier birth years, a far greater percentage of WM (20.1%) than of other groups (2.6%-6.7%) was deceased at the end of follow-up (Table 4-2). The percentage of unknown vital status among non-whites was about half that of whites.

Approximately 40% of decedents in the cohort died in the state of Idaho (Table 4-3). The other most common death states (in decreasing order of frequency) included Utah, California, Washington, Oregon, Arizona, Texas, Nevada, Colorado and Montana, indicating substantial out-migration from the INEEL region. These ten states together account for 83% of the INEEL cohort deaths occurring in a known locale. The death location information was missing for about 5% of the decedents. This information can assist in determining the likely success of a registry linkage to obtain data on incident cancers.

4.2 Cause of Death Information

The distribution of underlying COD in the full cohort is shown in Table 4-4. As expected, given the larger size and older age of the population, most causes of death occurred much more frequently among WM than among other groups. Exceptions included breast cancers, most common among WF. The most common major causes of death for all groups except WF were cardiovascular and cerebrovascular diseases combined (ICD-9 401-459) followed by cancers (ICD-9 140-208). For WF, this order was reversed. Numbers of specific types of cancers are of interest, because they show that, for most cancers, rate ratios reported in Chapter 5 were based on very small numbers of deaths for any specific group other than WM.

Table 4-2. INEEL cohort descriptive statistics.

	Total cohort	White males	White females	Non-white males	Non-white females
Total Number	63,561	50,775 (79.9%)	10,700 (16.8%)	1439 (2.3%)	647 (1.0%)
Year of Birth: median	1942	1939	1949	1950	1952
25th-75th percentile	1928-1953	1926-1951	1939-1958	1942-1956	1947-1960
N missing	128 (0.2%)	109 (0.2%)	18 (0.2%)	1 (0.1%)	0
Age at DLO*: median	54.4	56.4	48.3	48.8	46.6
25th-75th percentile	44.4-66.0	45.8-67.8	40.3-57.9	42.4-56.9	39.0-51.9
N missing	187 (0.3%)	157 (0.3%)	28 (0.3%)	1 (0.1%)	1 (0.2)
Length of FU*: median	21.1	21.5	19.4	19.4	19.6
25th-75th percentile	12.5-32.5	13.1-33.5	10.8-27.3	10.8-24.6	11.0-27.0
N missing	274	231	31	5	7
Year of hire: median	1975	1973	1979	1979	1979
25th-75th percentile	1960-1984	1959-1983	1967-1987	1974-1988	1972-1988
N missing	274 (0.43%)	231 (0.5%)	31 (0.3%)	5 (0.4%)	7 (1.1%)
Duration employed (years): median	3.33	3.41	3.20	2.43	2.57
25th-75th percentile	0.68-11.1	0.61-11.8	1.01-9.27	0.50-8.58	0.49-8.20
N missing	274 (0.4%)	231 (0.5%)	31 (0.3%)	5 (0.4%)	7 (1.1%)
Radiation-monitored On-site N (% of total)	36,449 (57.3%)	29,691 (58.5%)	5520 (51.6%)	805 (55.9%)	274 (42.4%)
On-site or off-site N (%)	36,531 (57.5%)	29,896 (58.9%)	5550 (51.9%)	809 (56.2%)	276 (42.7%)
Year first monitored on-site: median	1974	1972	1978	1979	1987
25th-75th percentile	1960-1984	1959-1984	1964-1986	1973-1987	1976-1993
Alive N (%)	48,821 (76.8%)	37,455 (73.8%)	9452 (88.3%)	1300 (90.3%)	614 (94.9%)
Deceased N (%)	10,906 (17.2%)	10,210 (20.1%)	582 (5.4%)	97 (6.7%)	17 (2.6%)
Unknown N (%)	3834 (6.0%)	3110 (6.1%)	666 (6.2%)	42 (2.9%)	16 (2.5%)

* DLO: date last observed; FU: follow-up.

Table 4-3. State of death distribution for 10,906 deceased INEEL study cohort members.

State of death (SOD)	Frequency (% of 10363 with known SOD)	State of death (SOD)	Frequency (% of 10363 with known SOD)
Unknown	543 (5.0% of total deceased)	Montana	204 (2.0%)
Alabama	20	Nebraska	36
Alaska	32	North Carolina	19
Arizona	290 (2.8%)	North Dakota	4
Arkansas	42	Nevada	237 (2.3%)
California	1051 (10.1%)	New Hampshire	5
Colorado	231 (2.2%)	New Jersey	37
Connecticut	128	New Mexico	78
Delaware	10	New York	56
District of Columbia	12	Ohio	46
Florida	161	Oklahoma	103
Georgia	43	Oregon	336 (3.2%)
Hawaii	12	Pennsylvania	105
Idaho	4136 (39.9%)	Rhode Island	29
Illinois	110	South Carolina	28
Indiana	25	South Dakota	14
Iowa	29	Tennessee	56
Kansas	34	Texas	277 (2.7%)
Kentucky	18	Utah	1101 (10.6%)
Louisiana	30	Vermont	3
Maine	9	Virginia	97
Maryland	47	Washington	711 (6.9%)
Massachusetts	31	West Virginia	10
Michigan	34	Wisconsin	25
Minnesota	28	Wyoming	96
Mississippi	16	Puerto Rico	1
Missouri	69	Other U.S. territory	1

Table 4-4. Distribution of underlying COD (ICD-9 codes) in the INEEL cohort, by sex and race.

ICD-9 code	Description	WM[1] deaths	WF[1] deaths	NWM[1] deaths	NWF[1] deaths	Total deaths
001-139	**Infectious diseases**	**101**	**6**	**4**	**2**	**113**
001-009	Intestinal infectious diseases	1	0	0	0	1
010-018	Tuberculosis	2	0	0	1	3
030-041	Non-zoonotic bacterial disease	43	3	0	1	47
042-044	Human immunodeficiency virus infection	21	0	4	1	26
046-049	Viral disease of the CNS (non-polio)	3	1	0	0	4
053-054	Herpes (zoster & simplex)	3	0	0	0	3
070, 079	Viral hepatitis & viral infection of unspecified site	9	1	0	0	10
083	Non-tick-borne rickettsioses	1	0	0	0	1
110-118	Mycoses	10	0	0	0	10
135-136	Sarcoidosis & other & unspecified infectious disease	6	0	0	0	6
137-139	Late effects of infectious & parasitic disease	1	1	0	0	2
140-208	**Malignant neoplasms (MN)**	**2677**	**197**	**26**	**3**	**2903**
140-145	MN of lip & oral cavity	24	4	0	0	28
146-149.0	MN of pharynx	16	2	1	0	19
150	MN of esophagus	67	2	1	0	70
151	MN of stomach	85	4	0	0	89
152-153	MN of small intestine & colon	215	13	2	0	230
154	MN of rectum & anus	44	2	1	0	47
155	MN of liver	36	4	1	1	42
156	MN of gall bladder	21	0	0	0	21
157	MN of pancreas	146	7	0	0	153
158	MN of peritoneum	5	0	0	0	5
159	MN of other & ill-defined digestive	9	1	0	0	10
160	MN of nasal cavities	6	0	0	0	6
161	MN larynx	25	2	0	0	27
162	MN trachea, bronchus & lung	863	36	8	1	908
163	MN pleura	9	0	0	0	9
164	MN thymus, heart & mediastinum	8	0	0	0	8
170	MN bone & articular cartilage	3	1	0	0	4
171	MN connective & soft tissue	14	1	0	0	15

Table 4-4. Distribution of underlying COD (ICD-9 codes) in the INEEL cohort, by sex and race.

ICD-9 code	Description	WM[1] deaths	WF[1] deaths	NWM[1] deaths	NWF[1] deaths	Total deaths
172	Malignant melanoma of skin	53	1	0	0	54
173	Other MN of skin	10	0	0	0	10
174-175	MN breast	5	49	0	0	54
180	MN uterine cervix	0	6	0	0	6
179, 182	MN other parts of uterus or part unspecified	0	6	0	0	6
183	MN ovary	0	8	0	0	8
185	MN prostate	281	0	4	0	285
186	MN testis	3	0	0	0	3
187	MN other male genital	2	0	0	0	2
188	MN bladder	70	2	1	0	73
189	MN kidney & other urinary	79	2	0	0	81
191	MN brain	87	11	0	0	98
192	MN other & unspecified parts of nervous system	3	0	0	0	3
193	MN thyroid gland	5	0	0	0	5
194	MN other endocrine glands	3	0	0	0	3
195	MN other & ill-defined sites	17	1	0	0	18
197	Secondary MN of liver	1	0	0	0	1
199	MN without specification of site	187	9	3	0	199
200, 202	Lymphosarcoma, reticulosarcoma & other MN of lymphoid tissue (non-Hodgkin lymphoma)	121	9	0	0	130
201	Hodgkin's disease	10	1	0	0	11
203.0	Multiple myeloma	37	3	1	0	41
204-208, ex 204.1	Non-CLL leukemia	88	10	3	1	102
204.1	Chronic lymphocytic leukemia	19	0	0	0	19
210-229	**Benign neoplasms (BN)**	7	0	0	0	7
211	BN of digestive system (non-oral cavity)	2	0	0	0	2
216	BN of skin	1	0	0	0	1
225	BN of nervous system	2	0	0	0	2
227.3	BN of pituitary gland	1	0	0	0	1
228.1	Lymphangioma	1	0	0	0	1
235-239	**Neoplasms of uncertain behavior (NUB) & neoplasms of unspecified nature (NUN)**	30	4	1	0	35
236	NUB of unspecified urinary organ	1	0	0	0	1
238.1	NUB of connective & other soft tissue	1	0	0	0	1

Table 4-4. Distribution of underlying COD (ICD-9 codes) in the INEEL cohort, by sex and race.

ICD-9 code	Description	WM[1] deaths	WF[1] deaths	NWM[1] deaths	NWF[1] deaths	Total deaths
238.4-238.7		10	0	0	0	10
239.1		1	1	0	0	2
239.6		10	3	1	0	14
239.0, 239.2-239.5, 239.7-239.9		7	0	0	0	7
240-259	Diseases of thyroid & other endocrine glands	188	17	2	0	207
242	Thyrotoxicosis	1	0	0	0	1
250	Diabetes mellitus	186	17	2	0	205
259	Other endocrine disorders	1	0	0	0	1
260-269		11	1	0	0	12
270-279	Other metabolic disorders & immunity disorders	32	5	1	0	38
280-289		26	2	0	0	28
284		5	0	0	0	5
285		6	0	0	0	6
286-289		15	2	0	0	17
290-294	Organic psychotic conditions	39	2	0	0	41
290	Senile & presenile organic psychotic conditions	34	2	0	0	36
291	Alcoholic psychoses	5	0	0	0	5
295-299		6	2	0	0	8
300-316	Neurotic disorders, personality disorders, & other non-psychotic mental disorders	60	5	0	0	65
300	Neurotic disorders	1	0	0	0	1
303	Alcohol dependence syndrome	37	1	0	0	38
304-305	Drug dependence & non-dependent use of drugs	13	1	0	0	14
306, 310	Other mental disorders	9	3	0	0	12
318.1		1	0	0	0	1
320-324	Inflammatory diseases of the CNS	5	1	0	0	6
330-337		144	11	0	0	155
331		56	5	0	0	61
332-333		53	2	0	0	55
334		1	0	0	0	1
335.2		34	4	0	0	38
340-349	Other disorders of the CNS	40	5	0	0	45
340-341	Multiple sclerosis & other demyelinating diseases of CNS	13	3	0	0	16

Table 4-4. Distribution of underlying COD (ICD-9 codes) in the INEEL cohort, by sex and race.

ICD-9 code	Description	WM[1] deaths	WF[1] deaths	NWM[1] deaths	NWF[1] deaths	Total deaths
344	Non-infantile paralytic syndromes	4	1	0	0	5
345	Epilepsy	7	0	0	0	7
348-349	Other & unspecified conditions of brain & nervous system	16	1	0	0	17
356-359		10	0	0	0	10
391	Rheumatic fever with heart involvement	2	1	0	0	3
393-398		46	7	0	0	53
401-404	Hypertensive disease	100	9	1	0	110
401	Essential hypertension	15	1	0	0	16
402	Hypertensive heart disease	59	6	0	0	65
403	Hypertensive renal disease	15	1	1	0	17
404	Hypertensive heart & renal disease	11	1	0	0	12
410-414		2426	63	16	1	2506
410		1483	46	10	1	1540
411-414		943	17	6	0	966
415-416	Acute & chronic pulmonary heart disease	54	5	1	0	60
420-429		642	33	5	2	682
420-424		58	5	0	0	63
425		71	2	1	1	75
426		4	1	0	0	5
427		158	10	1	1	170
428		110	8	1	0	119
429		241	7	2	0	250
430-438	Cerebrovascular diseases	463	41	6	1	511
430-432	Subarachnoid & intracran. hemorrhage	124	11	4	1	140
433-434	Occlusion & stenosis of pre-cerebral & cerebral arteries	63	6	0	0	69
436-438	Other & ill-defined cerebrovascular disease	276	24	2	0	302
440-447		209	7	1	0	217
440		62	0	1	0	63
441-442		121	4	0	0	125
443-447		26	3	0	0	29
451-459	Other diseases of circulatory system	15	0	1	1	17
460-478		2	0	0	0	2
480-487	Pneumonia & influenza	211	6	1	0	218

Table 4-4. Distribution of underlying COD (ICD-9 codes) in the INEEL cohort, by sex and race.

ICD-9 code	Description	WM[1] deaths	WF[1] deaths	NWM[1] deaths	NWF[1] deaths	Total deaths
490-496		623	22	1	0	646
490-491		23	1	0	0	24
492		158	2	0	0	160
493-495		27	2	0	0	29
496		415	17	1	0	433
500-508	Pneumoconioses & other lung diseases due to external agents	33	1	0	0	34
501	Asbestosis	10	0	0	0	10
502	Pneumoconiosis due to silica or silicates	2	0	0	0	2
507	Pneumoconiosis due to solids & liquids	21	1	0	0	22
510-519		80	6	0	0	86
530-537	Diseases of esophagus, stomach & duodenum	49	3	0	0	52
540-542		3	0	0	0	3
550-553	Hernia of abdominal cavity	3	0	0	0	3
555-569		82	6	1	0	89
570-579	Other diseases of digestive system	244	14	2	1	261
570	Acute & subacute necrosis of liver	2	0	0	0	2
571	Chronic liver disease & cirrhosis	170	8	2	1	181
572-573	Other disorders of liver	23	1	0	0	24
574-576	Disorders of gall bladder & biliary tract	13	0	0	0	13
577	Diseases of pancreas	18	4	0	0	22
578-579	Other diseases of digestive system	18	1	0	0	19
580-589		76	4	0	0	80
590-599	Other diseases of urinary system	24	5	0	0	29
600-602		4	0	0	0	4
682-686	Infections of skin & subcutaneous tissue	5	0	0	0	5
692		1	0	0	0	1
707	Chronic ulcer of skin	2	0	0	0	2
710-716		13	5	1	0	19
710.0		1	4	0	0	5
710.1-710.9		4	0	1	0	5
714		5	1	0	0	6
715-716		3	0	0	0	3
730, 733	Osteopathies	2	0	0	0	2

Table 4-4. Distribution of underlying COD (ICD-9 codes) in the INEEL cohort, by sex and race.

ICD-9 code	Description	WM[1] deaths	WF[1] deaths	NWM[1] deaths	NWF[1] deaths	Total deaths
740-759		12	1	0	0	13
780-789	Symptoms	7	0	0	0	7
797-799		93	8	2	0	103
800-807	Railway accidents	2	0	0	0	2
810-825		322	33	10	4	369
826-829	Other road vehicle accidents	1	1	0	0	2
830-838		19	0	1	0	20
840-844	Air transport accidents	57	1	0	0	58
846-848		2	0	0	0	2
850-869	Accidental poisoning	33	2	1	1	37
870-876		0	1	0	0	1
878-879	Surgical & medical procedures, without mention of misadventure	15	0	0	0	15
880-888		76	2	0	0	78
890-899	Accidents caused by fire	16	0	0	0	16
900-909		14	1	0	0	15
910-913	Accidents caused by submersion & suffocation	47	1	1	0	49
916-929		107	2	3	0	112
947	Drugs & medicaments causing adverse effects in medical use	1	1	0	0	2
950-959		294	13	4	1	312
960-969	Homicide	44	7	2	0	53
980-989		15	3	1	0	19
999	COD not available	242 (2.4%)	9 (1.5%)	1 (1.0%)	0	252 (2.3%)
		10,210	582	97	17	10,906

[1]WM: white males; WF: white females; NWM: Non-white males; NWF: Non-white females.

4.3 Covariates

Among all workers in the study, slightly more were from the local region (Idaho, Wyoming, Utah and Montana) than migrants from elsewhere (Table 4-5). SES could be assigned based on job title to nearly 80% of the cohort. Of those with known SES, the greatest percentage was assigned to the skilled manual worker category, followed by professionals, intermediate, skilled non-manual, partly skilled and unskilled workers. A small percentage of the cohort had an unknown employer type; the most common known type of employer was the subcontractor grouping, followed by workers who were employed by both "prime" (including DOE, NRF and ANL-W) and subcontractors, and lastly the group of "prime" contractor employees.

There were large differences among men and women and among whites and non-whites in the distribution of some of the attributes and work experiences of the cohort (Table 4-5). WF were much less likely and NWM much more likely to have been migrants to the INEEL region, compared to other groups in the cohort. Males were approximately three times as likely to have been classified as professional compared to females and were much less likely to have been skilled non-manual workers. Over half the entire group of women in the cohort, in contrast, was classified as skilled non-manual workers. The most prevalent socioeconomic group for WM was skilled manual employment, at approximately 25%. NWM consisted of a similar preponderance of professional, intermediate and skilled non-manual workers as WM; however, the proportion of skilled manual workers was about half that of WM. Female workers were less likely to have been of unknown SES than males, perhaps because of the improvements in record keeping and record availability in more recent time periods, during which women were more likely to have been employed.

WM were much more likely to have been employed by a subcontractor type than were other groups. WF, NWM and NWF were more likely to have been employed by a "prime" contractor (including NRF, DOE and ANL-W) than were WM. Duration of employment differed greatly by employer category. Prime employees worked three to four times as long on average as subcontractors and unknown contractor types, and persons who worked for multiple types of employers worked five to eleven times as long as subcontractors and unknown employers (Table 4-6).

4.4 Non-radiological Exposures (Subcohorts)

Nearly half the workers in the INEEL cohort were employed at some point as a construction or service work contractor, as identified by job title and contractor codes (Table 4-7). This observation primarily reflected the experiences of white male workers, over half of whom were ever employed as a construction or service worker. Only one-third of NWM and one-fifth of female INEEL employees were ever employed in this type of job.

Over 5000 workers (about 8% of cohort) ever were employed as a chemical worker, based on job title and contractor category (Table 4-7). A greater percentage of females than males was employed in the chemical worker category. In contrast, WM were five to ten times as likely to have been identified in a job related to asbestos exposure, compared to females and NWM. Almost 2700 WM were identified as asbestos workers in this cohort.

Truck and bus drivers were approximately 3% of the workers with known job descriptions, and WM were again more likely than others to have held these jobs (Table 4-7). Reactor workers were a fairly small group within the cohort; 2.6% of males and 0.5% of females ever held jobs of this type. About 2% of the cohort worked in a job related to security, and percentages were only slightly greater among males than females. Lastly, only a very small percentage of the workers in the cohort were painters, with males much more likely than females to have ever been employed in this category.

4.5 Radiological Exposures

Over half the cohort was monitored for external ionizing radiation exposure (including photons and neutrons) at the INEEL facility (Table 4-8). About 250 additional people were monitored only off-site for radiation exposure (presumably, from previous employment at another nuclear facility). Approximately one-third of the cohort was ever monitored for exposure to transuranic radionuclides, tritium, or fission products. Men were more likely to have been monitored for both external and internal exposure than were women, and whites were slightly more likely than non-whites to have been monitored for these exposures.

The percentage of the cohort with non-zero cumulative dose, as well as average cumulative dose, was fairly low in this cohort: just over one-third of the total cohort had a positive on-site dose (Table 4-8). This percentage was much higher for men than for women and for whites compared to non-whites. Although the percentages having a positive dose were slightly higher for on-site and off-site doses combined, patterns were very similar by sex and race.

The age distribution (as of the DLO) differed somewhat among the various dose categories used in the life table and Poisson regression analysis (Figure 4-1). The attained ages tended to be higher among those receiving more than 1 mSv of dose, compared to the unmonitored and those receiving no dose.

Among those monitored at the INEEL, the cohort-wide mean on-site and total doses were 12.7 mSv and 14.5 mSv, respectively. Men received about ten times the average dose of women, and WM had higher average doses than NWM (Table 4-8). The median dose for both WF and NWF was zero. When considering only those who ever had a positive dose, the mean on-site and total doses were 19.8 and 21.5, respectively. Patterns were similar by sex and race.

Radiation exposures also varied by migrant status (to the INEEL region) and SES. Migrants in general had lower external doses (both on-site and off-site) than locals did (Figure 4-2a). Those in the skilled manual and intermediate employment categories had the highest cumulative external doses, followed by partly skilled, unskilled and professional workers. Skilled non-manual workers had the lowest average doses (Figure 4-2a). This finding may be the reason that female workers had much lower doses than males did, since they were much more likely to have been employed in the skilled non-manual category.

The differences in dose by migrant category were largely attenuated after adjusting for duration of employment (Figure 4-2b); that is, local workers in general worked much longer periods of time than migrants to the region. The differences in cumulative dose by SES category persisted after adjustment for duration, especially for local workers. Similar observations were made when considering cumulative on-site and off-site doses combined by SES and migrant status, except that in each category average doses increased by approximately two mSv (data not shown).

Table 4-5. Distribution of covariates within the INEEL cohort: demographic variables and employer type.

Covariate	Value	Total	White Males	White Females	Non-White Males	Non-White Females
Migrant	From ID, UT, WY, MT	31,010 (48.8%)	23,638 (46.6%)	6561 (61.3%)	476 (33.1%)	335 (51.8%)
	From elsewhere	29,425 (46.3%)	24,316 (47.9%)	3834 (35.8%)	963 (66.9%)	312 (48.2%)
	Unknown	3126 (4.9%)	2821 (5.6%)	305 (2.9%)	0	0
SES	Professional	10,245 (16.1%)	9170 (18.1%)	735 (6.9%)	282 (19.6%)	58 (8.9%)
	Intermediate	9693 (15.3%)	7875 (15.5%)	1423 (13.3%)	280 (19.5%)	115 (17.8%)
	Skilled non-manual	7771 (12.2%)	1693 (3.3%)	5683 (53.1%)	60 (4.2%)	335 (51.8%)
	Skilled manual	13,145 (20.7%)	12,830 (25.3%)	134 (1.3%)	176 (12.2%)	5 (0.8%)
	Partly skilled	4912 (7.7%)	4268 (8.4%)	525 (4.9%)	103 (7.2%)	16 (2.5%)
	Unskilled	4344 (6.8%)	3873 (7.6%)	315 (2.9%)	136 (9.5%)	20 (3.1%)
	Unknown	13,451 (21.2%)	11,066 (21.8%)	1885 (17.6%)	402 (27.9%)	98 (15.2%)
Contractor types	Prime (DOE, NRF, ANL-W or prime site contractor)	16,036 (25.2%)	10,573 (20.8%)	4590 (42.9%)	546 (37.9%)	327 (50.5%)
	Subcontractor	24,696 (38.9%)	21,853 (43.0%)	2340 (21.9%)	429 (29.8%)	74 (11.4%)
	Multiple types	21,178 (33.3%)	16,918 (33.3%)	3618 (33.8%)	413 (28.7%)	229 (35.4%)
	Unknown	1651 (2.6%)	1431 (2.8%)	152 (1.4%)	51 (3.5%)	17 (2.6%)

Table 4-6. Average duration employed (in years) among contractor types, of workers with non-missing DOH and DOT.

Employer category	Males			Females		
	N	Mean duration	Median duration	N	Mean duration	Median duration
Subcontractor	21,650	3.02	1.02	2318	2.23	1.20
Prime contractor	11,078	8.86	5.54	4884	6.19	3.49
Unknown contractor	1074	1.86	1.00	92	1.62	1.60
Multiple contractors	17,293	13.3	11.0	3840	9.09	6.06

Table 4-7. Distribution of covariates within the INEEL cohort: exposure-related variables.

Covariate	Value	Total	White Males	White Females	Non-White Males	Non-White Females
Construction & service workers	Ever	29,631 (46.6%)	26,880 (52.9%)	2199 (20.6%)	464 (32.2%)	88 (13.6%)
	Never	32,647 (51.4%)	22,835 (45.0%)	8343 (78.0%)	927 (64.4%)	542 (83.8%)
	Unknown	1283 (2.0%)	1060 (2.1%)	158 (1.5%)	48 (3.3%)	17 (2.6%)
Chemical workers	Ever	5332 (8.4%)	3876 (7.6%)	1234 (11.5%)	135 (9.4%)	87 (13.5%)
	Never	57,915 (91.1%)	46,630 (91.8%)	9443 (88.3%)	1287 (89.4%)	542 (83.8%)
	Unknown	314 (0.49%)	269 (0.5%)	23 (0.2%)	17 (1.2%)	88 (13.6%)
Asbestos workers	Ever	2741 (4.3%)	2691 (5.3%)	26 (0.2%)	23 (1.6%)	1 (0.2%)
	Never	59,537 (93.7%)	47,024 (92.6%)	10,516 (98.3%)	1368 (95.1%)	629 (97.2%)
	Unknown	1283 (2.0%)	1060 (2.1%)	158 (1.5%)	48 (3.3%)	17 (2.6%)
Drivers	Ever	1947 (3.1%)	1850 (3.6%)	77 (0.7%)	20 (1.4%)	0
	Never	48,163 (75.8%)	37,859 (74.6%)	8738 (81.7%)	1017 (70.7%)	549 (84.9%)
	Unknown	13,451 (21.2%)	11,066 (21.8%)	1885 (17.6%)	402 (27.9%)	98 (15.2%)
Reactor workers	Ever	1440 (2.3%)	1346 (2.7%)	55 (0.5%)	37 (2.6%)	2 (0.3%)
	Never	48,670 (76.6%)	38,363 (75.6%)	8760 (81.9%)	1000 (69.5%)	547 (84.5%)
	Unknown	13,451 (21.2%)	11,066 (21.8%)	1885 (17.6%)	402 (27.9%)	98 (15.2%)
Security workers	Ever	1276 (2.0%)	1110 (2.2%)	133 (1.2%)	27 (1.9%)	6 (0.9%)
	Never	48,834 (76.8%)	38,599 (76.0%)	8682 (81.1%)	1010 (70.2%)	543 (83.9%)
	Unknown	13,451 (21.2%)	11066 (21.8%)	1885 (17.6%)	402 (27.9%)	98 (15.2%)
Painters	Ever	690 (1.1%)	671 (1.3%)	8 (0.1%)	11 (0.8%)	0
	Never	61,588 (96.9%)	49,044 (96.6%)	10,534 (98.5%)	1380 (95.9%)	630 (97.4%)
	Unknown	1283 (2.0%)	1060 (2.1%)	158 (1.5%)	48 (3.3%)	17 (2.6%)

Table 4-8. Dosimetry characteristics of INEEL cohort.					
	Total cohort	WM*	WF*	NWM*	NWF*
Monitored external, on-site (%)	36,290 (57.1%)	29,691 (58.5%)	5520 (51.6%)	805 (55.9%)	274 (42.4%)
Positive external, on-site (%)	23,280 (36.6%)	20,333 (40.0%)	2378 (22.2%)	467 (32.5%)	102 (15.8%)
Monitored external, on-site or off-site (%)	36,531 (57.5%)	29,896 (58.9%)	5550 (51.9%)	809 (56.2%)	276 (42.7%)
Positive external, on-site or off-site (%)	24,532 (38.6%)	21,444 (42.2%)	2480 (23.2%)	502 (34.9%)	106 (16.4%)
Monitored internal (%)	22,006 (34.6%)	17,940 (35.3%)	3399 (31.8%)	492 (34.2%)	175 (27.0%)
Monitored internal, some exposure (%)	9426 (14.8%)	8312 (16.4%)	977 (9.1%)	107 (7.4%)	30 (4.6%)
On-site doses only (in mSv)					
Among ever-monitored:					
Mean dose	12.72	15.12	1.29	6.87	0.69
Median dose	0.41	0.69	0.00	0.20	0.00
75th percentile	5.32	8.06	0.40	2.48	0.30
90th percentile	29.69	37.27	1.80	16.30	2.02
95th percentile	68.40	80.35	4.93	43.02	5.08
Among monitored with positive dose:					
Mean dose	19.84	22.08	2.99	11.83	1.86
Median dose	2.41	3.15	0.55	1.80	0.55
75th percentile	14.70	17.64	1.68	8.91	2.25
90th percentile	52.35	60.00	5.88	36.63	5.56
95th percentile	98.77	109.48	12.28	64.80	8.02
On-site and off-site doses combined					
Among ever-monitored:					
Mean dose	14.45	17.17	1.43	7.63	0.70
Median dose	0.58	0.99	0.00	0.32	0.00
75th percentile	7.20	10.84	0.45	3.47	0.31
90th percentile	35.75	45.23	1.95	18.80	2.20
95th percentile	76.29	89.40	5.40	47.10	5.08
Among monitored with positive dose:					
Mean dose	21.51	23.94	3.21	12.30	1.83
Median dose	2.90	3.80	0.58	1.88	0.55
75th percentile	17.17	20.18	1.71	9.28	2.25
90th percentile	58.15	65.40	6.35	35.39	5.56
95th percentile	106.70	116.63	13.31	66.91	8.02

*WM, White males; WF, White females; NWM, Non-White males; NWF, Non-White females.

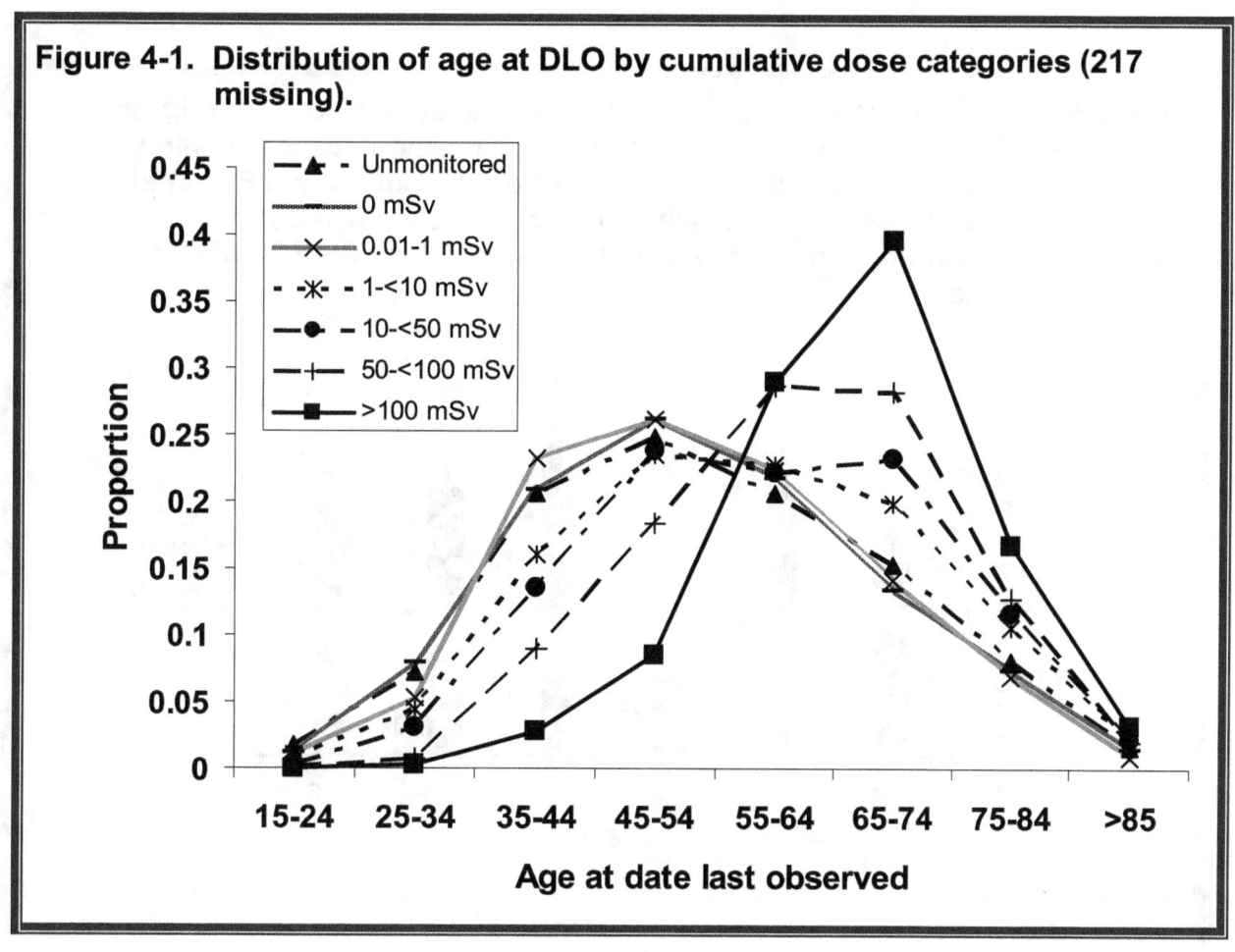

Figure 4-1. Distribution of age at DLO by cumulative dose categories (217 missing).

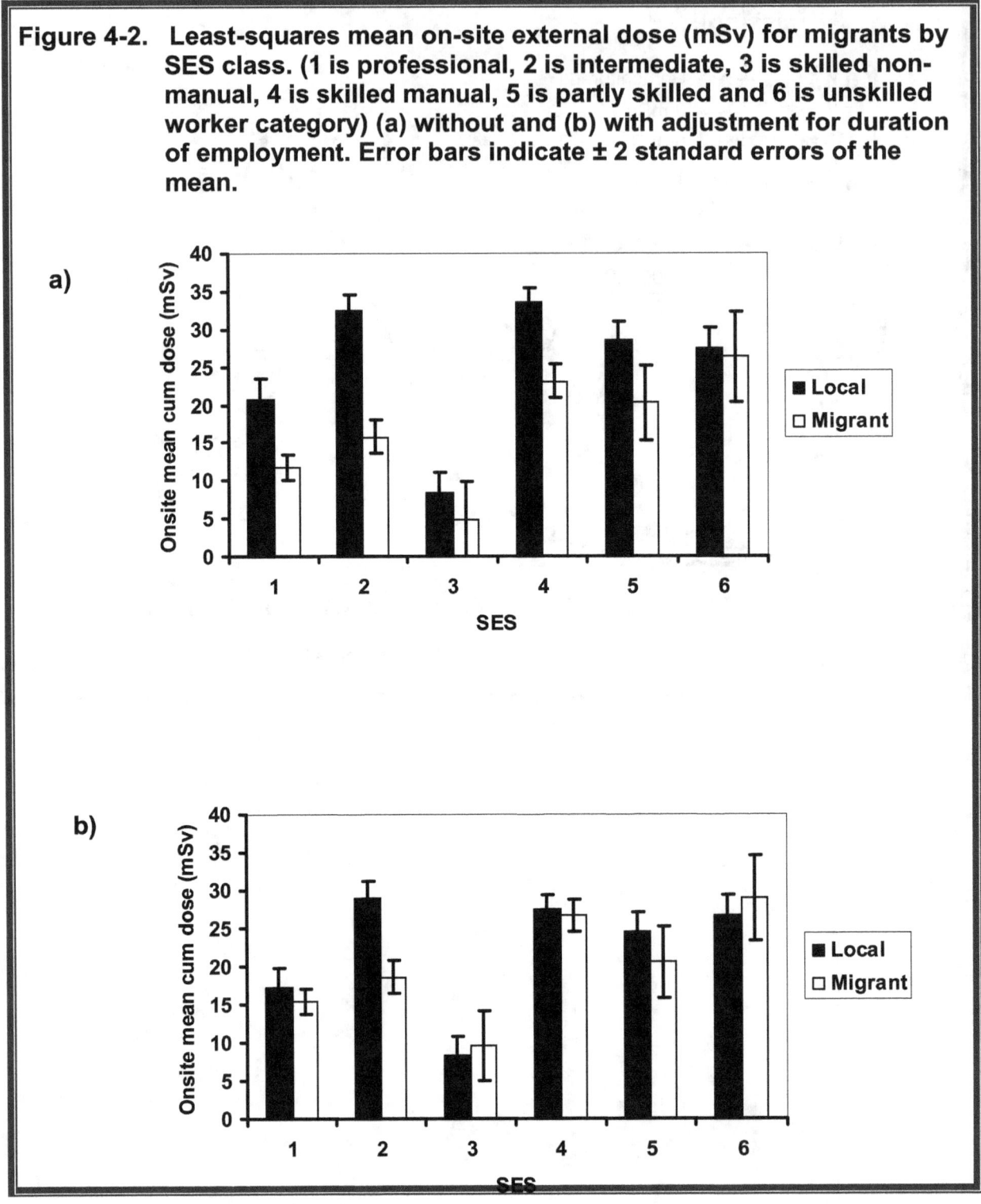

Figure 4-2. Least-squares mean on-site external dose (mSv) for migrants by SES class. (1 is professional, 2 is intermediate, 3 is skilled non-manual, 4 is skilled manual, 5 is partly skilled and 6 is unskilled worker category) (a) without and (b) with adjustment for duration of employment. Error bars indicate ± 2 standard errors of the mean.

5 Life Table Analysis Results

5.1 Required Information for Analysis

The life table requires for each individual several known dates for analysis, including DOB, entry date into the cohort (the DOH for each worker), and DLO. Of the 63,561 workers otherwise eligible for analysis, required information was available for a total of 63,129 workers. The numbers of workers missing the required information are shown by date type in Table 5-1. Thus, 432 otherwise-eligible workers were excluded from analysis. As described in §2.1, a number of additional workers were excluded from analyses involving regional population rates because these rates do not begin until 1960 and the corresponding worker's DLO was before 1960. Many of these workers were also excluded because they lacked a DLO, DOB or DOH.

Table 5-1. Distribution of INEEL cohort members by presence of dates required for life table analysis.

	Valid DOB			
	Yes		No	
	DLO*=Yes	DLO=No	DLO=Yes	DLO=No
Hire date=Yes	63,129	30	123	5
Hire date=No	215	59	0	0

* DLO: date last observed.

The demographic and employment characteristics of cohort members who were ineligible for inclusion in the life table analysis are shown in Table 5-2. The ineligible cohort members were similar to the eligible members in percentage deceased, distribution by race/ethnicity and sex, year of first positive dose, and median duration of employment. They were also older at the last observed date (having earlier years of birth), were less likely to have had long durations of employment and to have been monitored for exposure to radiation, and had much lower on-site dose than eligible workers. In addition, they were much more likely to have been lost to follow-up. The distribution of first DOH among the workers excluded from the analysis is skewed by the large proportion (63%) of workers missing hire dates, which was much more likely to have occurred among workers born in the early decades of the 20th century.

Table 5-2. Characteristics of workers ineligible for life table analysis, compared to eligible cohort.

	Workers ineligible	Workers eligible
Total Number	432	63,129
Male (%)	85.0%	82.1%
Female (%)	14.8%	17.9%
Unknown (%)	0.2%	0.02%
White (%)	93.4%	96.1%
Asian (%)	3.5%	2.2%
Black or African-American (%)	3.0%	1.2%
Other or Unknown (%)	0	0.5%
Year of Birth: median	1928	1940
25th-75th percentile	1918-1939	1928-1953
N missing	128 (29.6%)	0
Age at DLO: median	65.5	54.3
25th-75th percentile	56.7-74.3	44.2-65.9
N missing	222 (51.4%)	0
Year of hire: median	1980	1973
25th-75th percentile	1964-1989	1960-1984
N missing	274 (63.4%)	0
Duration employed (years): median	2.96	3.20
25th-75th percentile	0.39-3.02	0.65-11.0
N missing	313 (72.5%)	1019
Radiation-monitored		
On-site (%)	21.5%	57.3%
On-site or off-site (%)	25.9%	57.3%
Year first monitored on-site: median	1968	1974
25th-75th percentile	1964-1988	1960-1984
Year first positive dose: median	1972	1972
25th-75th percentile	1965-1981	1959-1982
On-site dose: Mean (mSv)	1.30	12.8
Collective (Sv)	0.12	461.5
On-site & off-site dose: Mean (mSv)	6.58	14.5
Collective (Sv)	0.74	527.3
Alive	208 (48.2%)	48,613 (77.0%)
Deceased	92 (21.3%)	10,814 (17.1%)
Unknown	132 (30.6%)	3702 (5.9%)

5.2 Full Cohort Analysis

5.2.1 Comparison to General U.S. Population

5.2.1.1 Combined Cohort, 1949-1999

A total of 1,432,653 person-years were accrued among INEEL cohort members over the entire follow-up period. The all-cause mortality rate, as well as mortality rates for most individual causes of death, was lower in the INEEL cohort than in the general U.S. population (Table 5-3). The all-cancer SMR was substantially below unity in the INEEL population, as well [0.81, 95% CI 0.78-0.84]. Individual cancer mortality rates tended to be much lower than in the general population, for example, cancers of the oral cavity, digestive system and respiratory system (Table 5-3). Exceptions were cancers of the peritoneum and "other respiratory" (including pleura), which were slightly or substantially elevated compared to the U.S. population. Death rates for many of the cancers of *a priori* interest with respect to radiation exposure (e.g., leukemia, bone, breast and thyroid) were lower than in the U.S. population.

Mortality rates for most non-malignant causes of death were also lower in the full cohort than among the age-, calendar year-, race- and sex-adjusted U.S. population (Table 5-3). For example, the SMR for deaths from heart disease was 0.68 (CI, 0.66-0.70) and for cerebrovascular disease was 0.74 (CI, 0.68-0.81). Deaths from pneumoconiosis and other non-malignant respiratory diseases and emphysema were slightly higher than in the general U.S. population. Mortality rates were much lower among the cohort for most lifestyle-related causes of death, including alcoholism and cirrhosis of liver (SMRs=0.65 and 0.55, respectively). The number of accidental deaths in the cohort was lower than expected, except for transportation accidents, which were slightly higher than expected. The rate of suicide was similar to the U.S. population, but the homicide rate was about half that in the U.S. population.

5.2.1.2 White Males, 1949-1999

White male workers contributed 1,173,851 person-years (82% of the total) to the study, over the entire follow-up period. The mortality experience of WM was very similar to results seen in the combined cohort overall, as expected since white male deaths constituted 94% of the total deaths in the cohort. For WM, the all-cause SMR was 0.79 (CI, 0.78-0.81) and the all-cancer SMR was 0.82 (CI, 0.79-0.85). Most individual cancers had SMR estimates well below 1.00, with the exception of cancers of the peritoneum (SMR=1.11; CI, 0.59-1.90), respiratory tract other than larynx, trachea, bronchus and lung (SMR=1.85; CI, 1.16-2.80), male breast (SMR=1.27; CI, 0.41-2.98), prostate (SMR=1.11; CI, 0.98-1.25), and brain (SMR=0.96; CI, 0.77-1.18).

The SMR for heart disease was even lower than all cancers among WM, at 0.69 (CI, 0.66-0.71). The SMRs of lifestyle-related disease, for example, alcoholism and cirrhosis of the liver were similarly low, at 0.69 (CI, 0.49-0.96) and 0.56 (CI, 0.48-0.65) respectively. In contrast, the SMRs for emphysema and for pneumoconiosis were elevated among WM (SMRs=1.15; CI, 0.97-1.34 and 1.11; CI, 1.02-1.21, respectively). Among external causes, suicide and accidental death rates were similar to the general U.S. population, and the homicide rate for WM was less than half that of U.S. population. Death from "other and unspecified" causes was elevated, as expected given the inclusion in this category of deaths for which no death certificate could be obtained.

5.2.1.3 White Females, 1949-1999

There were 218,983 person-years contributed by white female workers in the INEEL cohort (15% of the total). Mortality rates for most causes of death were low among WF at INEEL, compared to the U.S. population (Table 5-3). The SMR for all cancers combined was 0.75 (CI, 0.65-0.87) and for all causes was 0.76 (CI, 0.70-0.82). For most individual causes of death, CIs were very wide because of the low number of deaths in each category. Most individual cancers had SMRs below 1.00 as well. Exceptions included cancers of the oral cavity and pharynx, with an SMR of 2.08 (CI, 0.76-4.53).

Cardiovascular mortality rates were quite low for WF, as for WM, with an SMR of 0.60 (CI, 0.50-0.73). Mortality rates for lifestyle-related causes of death, such as alcoholism, cirrhosis of liver and emphysema, were also very low, as was the mortality rate for other non-malignant respiratory diseases (Table 5-3).

Accidental death rates were similar to the general population, with transportation accidents estimated at 38% above U.S. rates (CI on SMR, 0.96-1.92). Rates of suicide, homicide and other and unspecified causes of death were similar to the general U.S. population (Table 5-3).

5.2.1.4 Non-white Males

Non-white male workers contributed 27,410 person-years of follow-up (1.9% of the total). As noted in §4.2, there were very few deaths among NWM, due in part to the low numbers and relatively young age of this group within the cohort. Mortality rates for all causes combined (SMR=0.41; CI, 0.33-0.50) and all cancers combined (SMR=0.53, 0.34-0.77) were far below U.S. rates for NWM. No precisely estimated COD was substantially in excess in this group (Table 5-3), although the SMR for leukemia was estimated at 2.18 (CI, 0.45-6.37, N=3).

Death rates from specific cancers, including those of the digestive and respiratory systems, were substantially lower than the U.S. rates (Table 5-3). Mortality from cardiovascular disease and non-malignant respiratory disease was far below expectation. Similarly, death rates from lifestyle-related causes of death, including alcoholism, cirrhosis of liver and emphysema, were quite low for NWM. The homicide rate was only one-eighth that of the general U.S. population.

5.2.1.5 Non-white Females, 1949-1999

A very small person-year contribution (12,409 or 0.9%) was by non-white female workers in the cohort, and very few deaths occurred in the cohort among NWF (Table 4-4). Just one cancer death was observed, which was well below expectation (SMR=0.11; CI, 0.00-0.60), and the all-cause mortality SMR was also very low in this group at 0.40 (0.22-0.66). No precisely estimated COD was elevated among this group, although the point estimate for the accidental death SMR was high, particularly for transportation accidents (Table 5-3). Most other SMR point estimates were well below one, although they were unreliably estimated as they are based on very small numbers.

5.2.1.6 Comparison to U.S. Population, 1960-1999

A life table analysis was conducted using U.S. rates beginning in 1960 rather than at the start of follow-up for two reasons: to facilitate comparisons to analyses using state rates (which do not begin until 1960) and to more closely evaluate certain causes of death, such as non-Hodgkin lymphoma (NHL), multiple myeloma, malignant melanoma, and certain non-malignant respiratory diseases, such as asbestosis, for which rates do not begin until 1960. A relatively small percentage of overall person-time was lost by excluding those who died or were lost to follow-up before 1960 and excluding all pre-1960 person-time. The person-years for the full cohort decreased to 1,369,273 (a 4.4% decrease). Just 26 deaths were lost by excluding follow-up before 1960, of which five were due to cancer (only three of which were on the "radiogenic cancers" list given in Table 1-4). The summary and sex- and race-specific SMR estimates and CIs were very similar to the results obtained using the 1940-1999 rates but were slightly higher for most cancers (data not shown), because of the fact that a greater proportion of person-time than deaths was excluded by excluding follow-up before 1960. The inferences did not change regarding any causes found to have been in excess or in deficit because of this change in follow-up period.

The SMRs for causes of death not available using the 1940-1999 comparison rates are shown in Table 5-4. SMRs for most causes were estimated at below 1.00, although the NHL SMR was estimated to be very close to unity. The asbestosis rate was much higher than expected (SMR=3.70, CI 1.77-6.81 among WM), with a total of ten cases identified. All deaths from this cause occurred among WM. Deaths from other pneumoconioses were absent from the cohort; however, the "other respiratory disease" SMR was slightly elevated, with a 95% CI excluding one, among WM. This category consists of ICD-9 codes 470-478, 494-499, 504, and 506-519, including some acute respiratory infections, bronchiectasis and extrinsic allergic alveolitis, organic dust pneumoconiosis, and chronic airways obstruction not elsewhere classified. The distribution of the 543 deaths in these ICD codes is shown in Table 4-4. Most of the deaths in this category were attributable to the last category mentioned. Mortality from HIV was particularly below expectation when compared to the general U.S. population.

Table 5-3. Standardized mortality ratios (SMRs) for all workers, combined and by sex and race categories, compared to general U.S. population rates 1940-1999, adjusted for five-year age and calendar period intervals.

Cause of death	Total cohort	White male	White female	Non-white male	Non-white female
			SMR (95% CIs) N		
Tuberculosis	0.12 (0.02-0.35), 3	0.13 (0.03-0.38), 3	0 (0.76 expected)	0 (1.04 expected)	0 (0.12 expected)
MN Buccal cavity	0.57 (0.42-0.76), 47	0.52 (0.37-0.70), 40	2.08 (0.76-4.53), 6	0.47 (0.01-2.60), 1	0 (0.14 expected)
MN Pharynx	0.47 (0.28-0.74), 19	0.42 (0.24-0.69), 16	1.56 (0.19-5.62), 2	0.80 (0.02-4.45), 1	0
MN Digestive	0.77 (0.72-0.84), 658	0.79 (0.73-0.85), 620	0.68 (0.47-0.96), 33	0.36 (0.12-0.84), 5	0 (1.87 expected)
MN Esophagus	0.75 (0.59-0.95), 69	0.76 (0.59-0.97), 66	0.99 (0.12-3.56), 2	0.37 (0.01-2.03), 1	0
MN Stomach	0.79 (0.63-0.97), 87	0.80 (0.64-0.99), 83	0.87 (0.24-2.23), 4	0 (2.27 expected)	0
MN Intestine	0.74 (0.65-0.84), 228	0.75 (0.66-0.86), 213	0.62 (0.33-1.07), 13	0.56 (0.07-2.02), 2	0
MN Rectum	0.67 (0.49-0.89), 45	0.67 (0.48-0.90), 42	0.54 (0.07-1.94), 2	1.30 (0.03-7.22), 1	0
MN Liver & Gall Bladder	0.77 (0.57-1.02), 48	0.78 (0.57-1.04), 44	0.73 (0.15-2.14), 3	0.65 (0.02-3.64), 1	0
MN Liver Unspecified	0.58 (0.32-0.98), 14	0.59 (0.31-1.00), 13	0.81 (0.02-4.49), 1	0 (0.56 expected)	0
MN Pancreas	0.89 (0.75-1.04), 153	0.92 (0.77-1.08), 146	0.65 (0.26-1.33), 7	0 (2.30 expected)	0
MN Peritoneum & Other	1.09 (0.60-1.84), 14	1.11 (0.59-1.90), 13	1.07 (0.03-5.94), 1	0 (0.12 expected)	0
MN Respiratory	0.74 (0.69-0.79), 948	0.75 (0.70-0.80), 901	0.66 (0.46-0.90), 38	0.44 (0.19-0.86), 8	0.65 (0.02-3.59), 1
MN Larynx	0.64 (0.42-0.93), 27	0.62 (0.40-0.92), 25	2.37 (0.29-8.56), 2	0 (1.01 expected)	0 (0.05 expected)
MN Trach., Bronch., Lung	0.73 (0.69-0.78), 899	0.74 (0.69-0.79), 854	0.64 (0.45-0.88), 36	0.47 (0.20-0.92), 8	0.68 (0.02-3.76), 1
MN Other Respiratory	1.73 (1.09-2.62), 22	1.85 (1.16-2.80), 22	0 (0.57 expected)	0 (0.18 expected)	0 (0.02 expected)
MN Breast	0.83 (0.62-1.08), 54	1.27 (0.41-2.98), 5	0.84 (0.62-1.11), 49	0 (0.07 expected)	0 (2.48 expected)
MN Genital	See sex- and race-specific values	1.08 (0.96-1.21), 282	0.63 (0.38-0.97), 20	1.00 (0.27-2.55), 4	0 (1.29 expected)
Prostate (M) or Cervix (F)		1.11 (0.98-1.25), 277	0.72 (0.26-1.56), 6	1.03 (0.28-2.62), 4	0
MN Ovary (F)		—	0.48 (0.21-0.94), 8	—	0
MN Other Genital		0.45 (0.15-1.06), 5	0 (0.88 expected)	0 (0.12 expected)	0
MN Urinary Organs	0.89 (0.76-1.04), 154	0.90 (0.77-1.06), 149	0.62 (0.17-1.59), 4	0.67 (0.02-3.70), 1	0 (0.19 expected)
MN Kidney	0.89 (0.70-1.10), 79	0.92 (0.72-1.15), 77	0.48 (0.06-1.73), 2	0 (0.98 expected)	0
MN Bladder	0.90 (0.71-1.12), 75	0.89 (0.70-1.12), 72	0.88 (0.11-3.19), 2	1.90 (0.05-10.6), 1	0
MN Other & Unspecified Sites	0.89 (0.81-0.99), 409	0.91 (0.82-1.01), 382	0.73 (0.47-1.09), 24	0.55 (0.11-1.61), 3	0 (0.99 expected)
MN Skin	0.79 (0.61-1.02), 63	0.84 (0.64-1.07), 62	0.21 (0.01-1.15), 1	0 (0.38 expected)	0
MN Brain & Other Nervous System	0.98 (0.80-1.19), 101	0.96 (0.77-1.18), 90	1.38 (0.69-2.46), 11	0 (0.74 expected)	0
MN Thyroid	0.78 (0.25-1.83), 5	0.87 (0.28-2.04), 5	0 (0.61 expected)	0 (0.05 expected)	0
MN Bone	0.43 (0.12-1.11), 4	0.36 (0.07-1.04), 3	1.59 (0.04-8.83), 1	0 (0.13 expected)	0

Table 5-3. Standardized mortality ratios (SMRs) for all workers, combined and by sex and race categories, compared to general U.S. population rates 1940-1999, adjusted for five-year age and calendar period intervals.

Cause of death	Total cohort	SMR (95% CIs) N White male	White female	Non-white male	Non-white female
MN Connective Tissue		0.76 (0.42-1.28), 14	0.47 (0.01-2.59), 1	0 (0.33 expected)	0
MN Other & Unspec.		0.97 (0.84-1.11), 208	0.61 (0.29-1.13), 10	0.78 (0.16-2.29), 3	0
MN Lymphatic & Hematopoietic System	0.88 (0.79-0.99), 302	0.88 (0.77-0.99), 275	0.99 (0.63-1.49), 23	0.99 (0.27-2.54), 4	0 (0.75 expected)
Lymphosarcoma & Reticulosarcoma	0.87 (0.58-1.27), 27	0.90 (0.59-1.31), 26	0.60 (0.02-3.35), 1	0 (0.20 expected)	0
Hodgkin's Disease	0.50 (0.25-0.90), 11	0.50 (0.24-0.92), 10	0.61 (0.02-3.41), 1	0 (0.25 expected)	0
Leukemia	0.91 (0.76-1.09), 120	0.88 (0.73-1.07), 107	1.15 (0.55-2.12), 10	2.18 (0.45-6.37), 3	0
Other	0.91 (0.77-1.07), 144	0.91 (0.77-1.08), 132	0.99 (0.49-1.77), 11	0.45 (0.01-2.51), 1	0
Benign & Unspecified Neoplasms		0.97 (0.68-1.34), 37	1.12 (0.30-2.85), 4	1.78 (0.05-9.90), 1	0 (0.15 expected)
All Cancers	0.81 (0.78-0.84), 2878	0.82 (0.79-0.85), 2654	0.75 (0.65-0.87), 197	0.53 (0.34-0.77), 26	0.11 (0.00-0.60), 1
Diabetes Mellitus		0.80 (0.69-0.92), 185	0.77 (0.44-1.25), 16	0.36 (0.04-1.29), 2	0 (1.35 expected)
Diseases of Blood & Blood-Forming Organs	0.63 (0.42-0.91), 28	0.65 (0.43-0.96), 26	0.64 (0.08-2.32), 2	0 (1.09 expected)	0 (0.33 expected)
Non-pernicious & Unspecified Anemias	0.73 (0.36-1.30), 11	0.84 (0.42-1.50), 11	0 (1.14 expected)	0 (0.63 expected)	0
Alcoholism		0.69 (0.49-0.96), 37	0.50 (0.01-2.80), 1	0 (2.90 expected)	0 (0.27 expected)
Other Mental Disorders	0.98 (0.77-1.23), 76	0.98 (0.76-1.24), 68	1.42 (0.61-2.79), 8	0 (2.31 expected)	0 (0.33 expected)
Diseases of Nervous System & Sense Organs		1.08 (0.93-1.24), 196	1.08 (0.63-1.73), 17	0 (2.76 expected)	0 (0.59 expected)
Diseases of Heart	0.68 (0.66-0.70), 3292	0.69 (0.66-0.71), 3158	0.60 (0.50-0.73), 110	0.37 (0.23-0.57), 21	0.41 (0.09-1.21), 3
Ischemic Heart Disease	0.66 (0.64-0.69), 2485	0.67 (0.64-0.70), 2405	0.51 (0.39-0.65), 63	0.50 (0.29-0.82), 16	0.28 (0.01-1.56), 1
Hypertens. with Heart Dis.	0.65 (0.51-0.81), 76	0.66 (0.51-0.83), 69	1.06 (0.43-2.19), 7	0 (5.27 expected)	0 (0.89 expected)
Other Diseases of Circulatory System		0.78 (0.72-0.83), 768	0.81 (0.61-1.06), 54	0.56 (0.27-1.03), 10	0.56 (0.07-2.01), 2
Hypertens. w/o Heart Dis.		0.79 (0.53-1.12), 30	0.75 (0.09-2.71), 2	0.63 (0.02-3.48), 1	0 (0.29 expected)
Cerebrovascular Disease		0.73 (0.67-0.80), 460	0.89 (0.64-1.21), 41	0.50 (0.18-1.09), 6	0.42 (0.01-2.34), 1
Disease of Respiratory System	0.95 (0.89-1.01), 977	0.98 (0.91-1.04), 940	0.64 (0.44-0.89), 35	0.17 (0.02-0.60), 2	0 (1.82 expected)
Influenza	1.43 (0.71-2.56), 11	1.40 (0.67-2.57), 10	2.09 (0.05-11.6), 1	0 (0.05 expected)	0

Table 5-3. Standardized mortality ratios (SMRs) for all workers, combined and by sex and race categories, compared to general U.S. population rates 1940-1999, adjusted for five-year age and calendar period intervals.

Cause of death	Total cohort	White male	White female	Non-white male	Non-white female
			SMR (95% CIs) N		
Pneumonia	0.65 (0.57-0.75), 207	0.68 (0.59-0.78), 201	0.30 (0.10-0.70), 5	0.18 (0.00-1.03), 1	0
Chron. & Unspec. Bronch.	0.89 (0.57-1.33), 24	0.90 (0.57-1.36), 23	0.79 (0.02-4.41), 1	0 (0.16 expected)	0
Emphysema	1.11 (0.94-1.29), 158	1.15 (0.97-1.34), 156	0.34 (0.04-1.25), 2	0 (0.79 expected)	0
Asthma	0.92 (0.57-1.40), 21	1.01 (0.61-1.58), 19	0.66 (0.08-2.40), 2	0 (0.82 expected)	0
Pneumoconiosis & Other Respiratory Diseases	1.09 (1.00-1.18), 555	1.11 (1.02-1.21), 530	0.88 (0.56-1.30), 24	0.21 (0.01-1.16), 1	0
Diseases of Digestive System	0.67 (0.60-0.74), 378	0.68 (0.43-1.01), 23	0.24 (0.05-0.70), 3	0.50 (0.01-2.77), 1	
Cirrhosis of Liver		0.56 (0.48-0.65), 168	0.52 (0.22-1.02), 8	0.27 (0.03-0.98), 2	0.89 (0.02-4.97), 1
Diseases of Genitourinary System	0.65 (0.54-0.78), 113	0.66 (0.54-0.80), 104	0.80 (0.37-1.53), 9	0 (4.11 expected)	0 (0.82 expected)
Acute Glomerulonephritis & Acute Renal Failure	0.47 (0.20-0.92), 8	0.51 (0.22-1.01), 8	0 (0.97 expected)	0 (0.36 expected)	0
Chronic & Unspec. Nephritis & Renal Failure	0.85 (0.67-1.07), 72	0.88 (0.68-1.12), 68	0.80 (0.22-2.04), 4	0 (2.17 expected)	0
Diseases of Skin		0.83 (0.36-1.63), 8	0 (0.97 expected)	0 (0.31 expected)	0 (0.09 expected)
Diseases of Musculoskeletal & Connective Tissue	0.68 (0.42-1.05), 21	0.61 (0.34-1.00), 15	1.04 (0.34-2.43), 5	1.75 (0.04-9.71), 1	0 (0.50 expected)
Symptoms & Ill-defined Conditions		0.74 (0.60-0.91), 95	0.98 (0.42-1.93), 8	0.42 (0.05-1.51), 2	0 (0.77 expected)
Accidents	0.99 (0.92-1.06), 772	0.98 (0.91-1.06), 706	1.11 (0.81-1.49), 45	0.79 (0.45-1.29), 16	2.05 (0.66-4.79), 5
Transportation Accidents	1.06 (0.96-1.16), 452	1.03 (0.93-1.13), 402	1.38 (0.96-1.92), 35	1.12 (0.56-2.00), 11	3.05 (0.83-7.81), 4
Accidental Falls	0.97 (0.77-1.20), 81	1.00 (0.79-1.25), 79	0.59 (0.07-2.12), 2	0 (1.51 expected)	0 (0.10 expected)
Suicide		0.94 (0.85-1.06), 290	0.78 (0.42-1.34), 13	0.97 (0.26-2.48), 4	2.15 (0.05-12.0), 1
Homicide	0.41 (0.31-0.54), 53	0.43 (0.31-0.57), 44	1.02 (0.41-2.10), 7	0.12 (0.01-0.43), 2	0 (1.72 expected)
Other & Unspecified Causes		1.21 (1.09-1.33), 411	0.99 (0.64-1.46), 25	0.33 (0.13-0.69), 7	0.58 (0.07-2.09), 2
All Deaths	0.78 (0.77-0.80), 10,814	0.79 (0.78-0.81), 10,123	0.76 (0.70-0.82), 579	0.41 (0.33-0.50), 97	0.40 (0.22-0.66), 15

Table 5-4. SMRs for selected causes of death for INEEL cohort, compared to U.S. rates 1960-1999, adjusted for age and calendar period.

Cause of death	Total cohort	White male	SMR (95% CIs), N White female	Non-white male	Non-white female
MN Skin Melanoma	0.87 (0.65-1.14), 53	0.92 (0.69-1.21), 52	0.23 (0.01-1.30), 1	0 (0.09 expected)	0 (0.02 expected)
Other MN of Skin	0.58 (0.28-1.06), 10	0.61 (0.29-1.12), 10	0 (0.52 expected)	0 (0.29 expected)	0 (0.02 expected)
Non-Hodgkin Lymphoma	0.99 (0.83-1.17), 130	1.00 (0.83-1.20), 121	0.98 (0.45-1.86), 9	0 (1.36 expected)	0 (0.23 expected)
Hodgkin's Disease	0.55 (0.27-0.98), 11	0.55 (0.26-1.00), 10	0.64 (0.02-3.56), 1	0 (0.24 expected)	0 (0.05 expected)
Leukemia	0.93 (0.77-1.11), 119	0.90 (0.73-1.08), 106	1.17 (0.56-2.15), 10	2.19 (0.45-6.41), 3	0 (0.30 expected)
Myeloma	0.75 (0.53-1.01), 41	0.74 (0.52-1.01), 37	0.85 (0.18-2.49), 3	0.96 (0.02-5.35), 1	0 (0.18 expected)
Asbestosis	3.67 (1.76-6.75), 10	3.70 (1.77-6.81), 10	0 (0.01 expected)	0 (0.02 expected)	0 (0.00 expected)
Silicosis	1.02 (0.12-3.69), 2	1.04 (0.13-3.76), 2	0 (0.00 expected)	0 (0.03 expected)	0 (0.00 expected)
Other Pneumoconioses	0 (10.3 expected)	0 (10.3 expected)	0 (0.01 expected)	0 (0.02 expected)	0 (0.00 expected)
Other Respiratory Diseases	1.10 (1.01-1.20), 543	1.12 (1.03-1.23), 518	0.88 (0.56-1.31), 24	0.21 (0.01-1.18), 1	0 (0.63 expected)
HIV-Related	0.22 (0.14-0.32), 26	0.21 (0.13-0.32), 21	0 (2.72 expected)	0.31 (0.08-0.80), 4	0.65 (0.02-3.59), 1
All Cancers	0.82 (0.79-0.85), 2873	0.83 (0.80-0.86), 2649	0.76 (0.66-0.87), 197	0.53 (0.34-0.77), 26	0.11 (0.00-0.60), 1
All Deaths	0.80 (0.78-0.81), 10788	0.81 (0.79-0.82), 10097	0.76 (0.70-0.83), 579	0.41 (0.33-0.50), 97	0.40 (0.23-0.67), 15

5.2.2 Comparison to Regional Population (ID, WY, MT combined) 1960-1999

When comparing INEEL death rates to the regional population, SMRs were in general higher than those resulting from comparisons to the U.S. (Table 5-5). The all-cause SMR was still below unity (SMR=0.96; CI, 0.94-0.97) but was much higher than that derived from comparison to the U.S. population. The all-cancer SMR was slightly elevated (SMR=1.07; CI, 1.03-1.11). The findings for respiratory cancer are particularly notable: in the full cohort, the precisely estimated rate of mortality from respiratory cancers was lower than that of the U.S. population but was higher than the regional population.

The elevation in oral cavity cancers among WF was exacerbated when local rates were used for comparison but was based on very small numbers (SMR=2.57, CI 0.94-5.60, N=6). Brain cancer death rates, including neoplasms of unspecified nature (NUN) of nervous system, were also elevated but with wide CIs, among WF (Table 5-5).

Lymphopoietic cancer rates were imprecisely elevated in comparison to the local rates (Table 5-5). Most of this elevation appears attributable to NHL, which was elevated among the full cohort (SMR=1.26, CI 1.05-1.50) and WM (SMR=1.28, CI 1.06-1.53). The SMR for leukemia was also slightly higher when regional rates were used for comparison, and was elevated particularly among NWM, though based on only 3 deaths (SMR=3.98; CI, 0.82-11.6). Non-nervous system benign and unspecified nature neoplasms were elevated among WM (SMR=1.67; CI, 1.07-2.48, N=24). Ten of these deaths were NUN of the lymphatic and hematopoietic system. WF and NWM showed imprecise elevations in NUN of nervous system. The mortality rate for the subcategory of malignant neoplasms of "other and unspecified" sites was elevated (SMR=1.17; CI, 1.02-1.34). This category primarily comprises malignant neoplasms of unspecified site (N=187 out of 208; Table 4-4).

Cancers related to asbestos exposure showed elevated rates among WM in the cohort when compared to regional rates (Table 5-4). The SMR for "other respiratory" cancer was 2.34 (CI, 1.46-3.54) and for cancer of peritoneum was 1.26 (CI, 0.67-2.16). The SMR for asbestosis was substantially elevated, at 3.26 (CI, 1.56-5.99) for WM. This SMR was somewhat lower than that calculated using U.S. population rates, which likely resulted from higher regional rates, rather than a larger contribution of INEEL cohort deaths within the comparison population rates when local state rates were used. Only 3 of the 10 asbestosis deaths among cohort members occurred in Idaho, Montana, or Wyoming.

The number of deaths from nervous system diseases other than multiple sclerosis was greater than expected, particularly among WM (Table 5-5). This category was dominated by Parkinson's and Alzheimer's diseases in the INEEL cohort (Table 4-4). Heart disease death rates as a group were lower than expected, based on regional comparisons; however, hypertensive heart disease rates were elevated, particularly among WM (SMR=1.74, CI 1.36-2.20) and WF (SMR=2.48; CI, 1.00-5.12).

SMRs for accidents were lower than those calculated using U.S. rates. SMRs for both suicide and homicide were substantially below unity, although the SMR for suicide was lower, and for homicide was higher, than those calculated using U.S. rates. The mortality rate from "other and unspecified" causes of death was also substantially elevated at approximately double that of the regional population.

Numbers of several causes of death were much lower than expected among NWM. No deaths from alcoholism were observed. Mortality rates from cirrhosis of liver, alcoholism, accidents, suicide and homicide were one-tenth to one-third of the regional rates (Table 5-5). NWF showed similarly low mortality rates for most causes of death, when compared to regional rates. The overall SMR among this group was 0.38 (CI, 0.21-0.62).

Analysis of mortality by employment duration categories shows that all-cause mortality rates for the full cohort, and specifically for WM and WF, were elevated compared to the regional population for employment durations of 0-2 years. Mortality rates were lower than in the regional population for employment durations of greater than two years, and the all-cause SMR estimates tended to decrease with increasing duration of exposure categories (Table 5-6a).

All-cancer SMRs were also elevated for the total cohort, and for WM and WF, in the 0-2 year exposure category. The all-cancer SMRs also tended to decrease with increasing exposure duration categories (Table 5-6b).

Total brain tumors (MN, BN and NUB) were compared to expected based on regional rates through the creation of a combined rate file, as described in §2.6.2.2. The overall SMR for all brain tumors decreased for WM (Table 5-7) compared to the rate for brain cancers. No evidence was seen of an excess compared to the regional rates for WM or NWM. The SMR was elevated for WF, but with CIs that overlap unity.

SRRs for total brain tumors were elevated among WM who were employed between two and ten years, compared to those who worked less than two years (Table 5-7). However, rate ratios were not elevated for white male workers employed for durations of longer than ten years. For WF, rate ratios were elevated (although with very wide CIs) only for those who worked between twenty and thirty years (Table 5-7).

Table 5-5. Standardized mortality ratios (SMRs) for cohort, combined and by sex and race categories, compared to combined Idaho, Montana and Wyoming population 1960-1999, adjusted for age and calendar period.

Cause of death	Total cohort	SMR (95% CIs), N			
		White male	White female	Non-white male	Non-white female
MN Buccal cavity	0.89 (0.65-1.18), 47	0.80 (0.57-1.09), 40	2.57 (0.94-5.60), 6	2.19 (0.06-12.1), 1	0 (0.07 expected)
MN Pharynx	0.77 (0.46-1.20), 19	0.68 (0.39-1.11), 16	1.86 (0.23-6.73), 2	3.44 (0.09-19.0), 1	0
MN Digestive	1.04 (0.96-1.12), 658	1.06 (0.97-1.14), 620	0.91 (0.63-1.28), 33	0.60 (0.19-1.39), 5	0 (1.23 expected)
MN Esophagus	1.05 (0.82-1.33), 69	1.05 (0.81-1.33), 66	1.36 (0.16-4.90), 2	1.02 (0.03-5.69), 1	0
MN Stomach	1.03 (0.82-1.27), 87	1.04 (0.83-1.29), 83	1.15 (0.31-2.93), 4	0 (1.51 expected)	0
MN Intestine	1.04 (0.91-1.18), 228	1.05 (0.91-1.20), 213	0.86 (0.46-1.47), 13	1.23 (0.15-4.44), 2	0
MN Rectum	0.88 (0.64-1.18), 45	0.88 (0.64-1.19), 42	0.79 (0.10-2.86), 2	1.31 (0.03-7.30), 1	0
MN Liver & Gall Bladder	1.12 (0.83-1.49), 48	1.16 (0.84-1.56), 44	0.99 (0.20-2.89), 3	0.65 (0.02-3.62), 1	0
MN Liver Unspecified	0.92 (0.50-1.55), 14	0.94 (0.50-1.60), 13	1.10 (0.03-6.13), 1	0 (0.37 expected)	0
MN Pancreas	1.08 (0.91-1.26), 153	1.11 (0.93-1.30), 146	0.81 (0.32-1.66), 7	0 (1.47 expected)	0
MN Peritoneum & Other	1.23 (0.67-2.06), 14	1.26 (0.67-2.16), 13	1.12 (0.03-6.21), 1	0 (0.15 expected)	0
MN Respiratory	1.10 (1.04-1.18), 948	1.12 (1.04-1.19), 901	0.94 (0.66-1.29), 38	0.83 (0.36-1.64), 8	0.98 (0.02-5.44), 1
MN Larynx	1.03 (0.68-1.49), 27	0.99 (0.64-1.46), 25	2.81 (0.34-10.2), 2	0 (0.29 expected)	0 (0.01 expected)
MN Trachea, Bronchus, Lung	1.09 (1.02-1.17), 899	1.11 (1.03-1.18), 854	0.92 (0.64-1.27), 36	0.87 (0.38-1.72), 8	0.99 (0.02-5.48), 1
MN Other Respiratory	2.19 (1.37-3.32), 22	2.34 (1.46-3.54), 22	0 (0.50 expected)	0 (0.13 expected)	0 (0.00 expected)
MN Breast	1.06 (0.80-1.38), 54	1.81 (0.59-4.22), 5	1.05 (0.77-1.38), 49	0 (0.00 expected)	0 (1.34 expected)
MN Genital		1.09 (0.97-1.23), 279	0.79 (0.48-1.23), 20	2.26 (0.62-5.79), 4	0 (0.86 expected)
Prostate (M) or Cervix (F)	See sex- and race-specific values	1.12 (0.99-1.26), 277	0.98 (0.36-2.13), 6	2.48 (0.68-6.34), 4	
Testis (M) or Ovary (F)		0.25 (0.03-0.90), 2	0.57 (0.25-1.13), 8	0 (0.15 expected)	0 (0.15 expected)
MN Other Uterine		--	1.35 (0.49-2.93), 6	--	
MN Urinary Organs	1.10 (0.93-1.29), 154	1.12 (0.94-1.31), 149	0.87 (0.24-2.24), 4	0.62 (0.02-3.46), 1	0 (0.15 expected)
MN Kidney	1.09 (0.86-1.36), 79	1.13 (0.89-1.41), 77	0.67 (0.08-2.42), 2	0 (1.22 expected)	0
MN Bladder	1.11 (0.88-1.40), 75	1.10 (0.86-1.39), 72	1.25 (0.15-4.53), 2	2.57 (0.07-14.3), 1	0
MN Other & Unspecified Sites	1.08 (0.98-1.20), 408	1.10 (1.00-1.22), 381	0.86 (0.55-1.28), 24	1.12 (0.23-3.26), 3	0 (0.65 expected)
MN Skin Melanoma	1.08 (0.81-1.41), 53	1.15 (0.86-1.50), 52	0.29 (0.01-1.61), 1	0 (0.19 expected)	0
Other MN of Skin	0.88 (0.42-1.61), 10	0.91 (0.44-1.68), 10	0 (0.39 expected)	0 (0.07 expected)	0
MN Brain & Other NS	1.12 (0.91-1.36), 99	1.08 (0.87-1.33), 88	1.60 (0.80-2.87), 11	0 (0.55 expected)	0
MN Thyroid	1.10 (0.35-2.56), 5	1.23 (0.40-2.88), 5	0 (0.46 expected)	0 (0.03 expected)	0
MN Bone	0.50 (0.14-1.28), 4	0.40 (0.08-1.17), 3	2.05 (0.05-11.4), 1	0 (0.01 expected)	0

Table 5-5. Standardized mortality ratios (SMRs) for cohort, combined and by sex and race categories, compared to combined Idaho, Montana and Wyoming population 1960-1999, adjusted for age and calendar period.

Cause of death	Total cohort	SMR (95% CIs), N White male	White female	Non-white male	Non-white female
MN Connective Tissue	0.84 (0.47-1.39), 15	0.91 (0.49-1.52), 14	0.50 (0.01-2.75), 1	0 (0.30 expected)	0
MN Other & Unspecified	1.14 (0.99-1.30), 222	1.17 (1.02-1.34), 209	0.71 (0.34-1.30), 10	1.95 (0.40-5.71), 3	0
MN Lymphatic & Hematopoietic	1.07 (0.95-1.19), 301	1.05 (0.93-1.18), 274	1.24 (0.78-1.86), 23	1.59 (0.43-4.05), 4	0 (0.46 expected)
Non-Hodgkin Lymphoma	1.26 (1.05-1.50), 130	1.28 (1.06-1.53), 121	1.22 (0.56-2.33), 9	0 (1.04 expected)	0
Hodgkin's Disease	0.64 (0.32-1.14), 11	0.62 (0.30-1.14), 10	0.98 (0.02-5.42), 1	0 (0.10 expected)	0
Leukemia	1.07 (0.89-1.28), 119	1.03 (0.84-1.25), 106	1.38 (0.66-2.53), 10	3.98 (0.82-11.6), 3	0
Myeloma	0.81 (0.58-1.10), 41	0.79 (0.55-1.08), 37	1.02 (0.21-2.98), 3	1.59 (0.04-8.82), 1	0
Benign & Unspecified Neoplasms	1.14 (0.82-1.54), 42	1.11 (0.78-1.53), 37	1.28 (0.35-3.28), 4	8.24 (0.21-45.8), 1	0 (0.20 expected)
Eye, Brain, Other NS Unspecified Nature	0.89 (0.50-1.46), 15	0.71 (0.35-1.27), 11	2.41 (0.50-7.04), 3	14.6 (0.37-80.9), 1	0
Other Benign & Unspec.	1.58 (1.02-2.33), 25	1.67 (1.07-2.48), 24	0.76 (0.02-4.21), 1	0 (0.04 expected)	0
All Cancers	1.07 (1.03-1.11), 2873	1.08 (1.04-1.13), 2649	0.97 (0.84-1.12), 197	0.96 (0.63-1.41), 26	0.17 (0.00-0.96), 1
Diabetes Mellitus	1.02 (0.88-1.17), 203	1.06 (0.91-1.22), 185	0.94 (0.54-1.53), 16	0.34 (0.04-1.22), 2	0 (1.57 expected)
Diseases of Blood & Blood-Forming Organs	0.79 (0.52-1.14), 28	0.80 (0.52-1.17), 26	0.82 (0.10-2.97), 2	0 (0.50 expected)	0 (0.20 expected)
Non-Pernicious & Unspecified Anemias	0.93 (0.46-1.66), 11	1.02 (0.51-1.83), 11	0 (0.84 expected)	0	0
Alcoholism	0.56 (0.40-0.77), 38	0.64 (0.45-0.88), 37	0.49 (0.01-2.71), 1	0 (6.90 expected)	0 (1.31 expected)
Other Mental Disorders	1.11 (0.87-1.39), 76	1.12 (0.87-1.42), 68	1.61 (0.69-3.16), 8	0 (2.31 expected)	0 (0.39 expected)
Diseases of Nervous System & Sense Organs	1.11 (0.96-1.27), 213	1.13 (0.98-1.30), 196	1.04 (0.61-1.67), 17	0 (2.84 expected)	0 (0.58 expected)
Multiple Sclerosis	0.64 (0.34-1.10), 13	0.61 (0.29-1.12), 10	0.83 (0.17-2.44), 3	0	0
Other	1.16 (1.01-1.33), 200	1.19 (1.02-1.37), 186	1.10 (0.60-1.85), 14	0	0
Diseases of Heart	0.91 (0.88-0.94), 3285	0.92 (0.89-0.95), 3151	0.88 (0.72-1.06), 110	0.48 (0.30-0.73), 21	0.72 (0.15-2.09), 3
Ischemic Heart Disease	0.93 (0.89-0.96), 2702	0.94 (0.90-0.97), 2613	0.80 (0.62-1.01), 70	0.56 (0.33-0.88), 18	0.41 (0.01-2.26), 1
Hypertension with Heart Disease	1.77 (1.40-2.22), 76	1.74 (1.36-2.20), 69	2.48 (1.00-5.12), 7	0 (0.42 expected)	0 (0.05 expected)
Other Diseases of Circulatory System	0.92 (0.86-0.99), 831	0.92 (0.86-0.99), 765	0.93 (0.70-1.21), 54	0.92 (0.44-1.70), 10	0.91 (0.11-3.29), 2
Hypertens. w/o Heart Dis.	1.09 (0.75-1.54), 32	1.09 (0.73-1.56), 29	1.02 (0.12-3.67), 2	1.73 (0.04-9.63), 1	0 (0.11 expected)

Table 5-5. Standardized mortality ratios (SMRs) for cohort, combined and by sex and race categories, compared to combined Idaho, Montana and Wyoming population 1960-1999, adjusted for age and calendar period.

Cause of death	Total cohort	SMR (95% CIs), N White male	White female	Non-white male	Non-white female
Cerebrovascular Disease	0.92 (0.84-1.01), 507	0.91 (0.83-1.00), 459	1.05 (0.75-1.42), 41	0.88 (0.32-1.91), 6	0.70 (0.02-3.87), 1
Disease of Respiratory System	0.98 (0.92-1.04), 977	1.01 (0.94-1.07), 940	0.71 (0.49-0.98), 35	0.18 (0.02-0.65), 2	0 (1.68 expected)
Chronic & Unspecified Bronchitis	0.88 (0.56-1.31), 24	0.88 (0.56-1.33), 23	0.94 (0.02-5.20), 1	0 (0.09 expected)	0
Emphysema	1.00 (0.85-1.17), 158	1.03 (0.88-1.21), 156	0.36 (0.04-1.29), 2	0 (0.56 expected)	0
Asthma	0.77 (0.47-1.17), 21	0.79 (0.48-1.23), 19	0.69 (0.08-2.50), 2	0 (0.30 expected)	0
Asbestosis	3.25 (1.56-5.98), 10	3.26 (1.56-5.99), 10	0 (0.00 expected)	0 (0.00 expected)	0
Silicosis	0.58 (0.07-2.09), 2	0.59 (0.07-2.11), 2	0 (0.00 expected)	0 (0.03 expected)	0
Other Respiratory Diseases	1.04 (0.96-1.14), 543	1.06 (0.97-1.16), 518	0.90 (0.58-1.34), 24	0.24 (0.01-1.31), 1	0
Diseases of Digestive System	0.86 (0.78-0.95), 404	0.92 (0.83-1.02), 377	0.81 (0.52-1.22), 23	0.12 (0.03-0.35), 3	0.15 (0.00-0.83), 1
Cirrhosis of Liver	0.81 (0.70-0.94), 179	0.91 (0.78-1.06), 168	0.68 (0.29-1.34), 8	0.11 (0.01-0.40), 2	0.20 (0.01-1.10), 1
Diseases of Genitourinary System	0.93 (0.77-1.12), 113	0.94 (0.76-1.13), 104	1.26 (0.58-2.39), 9	0 (2.61 expected)	0 (0.82 expected)
Acute Glomerulonephritis & Acute Renal Failure	0.67 (0.29-1.32), 8	0.71 (0.31-1.40), 8	0 (0.57 expected)	0	0
Chronic & Unspec. Nephritis & Renal Failure	1.16 (0.91-1.46), 72	1.20 (0.93-1.52), 68	1.19 (0.32-3.04), 4	0	0
Diseases of Skin	1.19 (0.51-2.34), 8	1.40 (0.60-2.76), 8	0 (0.75 expected)	0 (0.27 expected)	0 (0.03 expected)
Symptoms & Ill-def. Cond.	0.88 (0.72-1.06), 105	0.85 (0.69-1.04), 95	1.41 (0.61-2.79), 8	0.95 (0.12-3.44), 2	0 (0.37 expected)
Accidents	0.73 (0.68-0.78), 767	0.74 (0.69-0.80), 701	0.91 (0.67-1.22), 45	0.34 (0.19-0.55), 16	0.61 (0.20-1.41), 5
Transportation Accidents	0.73 (0.67-0.80), 449	0.74 (0.67-0.81), 399	1.00 (0.70-1.39), 35	0.37 (0.18-0.66), 11	0.63 (0.17-1.61), 4
Accidental Falls	0.93 (0.74-1.16), 80	0.98 (0.77-1.22), 78	0.60 (0.07-2.15), 2	0 (2.83 expected)	0 (0.07 expected)
Other Accidents	0.64 (0.55-0.74), 183	0.66 (0.56-0.76), 175	0.62 (0.17-1.58), 4	0.33 (0.09-0.83), 4	0 (1.42 expected)
Suicide	0.76 (0.68-0.85), 307	0.77 (0.68-0.86), 289	0.79 (0.42-1.36), 13	0.38 (0.10-0.98), 4	0.78 (0.02-4.35), 1
Homicide	0.65 (0.48-0.84), 53	0.66 (0.48-0.89), 44	1.18 (0.47-2.44), 7	0.24 (0.03-0.86), 2	0 (1.37 expected)
Other & Unspecified	2.09 (1.89-2.30), 415	2.22 (2.00-2.45), 386	1.46 (0.95-2.16), 25	0.52 (0.11-1.53), 3	0.63 (0.02-3.50), 1
All Deaths	0.96 (0.94-0.97), 10,788	0.97 (0.95-0.99), 10,097	0.94 (0.86-1.02), 579	0.45 (0.37-0.55), 97	0.38 (0.21-0.62), 15

Table 5-6. Standardized mortality ratios by employment duration (compared to regional population of ID, MT, WY 1960-1999) for (a) all deaths combined and (b) cancer.

a) All deaths

SMRs by duration employed	Total cohort	White males	White females	Non-white males	Non-white females
0-<2 yr (95% CI), N	1.09 (1.06-1.12), 4940	1.11 (1.07-1.14), 4654	1.04 (0.91-1.19), 229	0.51 (0.38-0.68), 48	0.49 (0.23-0.97), 9
2-<5 yr (95% CI), N	0.90 (0.85-0.94), 1620	0.91 (0.87-0.96), 1486	0.87 (0.71-1.05), 108	0.54 (0.34-0.80), 24	0.27 (0.03-1.19), 2
5-<10 yr (95% CI), N	0.89 (0.85-0.94), 1365	0.91 (0.86-0.96), 1269	0.88 (0.71-1.09), 87	0.21 (0.08-0.47), 6	0.50 (0.10-1.65), 3
10-<20 yr (95% CI), N	0.85 (0.81-0.90), 1592	0.87 (0.82-0.91), 1485	0.87 (0.70-1.06), 96	0.32 (0.16-0.59), 11	0 (5.11 expected)
20-<30 yr (95% CI), N	0.83 (0.77-0.88), 866	0.82 (0.77-0.88), 810	1.00 (0.74-1.33), 48	0.76 (0.30-1.63), 7	0.36 (0.00-2.90), 1
≥30 yr (95% CI), N	0.80 (.72-0.88), 405	0.81 (0.73-0.89), 393	0.68 (0.34-1.24), 11	0.26 (0.00-2.16), 1	0 (0.23 expected)

b) Cancer*

SMRs by duration employed	Total cohort	White males	White females	Non-white males
0-<2 yr (95% CI), N	1.21 (1.14-1.28), 1273	1.22 (1.15-1.29), 1182	1.13 (0.90-1.41), 81	0.98 (0.47-1.84), 10
2-<5 yr (95% CI), N	1.04 (0.95-1.14), 435	1.06 (0.96-1.18), 394	0.87 (0.61-1.21), 36	0.96 (0.31-2.38), 5
5-<10 yr (95% CI), N	1.02 (0.92-1.13), 354	1.05 (0.94-1.17), 326	0.80 (0.53-1.19), 26	0.29 (0.00-2.38), 1
10-<20 yr (95% CI), N	0.96 (0.87-1.06), 415	0.97 (0.88-1.08), 378	0.88 (0.60-1.26), 31	1.11 (0.40-2.52), 6
20-<30 yr (95% CI), N	0.90 (0.79-1.02), 246	0.87 (0.76-1.00), 224	1.22 (0.73-1.92), 19	1.68 (0.34-5.49), 3
≥30 yr (95% CI), N	0.97 (0.82-1.13), 150	0.98 (0.83-1.15), 145	0.66 (0.18-1.83), 4	1.03 (0.01-8.41), 1

*Only one cancer death occurred among non-white females.

Table 5-7. Standardized mortality ratio (compared to regional population) and standardized rate ratio (for employment duration categories) for combined tumors of brain, including malignant, benign and tumors of unspecified nature*.

Rate ratios by duration employed	Total cohort	White males	White females
Overall SMR (95% CI), N	0.98 (0.80-1.18), 110	0.93 (0.75-1.13), 95	1.51 (0.82-2.53), 14
SMR 0-2 yr (95% CI), N	0.77 (0.53-1.08), 34	0.69 (0.46-1.01), 28	1.43 (0.46-3.54), 5
SRR 2-5 yr (95% CI), N	1.89 (1.14-3.15), 27	2.12 (1.23-3.67), 24	1.07 (0.25-4.52), 3
SRR 5-10 yr (95% CI), N	1.68 (0.94-2.99), 18	1.88 (1.01-3.49), 16	0.98 (0.19-5.09), 2
SRR 10-20 yr (95% CI), N	0.93 (0.48-1.83), 12	1.06 (0.50-2.15), 11	0.48 (0.06-4.14), 1
SRR 20-30 yr (95% CI), N	1.18 (0.60-2.31), 12	1.06 (0.50-2.29), 9	2.00 (0.45-8.77), 3
SRR ≥30 yr (95% CI), N	0.73 (0.32-1.67), 7	0.90 (0.39-2.08), 7	0 (0.24 expected)

*Only one case was observed among non-whites.

5.3 Badged and Unbadged Worker Subcohorts

5.3.1 White Males

Workers who were externally monitored as a group showed very different mortality rates than did unbadged workers (Table 5-8). The SMRs for many lifestyle-related and other causes of death (particularly smoking-related cancers, emphysema and other respiratory disease, ischemic heart disease, genitourinary system diseases, alcoholism, cirrhosis of liver, and accidents including transportation accidents) were lower among badged workers than in unbadged workers, with SRR point estimates below unity for many causes of death. However, for most specific causes of death, CIs of the SRR for badged compared to unbadged workers overlap 1.00. Exceptions included lung cancer, ischemic heart disease, cerebrovascular disease, emphysema, other respiratory disease, cirrhosis of liver, accidents and homicide, which were substantially lower among badged workers compared to unbadged workers (Table 5-8).

Causes of death for which badged workers had at least 10% higher mortality rates than unbadged workers included diabetes, bone cancer, connective tissue cancer, cancer of larynx, and Hodgkin's disease. Thyroid cancer, "other respiratory" cancers (including pleura), NHL and non-malignant diseases of skin were elevated both compared to the general regional population (among badged workers) and for badged compared to unbadged workers, although CIs on these SRRs generally were wide. Asbestosis rates were similarly elevated among badged and unbadged workers compared to the general regional population (Table 5-8).

White male workers receiving positive dose had generally lower risk of many lifestyle-related diseases than unbadged workers; for example, rates of death from cancers of lung and esophagus, ischemic heart disease, cerebrovascular disease, emphysema, cirrhosis of liver and transportation accidents were all lower among workers who received positive dose compared to unbadged workers (see SRRs in Table 5-9). For many causes of death, the CIs were very wide; however, mortality rate ratio estimates for cancers of thyroid and connective tissue, Hodgkin's disease and non-malignant skin disease were 20% or more higher among worker with positive dose, compared to unbadged workers. The death rate from asbestosis was also very highly elevated among the badged workers with positive dose, compared to the general regional population, and was slightly though imprecisely elevated compared to unbadged workers (Table 5-9).

Badged workers who received no dose consisted of a relatively small group. SRRs comparing mortality to unbadged workers were not greatly different from unity for most causes of death (Table 5-9) although the ischemic heart disease rate ratio was substantially reduced (SRR=0.77, CI 0.69-0.87). Table 5-9 also shows that, in general, the SRRs for badged workers with positive dose were substantially lower than the SRRs for badged workers with zero dose. Exceptions to this pattern included cancers of pancreas, peritoneum and other parts of digestive tract, breast, bladder, skin, connective tissue, and most lymphopoietic neoplasms, for which the SRRs for workers with positive dose were similar to or higher than those for badged workers with zero dose. SRRs for some non-cancer death categories, such as alcoholism, ischemic heart disease, emphysema, asbestosis and homicide were higher among badged workers with positive dose than among those with zero dose. However, CIs on the latter were wide.

Table 5-8. Standardized mortality ratios (SMRs) for white male radiation-badged and -unbadged workers, and standardized rate ratios (SRRs) for badged compared to unbadged workers. Comparison population for SMR analysis was combined Idaho, Montana and Wyoming.

Cause of death	SMR Unbadged (N)	SMR Badged (N)	SRR badged/ unbadged (95% CI)
MN Buccal cavity	0.93 (18)	0.72 (22)	0.76 (0.41-1.42)
MN Pharynx	0.78 (7)	0.63 (9)	0.85 (0.32-2.28)
MN Digestive	1.07 (245)	1.05 (375)	0.96 (0.82-1.13)
MN Esophagus	1.28 (31)	0.90 (35)	0.69 (0.43-1.13)
MN Stomach	1.02 (32)	1.06 (51)	1.00 (0.65-1.57)
MN Intestine	1.00 (79)	1.08 (134)	1.07 (0.81-1.41)
MN Rectum	0.86 (16)	0.90 (26)	1.03 (0.55-1.93)
MN Liver & Gall Bladder	1.22 (18)	1.12 (26)	0.91 (0.50-1.66)
MN Liver Unspecified	0.92 (5)	0.95 (8)	1.14 (0.37-3.50)
MN Pancreas	1.09 (56)	1.12 (90)	1.01 (0.72-1.41)
MN Peritoneum & Other	1.97 (8)	0.80 (5)	0.39 (0.13-1.19)
MN Respiratory	1.26† (392)	1.03 (509)	0.81 (0.71-0.92)
MN Larynx	0.92 (9)	1.03 (16)	1.12 (0.49-2.54)
MN Trachea, Bronchus, Lung	1.26† (376)	1.01 (478)	0.79 (0.69-0.91)
MN Other Respiratory	1.93 (7)	2.59† (15)	1.34 (0.55-3.31)
MN Breast	1.86 (2)	1.78 (3)	0.83 (0.14-4.97)
MN Male Genital	1.20 (122)	1.02 (157)	0.85 (0.67-1.08)
MN Prostate	1.23* (121)	1.05 (156)	0.85 (0.67-1.08)
MN Testis	0.33 (1)	0.20 (1)	0.58 (0.04-9.27)
MN Urinary Organs	1.17 (61)	1.08 (88)	0.90 (0.65-1.25)
MN Kidney	1.14 (30)	1.12 (47)	0.98 (0.62-1.55)
MN Bladder	1.20 (31)	1.04 (41)	0.83 (0.52-1.32)
MN Other & Unspec. Sites	1.08 (144)	1.12 (237)	1.03 (0.83-1.26)
MN Skin Melanoma	1.21 (21)	1.11 (31)	0.91 (0.52-1.59)
MN Brain & Other Nerv. Syst.	1.13 (35)	1.06 (53)	0.94 (0.61-1.44)
MN Thyroid	0.64 (1)	1.60 (4)	2.85 (0.32-25.6)
MN Bone	0.34 (1)	0.44 (2)	1.19 (0.11-13.2)
MN Connective Tissue	0.50 (3)	1.16 (11)	2.21 (0.62-7.91)
MN Other & Unspecified	1.11 (77)	1.20* (132)	1.07 (0.81-1.42)
MN Lymphatic & Hematopoietic	1.06 (107)	1.05 (167)	0.98 (0.77-1.25)
Non-Hodgkin Lymphoma	1.15 (42)	1.36† (79)	1.18 (0.81-1.71)
Hodgkin's Disease	0.33 (2)	0.80 (8)	2.53 (0.54-11.9)
Leukemia	1.10 (44)	0.99 (62)	0.89 (0.60-1.31)
Myeloma	1.04 (19)	0.63 (18)	0.59 (0.31-1.13)
Benign & Unspecified Neoplasms	1.46 (19)	0.89 (18)	0.58 (0.30-1.10)
Benign Neoplasms of Eye, Brain, Other Nervous System	1.52 (2)	0 (2.1 expected)	--
Neoplasms of Unspecified Nature of Nervous System	0.83 (5)	0.63 (6)	0.74 (0.22-2.42)
Other Benign & Unspecified Nature Neoplasms	2.11* (12)	1.38 (12)	0.62 (0.28-1.37)
All Cancers	1.15† (1091)	1.04 (1558)	Not available

Table 5-8. Standardized mortality ratios (SMRs) for white male radiation-badged and -unbadged workers, and standardized rate ratios (SRRs) for badged compared to unbadged workers. Comparison population for SMR analysis was combined Idaho, Montana and Wyoming.

Cause of death	SMR Unbadged (N)	SMR Badged (N)	SRR badged/unbadged (95% CI)
Diabetes Mellitus	0.92 (63)	1.15 (122)	1.22 (0.90-1.65)
Diseases of Blood & Blood-Forming Organs	1.01 (13)	0.66 (13)	0.60 (0.28-1.30)
Non-pernicious & Unspecified Anemias	*1.39 (6)*	*0.78 (5)*	*0.50 (0.15-1.63)*
Alcoholism	0.78 (17)	0.56† (20)	0.72 (0.38-1.38)
Other Mental Disorders	1.17 (28)	1.09 (40)	0.93 (0.57-1.51)
Diseases of Nervous System & Sense Organs	1.29* (87)	1.03 (109)	0.79 (0.60-1.05)
Diseases of Heart	1.04 (1410)	0.84† (1741)	0.80 (0.74-0.86)
Ischemic Heart Disease	*1.08† (1192)*	*0.84† (1421)*	*0.77 (0.71-0.83)*
Hypertension with Heart Dis.	*1.67* (27)*	*1.79† (42)*	*1.00 (0.61-1.62)*
Other Diseases of Circulatory System	1.01 (340)	0.86† (425)	0.83 (0.72-0.96)
Hypertension w/o Heart Dis.	*1.13 (12)*	*1.06 (17)*	*0.93 (0.44-1.95)*
Cerebrovascular Disease	*1.04 (213)*	*0.83† (246)*	*0.77 (0.64-0.93)*
Disease of Respiratory System	1.20† (445)	0.88† (495)	0.72 (0.63-0.82)
Chron. & Unspec. Bronchitis	*0.94 (10)*	*0.84 (13)*	*0.86 (0.38-1.97)*
Emphysema	*1.39† (85)*	*0.79 (71)*	*0.56 (0.41-0.76)*
Asthma	*0.95 (9)*	*0.69 (10)*	*0.75 (0.30-1.86)*
Asbestosis	*3.32 (4)*	*3.21* (6)*	*0.96 (0.27-3.43)*
Silicosis	*0.69 (1)*	*0.51 (1)*	*0.59 (0.04-9.45)*
Other Respiratory Disease	*1.29† (245)*	*0.92 (273)*	*0.70 (0.59-0.83)*
Diseases of Digestive System	1.16* (186)	0.76† (191)	0.65 (0.53-0.79)
Cirrhosis of Liver	*1.30* (92)*	*0.66† (76)*	*0.51 (0.38-0.69)*
Diseases of Genitourinary System	1.12 (50)		0.70 (0.48-1.03)
Acute Glomerulonephritis & Acute Renal Failure	*1.37 (6)*		*0.21 (0.04-1.02)*
Chronic & Unspec. Nephritis, Renal Failure & Other Renal Sclerosis	*1.25 (28)*	*1.17 (40)*	*0.90 (0.56-1.47)*
Diseases of Skin	0.87 (2)	1.76 (6)	1.96 (0.39-9.71)
Diseases of Musculoskeletal & Connective Tissue	0.56 (6)	0.54 (9)	0.89 (0.32-2.51)
Accidents	0.93 (340)	0.62† (361)	0.67 (0.58-0.78)
Transportation Accidents	*0.89 (185)*	*0.64† (214)*	*0.72 (0.59-0.88)*
Accidental Falls	*1.30 (41)*	*0.77 (37)*	*0.60 (0.38-0.94)*
Suicide	0.83* (121)	0.73† (168)	0.89 (0.70-1.12)
Homicide	0.94 (24)	0.49† (20)	0.53 (0.29-0.96)
HIV-related	1.53 (13)	0.66 (8)	0.42 (0.17-1.02)
Other & Unspecified	2.48† (169)	2.05† (217)	0.85 (0.69-1.04)
All Deaths	1.09† (4464)	0.89† (5633)	Not available

*95% CI of SMR excludes 1.00. SRRs are not flagged.
†99% CI of SMR excludes 1.00. SRRs are not flagged.

Table 5-9. SRRs for white males who were badged with zero dose and badged with positive dose, compared to unbadged workers.

Cause of death	SMR Badged-Zero dose (N)	SMR Badged-positive dose (N)	SRR badged-zero/unbadged (95% CI)	SRR badged-pos/unbadged (95% CI)
MN Buccal cavity	**1.17 (10)**	**0.54* (12)**	**1.22 (0.56-2.66)**	**0.57 (0.28-1.20)**
MN Pharynx	*1.26 (5)*	*0.38* (4)*	*1.59 (0.50-5.02)*	*0.57 (0.17-1.94)*
MN Digestive	**1.20* (122)**	**0.99 (253)**	**1.13 (0.91-1.40)**	**0.90 (0.75-1.07)**
MN Esophagus	*1.22 (13)*	*0.78 (22)*	*0.94 (0.49-1.80)*	*0.58 (0.34-1.01)*
MN Stomach	*1.30 (18)*	*0.96 (33)*	*1.25 (0.70-2.24)*	*0.91 (0.56-1.48)*
MN Intestine	*1.08 (38)*	*1.09 (96)*	*1.10 (0.74-1.62)*	*1.07 (0.79-1.44)*
MN Rectum	*1.32 (11)*	*0.73 (15)*	*1.61 (0.74-3.48)*	*0.83 (0.41-1.69)*
MN Liver & Gall Bladder		*1.02 (17)*	*1.12 (0.50-2.50)*	*0.80 (0.41-1.56)*
MN Liver Unspec.	*1.66 (4)*	*0.67 (4)*	*2.11 (0.57-7.88)*	*0.80 (0.21-2.98)*
MN Pancreas	*1.28 (29)*	*1.05 (61)*	*1.15 (0.73-1.80)*	*0.94 (0.65-1.36)*
MN Peritoneum & Other Digestive	*0 (1.81 exp.)*		--	*0.54 (0.18-1.64)*
MN Respiratory	**1.18* (162)**	**0.97 (347)**	**0.94 (0.78-1.13)**	**0.76 (0.66-0.88)**
MN Larynx	*1.17 (5)*	*0.97 (11)*	*1.35 (0.45-4.04)*	*1.09 (0.45-2.65)*
MN Trachea, Bronchus & Lung	*1.14 (150)*	*0.96 (328)*	*0.91 (0.75-1.10)*	*0.75 (0.65-0.87)*
MN Other Respiratory	**4.27† (7)**	*1.93 (8)*	*2.23 (0.78-6.38)*	*0.99 (0.36-2.75)*
MN Breast	**0 (0.47 exp.)**	**2.46 (3)**	--	**1.15 (0.19-6.91)**
MN Male Genital	**1.10 (50)**	**0.99 (107)**	**0.95 (0.68-1.32)**	**0.83 (0.64-1.08)**
MN Prostate	*1.14 (50)*	*1.01 (106)*	*0.96 (0.69-1.34)*	*0.83 (0.64-1.08)*
MN Testis	*0 (1.46 exp.)*	*0.28 (1)*	--	*0.76 (0.05-12.2)*
MN Urinary Organs	**0.90 (21)**	**1.15 (67)**	**0.81 (0.49-1.33)**	**0.98 (0.69-1.38)**
MN Kidney	*1.02 (12)*	*1.16 (35)*	*0.97 (0.50-1.91)*	*1.01 (0.62-1.65)*
MN Bladder	*0.78 (9)*	*1.15 (32)*	*0.65 (0.31-1.37)*	*0.94 (0.57-1.55)*
MN Other & Unspecified Sites	**1.15 (68)**	**1.11 (169)**	**1.04 (0.78-1.40)**	**1.02 (0.81-1.27)**
MN Skin Melanoma	*0.77 (6)*	*1.24 (25)*	*0.65 (0.26-1.62)*	*1.05 (0.58-1.88)*
MN Brain & Other Nervous System	*1.29 (18)*	*0.97 (35)*	*1.18 (0.67-2.09)*	*0.86 (0.53-1.37)*
MN Thyroid	*2.91 (2)*	*1.11 (2)*	*4.72 (0.43-52.1)*	*2.04 (0.18-22.5)*
MN Bone	*0.75 (1)*	*0.31 (1)*	*1.47 (0.09-23.5)*	*0.80 (0.05-12.8)*
MN Connective Tissue	*0.75 (2)*	*1.32 (9)*	*1.27 (0.21-7.60)*	*2.43 (0.66-8.98)*
MN Other & Unspec.	*1.24 (38)*	*1.19 (94)*	*1.09 (0.73-1.61)*	*1.06 (0.78-1.43)*
MN Lymphatic & Hematopoietic	**1.04 (47)**	**1.05 (120)**	**0.95 (0.67-1.34)**	**1.01 (0.76-1.34)**
Non-Hodgkin Lymphoma	*1.59* (26)*	*1.27 (53)*	*1.32 (0.81-2.16)*	*1.06 (0.70-1.59)*
Hodgkin's Disease	*0.35 (1)*	*0.98 (7)*	*1.00 (0.09-11.1)*	*3.13 (0.65-15.1)*
Leukemia	*1.00 (18)*	*0.98 (44)*	*0.89 (0.51-1.55)*	*0.98 (0.60-1.60)*

Table 5-9. SRRs for white males who were badged with zero dose and badged with positive dose, compared to unbadged workers.

Cause of death	SMR Badged-Zero dose (N)	SMR Badged-positive dose (N)	SRR badged-zero/unbadged (95% CI)	SRR badged-pos/unbadged (95% CI)
Myeloma	0.25* (2)	0.77 (16)	0.26 (0.06-1.14)	0.73 (0.38-1.43)
Benign & Unspec. Nature Neoplasms	0.68 (4)	0.97 (14)	0.42 (0.14-1.25)	0.61 (0.31-1.23)
Benign of Nervous System	0 (0.60 exp.)	0 (1.50 exp.)	--	--
Unspecified Nature of Nervous System	0.37 (1)	0.74 (5)	0.38 (0.04-3.28)	0.88 (0.25-3.08)
Other Benign & Unspecified	1.17 (3)	1.47 (9)	0.52 (0.15-1.85)	0.62 (0.26-1.47)
All Cancers	1.14† (480)	1.01 (1078)	Not available	
Diabetes Mellitus	1.28 (39)	1.09 (83)	1.38 (0.93-2.07)	1.14 (0.82-1.59)
Dis. of Blood & Blood-Forming Organs	0.69 (4)	0.65 (9)	0.62 (0.20-1.89)	0.56 (0.24-1.31)
Non-pernicious Anemia	1.01 (2)	0.67 (3)	0.66 (0.13-3.28)	0.40 (0.10-1.61)
Alcoholism	0.20† (2)	0.70 (18)	0.26 (0.06-1.13)	0.91 (0.47-1.77)
Other Mental Disorders	1.10 (12)	1.08 (28)	0.92 (0.46-1.80)	0.91 (0.54-1.54)
Dis. of Nervous System & Sense Organs	1.32 (40)	0.92 (69)	1.02 (0.70-1.48)	0.71 (0.51-0.97)
Diseases of Heart	0.87† (523)	0.83† (1218)	0.83 (0.75-0.91)	0.79 (0.73-0.85)
Ischemic Heart Disease	0.84† (411)	0.85† (1010)	0.77 (0.69-0.86)	0.77 (0.71-0.84)
Hypertension with Heart Disease	1.83* (13)		1.10 (0.56-2.14)	0.99 (0.59-1.68)
Other Diseases of Circulatory System	0.98 (147)	0.81† (278)	0.96 (0.79-1.17)	0.78 (0.66-0.91)
Hypertension w/o Heart Disease	1.28 (6)	0.97 (11)	1.22 (0.46-3.28)	0.84 (0.37-1.92)
Cerebrovascular Disease	0.94 (86)	0.78† (160)	0.88 (0.68-1.13)	0.71 (0.58-0.87)
Disease of Respiratory System	1.05 (173)		0.87 (0.73-1.04)	0.67 (0.58-0.77)
Pneumonia	1.00 (40)	0.83 (75)	1.02 (0.70-1.49)	0.85 (0.62-1.16)
Chronic & Unspec. Bronchitis	0.42 (2)		0.45 (0.10-2.10)	1.06 (0.45-2.50)
Emphysema	0.70 (19)	0.83 (52)	0.50 (0.30-0.83)	0.59 (0.42-0.84)
Asthma	0.71 (3)	0.68 (7)	0.68 (0.18-2.52)	0.76 (0.28-2.06)
Asbestosis	1.87 (1)	3.76* (5)	0.48 (0.05-4.30)	1.19 (0.32-4.49)
Silicosis	1.57 (1)	0 (1.34 exp.)	2.45 (0.15-39.2)	--
Other Resp. Disease	1.27* (107)	0.78† (166)	0.98 (0.78-1.23)	0.60 (0.49-0.73)

Table 5-9. SRRs for white males who were badged with zero dose and badged with positive dose, compared to unbadged workers.

Cause of death	SMR Badged-Zero dose (N)	SMR Badged-positive dose (N)	SRR badged-zero/unbadged (95% CI)	SRR badged-pos/unbadged (95% CI)
Diseases of Digestive System	0.95 (68)	0.69† (123)	0.82 (0.62-1.09)	0.58 (0.46-0.73)
Cirrhosis of Liver	0.85 (27)	0.59† (49)	0.68 (0.44-1.04)	0.45 (0.32-0.64)
Diseases of Genito-Urinary System	0.85 (17)	0.79 (37)	0.75 (0.43-1.30)	0.71 (0.46-1.08)
Acute Glomerulo-nephritis & Acute Renal Failure	0.51 (1)	0.20 (1)	0.44 (0.05-3.63)	0.15 (0.02-1.23)
Chronic & Unspec. Nephritis, Renal Failure & Other Renal Sclerosis	0.99 (10)	1.24 (30)	0.77 (0.37-1.59)	0.99 (0.59-1.66)
Diseases of Skin	0.97 (1)	2.10 (5)	1.10 (0.10-12.2)	2.24 (0.43-11.6)
Diseases of Musculo-skeletal System	0.21 (1)	0.68 (8)	0.34 (0.04-2.83)	1.11 (0.38-3.22)
Symptoms & Ill-Defined Conditions	0.60 (12)	0.94 (44)	0.68 (0.35-1.31)	1.03 (0.67-1.59)
Accidents	0.67† (114)	0.60† (247)	0.73 (0.59-0.91)	0.65 (0.55-0.77)
Transportation Accidents	0.76* (74)	0.59† (140)	0.87 (0.66-1.14)	0.67 (0.54-0.84)
Accidental Falls	0.84 (12)	0.74 (25)	0.64 (0.34-1.23)	0.58 (0.35-0.96)
Suicide	0.75* (50)	0.72† (118)	0.93 (0.67-1.29)	0.89 (0.69-1.15)
Homicide	0.25† (3)	0.59* (17)	0.30 (0.09-0.98)	0.63 (0.34-1.18)
HIV-related	0.50 (2)	0.74 (6)	0.36 (0.08-1.62)	0.51 (0.19-1.36)
Other & Unspecified	2.24† (69)	1.97† (148)	0.91 (0.69-1.21)	0.84 (0.66-1.08)
All Deaths	0.96 (1762)	0.86† (3871)	Not available	

*95% CI of SMR excludes 1.00. SRRs are not flagged.
†99% CI of SMR excludes 1.00. SRRs are not flagged.

5.3.2 White Females

For WF, very small numbers of deaths made it difficult to differentiate mortality rates among workers badged and not badged for external radiation (Table 5-10). The SMR for cancer of larynx was substantially elevated but was based on small numbers of deaths. The sole COD that was substantially elevated among badged female workers (compared to unbadged workers) is accidents, particularly transportation accidents (SRR=2.18; CI, 1.04-4.58). No precisely estimated causes of death showed lower rates among badged white female workers. Leukemia risk was elevated with an SRR of 2.16 (CI, 0.56-8.39) among badged compared to unbadged white female workers. The SRRs were very similarly elevated for leukemia among both workers who were badged with positive dose and badged with zero dose compared to unbadged workers (Table 5-11). Mortality rates for cancer of pancreas, diseases of musculoskeletal and connective tissue, and BN and NUN were elevated with very wide CIs for badged workers compared to unbadged workers (Table 5-10). The excess in the benign neoplasm group was confined to badged workers with no dose (Table 5-11). Brain cancer SRRs were also elevated, but with very wide CIs, among these workers. The cancer of larynx SMR was highly elevated in monitored white female workers with positive dose, but only two cases occurred in this group.

Unlike white male workers, a lower mortality rate was not observed in some lifestyle-related causes of death (such as ischemic heart disease and emphysema) for badged white female workers (Table 5-10). However, others such as cancers of oral cavity, esophagus and lung, cirrhosis of liver, and alcoholism showed lower mortality rates among badged compared to unbadged workers, although CIs were very wide for SRRs for these causes of death. Breast cancer mortality rates were very similar among badged (and among those receiving a positive dose) compared to unbadged workers.

SRRs for badged female workers with positive dose did not differ markedly from those of badged workers with zero dose, with the exceptions noted above (Table 5-11) and the following: badged workers with positive dose showed elevated but imprecisely estimated SRRs (compared to unbadged workers) for cancers of pancreas and intestine, but badged-zero-dose workers did not. The mortality rate ratio from transportation accidents was highly elevated among badged workers with zero dose only (compared to unbadged workers) SRR=2.94; CI, 1.35-6.42.

5.3.3 Non-white Males and Females

Very few deaths occurred among badged subcohorts for non-white male workers. Among the causes of death with sufficient numbers to evaluate differences, lung cancers were highly elevated particularly among the badged workers receiving a positive dose (SRR=8.07; CI, 0.97-67.3, N=6) compared to unbadged workers. Prostate cancer was elevated, but with very wide CIs, for badged workers with zero dose (SRR=2.79; CI, 0.39-20.2, N=2). Of the three leukemia cases among NWM, two occurred in unbadged workers and one among badged workers with zero dose, leading to an SRR of 1.16 (CI, 0.10-13.1). No analyses were conducted among NWF based on badging status, because of the very low number of total deaths in the cohort.

Table 5-10. Standardized mortality ratios (SMRs) and standardized rate ratios (SRRs) for white female workers. Badged compared to unbadged workers person-time. Comparison population for SMR analysis was combined Idaho, Montana and Wyoming 1960-1999.

Cause of death	SMR Unbadged (N)	SMR Badged (N)	SRR badged/unbadged (95% CI)
MN Buccal cavity	4.03† (4)	1.49 (2)	0.39 (0.07-2.14)
MN Digestive	0.91 (14)	0.91 (19)	1.01 (0.50-2.02)
MN Esophagus	1.58 (1)	1.19 (1)	0.88 (0.06-14.1)
MN Stomach	2.72 (4)	0 (2.0 exp.)	--*
MN Intestine	0.62 (4)	1.03 (9)	1.71 (0.52-5.56)
MN Rectum	0.94 (1)	0.69 (1)	0.98 (0.06-15.6)
MN Liver & Gall Bladder	1.55 (2)	0.57 (1)	0.34 (0.03-3.76)
MN Liver Unspecified Nature	0 (0.38 exp.)	1.90 (1)	--
MN Pancreas	0.54 (2)	1.00 (5)	1.87 (0.36-9.66)
MN Peritoneum & Other	0 (0.37 exp.)	1.92 (1)	--
MN Respiratory	1.15 (20)	0.78 (18)	0.70 (0.37-1.33)
MN Larynx	0 (0.30 exp.)	4.87 (2)	--
MN Trachea, Bronchus, Lung	1.19 (20)	0.71 (16)	0.63 (0.32-1.21)
MN Breast	1.09 (22)	1.01 (27)	0.95 (0.54-1.66)
MN Female Genital	0.74 (8)	0.83 (12)	1.15 (0.47-2.83)
MN Cervix, Uterine Organs	1.53 (4)	0.57 (2)	0.36 (0.07-1.97)
MN Ovary, Fallopian Tube, Broad Ligament	0.50 (3)	0.62 (5)	1.29 (0.31-5.42)
MN Urinary Organs	0 (1.95 exp.)	1.52 (4)	--
MN Kidney	0 (1.27 exp.)	1.17 (2)	--
MN Bladder	0 (0.68 exp.)	2.18 (2)	--
MN Other & Unspecified Sites	0.91 (11)	0.82 (13)	0.93 (0.42-2.08)
MN Skin Melanoma	0.66 (1)	0 (1.93 exp.)	--
MN Brain & Other Nervous System	1.34 (4)	1.80 (7)	1.38 (0.40-4.71)
MN Bone	4.76 (1)	0 (0.28 exp.)	--
MN Connective Tissue	0 (0.89 exp.)	0.88 (1)	--
MN Other & Unspecified Site	0.83 (5)	0.62 (5)	0.78 (0.23-2.70)
MN Lymphatic & Hematopoietic	1.00 (8)	1.41 (15)	1.50 (0.63-3.55)
Non-Hodgkin Lymphoma	1.27 (4)	1.19 (5)	0.93 (0.25-3.46)
Hodgkin's Disease	0 (0.43 exp.)	1.67 (1)	--
Leukemia	0.96 (3)	1.69 (7)	2.16 (0.56-8.39)
Myeloma	0.80 (1)	1.18 (2)	1.50 (0.14-16.6)
Benign & Unspecified Neoplasms	0.76 (1)	1.67 (3)	2.52 (0.26-24.3)
Neoplasms of Unspecified Nature of Nervous System	1.90 (1)	2.78 (2)	1.74 (0.16-19.2)
All Cancers	1.01 (87)	0.95 (110)	Not available
Diabetes Mellitus	1.09 (8)	0.82 (8)	0.81 (0.30-2.16)
Diseases of Blood & Blood-Forming Organs	0.96 (1)	0.72 (1)	0.64 (0.04-10.2)
Alcoholism	1.12 (1)	0 (1.16 exp.)	--

Table 5-10. Standardized mortality ratios (SMRs) and standardized rate ratios (SRRs) for white female workers. Badged compared to unbadged workers person-time. Comparison population for SMR analysis was combined Idaho, Montana and Wyoming 1960-1999.

Cause of death	SMR Unbadged (N)	SMR Badged (N)	SRR badged/unbadged (95% CI)
Other Mental Disorders	1.83 (4)	1.43 (4)	0.76 (0.19-3.07)
Diseases of Nervous System & Sense Organs	1.43 (10)	0.75 (7)	0.50 (0.19-1.33)
Diseases of Heart	0.83 (44)	0.91 (66)	1.14 (0.78-1.67)
Ischemic Heart Disease	*0.79 (29)*	*0.81 (41)*	*1.07 (0.66-1.72)*
Hypertension with Heart Disease	*2.51 (3)*	*2.47 (4)*	*1.02 (0.23-4.64)*
Other Diseases of Circulatory System	1.14 (28)	0.78 (26)	0.71 (0.42-1.22)
Hypertension w/o Heart Disease	*1.22 (1)*	*0.87 (1)*	*0.57 (0.04-9.17)*
Cerebrovascular Disease	*1.27 (21)*	*0.89 (20)*	*0.72 (0.39-1.33)*
Disease of Respiratory System	0.62 (13)	0.77 (22)	1.27 (0.64-2.54)
Influenza	*0 (0.34 exp.)*	*2.05 (1)*	*--*
Pneumonia	*0.96 (5)*	*0‡ (6.97 exp.)*	*--*
Chronic & Unspecified Bronchitis	*0 (0.45 exp.)*	*1.61 (1)*	*--*
Emphysema	*0 (2.35 exp.)*	*0.62 (2)*	*--*
Asthma	*0 (1.24 exp.)*	*1.21 (2)*	*--*
Other Respiratory Disease	*0.70 (8)*	*1.05 (16)*	*1.47 (0.63-3.44)*
Diseases of Digestive System	1.00 (12)	0.68 (11)	0.71 (0.31-1.60)
Cirrhosis of Liver	*1.00 (5)*	*0.44 (3)*	*0.45 (0.11-1.88)*
Diseases of Genitourinary System	1.35 (4)	1.20 (5)	0.95 (0.25 -3.55)
Chronic & Unspecified Nephritis, Renal Failure, Other Renal Dis.	*2.13 (3)*	*0.51 (1)*	*0.22 (0.02-2.15)*
Diseases of Musculoskeletal & Connective Tissue	0.51 (1)	1.52 (4)	3.46 (0.39-31.0)
Symptoms & Ill-Def. Conditions	0.83 (2)	1.85 (6)	2.58 (0.52-12.8)
Accidents	0.59 (13)	1.18 (32)	2.13 (1.11-4.08)
Transportation Accidents	*0.63 (10)*	*1.31 (25)*	*2.18 (1.04-4.58)*
Accidental Falls	*0 (1.43 exp.)*	*1.04 (2)*	*--*
Suicide	0.96 (7)	0.66 (6)	0.70 (0.23-2.10)
Homicide	1.12 (3)	1.24 (4)	1.02 (0.22-4.70)
Other & Unspecified	1.48 (11)	1.44 (14)	0.99 (0.45-2.20)
All Deaths	0.95 (250)	0.93 (329)	Not available

* SRR not computed if 0 deaths occurred in baseline or comparison category
†95% CI of SMR excludes 1.00. SRRs are not flagged.
‡99% CI of SMR excludes 1.00. SRRs are not flagged.

Table 5-11. SRRs for white females who were badged with zero dose and badged with positive dose, compared to unbadged workers.

Cause of death	SMR Badged-Zero dose (N)	SMR Badged-positive dose (N)	SRR badged-zero/unbadged (95% CI)	SRR badged-pos/unbadged (95% CI)
MN Buccal cavity	0 (0.63 exp.)	2.80 (2)	--*	0.73 (0.13-3.99)
MN Digestive	0.63 (6)	1.15 (13)	0.68 (0.26-1.78)	1.34 (0.62-2.88)
MN Esophagus	2.61 (1)	0 (0.46 exp.)	1.33 (0.08-21.3)	--
MN Intestine	0.25 (1)	1.68 (8)	0.41 (0.05-3.67)	2.85 (0.85-9.53)
MN Rectum	1.51 (1)	0 (0.79 exp.)	2.14 (0.13-34.2)	--
MN Liver & Gall Bladder	1.23 (1)	0 (0.94 exp.)	0.70 (0.06-7.68)	--
MN Pancreas	0.44 (1)	1.47 (4)	0.87 (0.08-9.60)	3.24 (0.59-17.8)
MN Peritoneum & Other Digestive	4.33 (1)	0 (0.29 exp.)	--	--
MN Respiratory	0.74 (8)	0.81 (10)	0.67 (0.30-1.53)	0.70 (0.33-1.51)
MN Larynx	0 (0.19 exp.)	8.96† (2)	--	--
MN Trachea, Bronchus & Lung	0.76 (8)	0.67 (8)	0.67 (0.30-1.53)	0.56 (0.25-1.28)
MN Breast	0.99 (13)	1.03 (14)	0.95 (0.48-1.90)	0.99 (0.50-1.97)
MN Female Genital	1.02 (7)	0.66 (5)	1.43 (0.52-3.98)	0.90 (0.29-2.82)
MN Cervix Uteri	1.15 (2)	0 (1.77 exp.)	0.85 (0.15-4.64)	--
MN Other Uterine	0.86 (1)	2.84 (4)	1.95 (0.12-31.1)	6.56 (0.72-59.5)
MN Ovary	1.06 (4)	0.24 (1)	2.09 (0.46-9.45)	0.38 (0.04-3.69)
MN Urinary Organs	0 (1.20 exp.)	2.80 (4)	--	--
MN Kidney	0 (0.79 exp.)	2.16 (2)	--	--
MN Bladder	0 (0.41 exp.)	3.99 (2)	--	--
MN Other & Unspecified Sites	1.17 (9)	0.49 (4)	1.33 (0.55-3.23)	0.53 (0.17-1.67)
MN Brain & Other Nervous System	2.57 (5)	1.03 (2)	1.92 (0.51-7.19)	0.64 (0.12-3.54)
MN Connective Tissue	1.74 (1)	0 (0.56 exp.)	--	--
MN Other & Unspec.	0.79 (3)	0.47 (2)	1.12 (0.27-4.71)	0.65 (0.13-3.36)
MN Lymphatic & Hematopoietic	1.38 (7)	1.44 (8)	1.50 (0.54-4.16)	1.50 (0.56-4.02)
Non-Hodgkin Lymphoma	1.51 (3)	0.91 (2)	1.19 (0.26-5.35)	0.64 (0.12-3.56)
Hodgkin's Disease	0 (0.30 exp.)	3.32 (1)	--	--
Leukemia	1.49 (3)	1.88 (4)	1.97 (0.40-9.78)	2.37 (0.53-10.6)
Myeloma	1.31 (1)	1.08 (1)	1.61 (0.10-25.7)	1.45 (0.09-23.2)
Benign & Unspec. Nature Neoplasms	3.56 (3)	0 (0.95 exp.)	5.66 (0.59-54.7)	--
Unspecified of Nervous System	5.79 (2)	0 (0.37 exp.)	--	--
All Cancers	0.91 (50)	0.99 (60)	Not available	
Diabetes Mellitus	0.22 (1)	1.35 (7)	0.21 (0.03-1.72)	1.31 (0.47-3.62)
Diseases of Blood & Blood-Forming Organs	1.55 (1)	0 (0.75 exp.)	1.67 (0.10-26.7)	--
Other Mental Disorders	0.84 (1)	1.88 (3)	0.49 (0.05-4.42)	0.95 (0.21-4.27)

Table 5-11. SRRs for white females who were badged with zero dose and badged with positive dose, compared to unbadged workers.

Cause of death	SMR Badged-Zero dose (N)	SMR Badged-positive dose (N)	SRR badged-zero/unbadged (95% CI)	SRR badged-pos/unbadged (95% CI)
Diseases of Nervous System & Sense Organs	0.69 (3)	0.81 (4)	0.53 (0.15-1.92)	0.59 (0.18-1.87)
Diseases of Heart	1.05 (33)	0.80 (33)	1.37 (0.87-2.16)	0.97 (0.62-1.53)
Ischemic Heart Disease	0.96 (21)	0.69 (20)	1.27 (0.72-2.23)	0.92 (0.52-1.63)
Hypertension with Heart Disease	1.48 (1)	3.18 (3)	0.64 (0.07-6.17)	1.25 (0.25-6.24)
Other Diseases of Circulatory System	0.75 (11)	0.80 (15)	0.73 (0.36-1.49)	0.79 (0.41-1.49)
Hypertension w/o Heart Disease	2.04 (1)	0 (0.66 exp.)	1.43 (0.09-22.9)	--
Cerebrovascular Disease	0.71 (7)	1.02 (13)	0.61 (0.26-1.47)	0.86 (0.42-1.74)
Disease of Respiratory System	0.79 (10)	0.76 (12)	1.42 (0.61-3.29)	1.40 (0.62-3.17)
Chronic & Unspecified Bronchitis	3.56 (1)	0 (0.34 exp.)	--	--
Emphysema	0.69 (1)	0.56 (1)	--	--
Asthma	1.28 (1)	1.15 (1)	--	--
Other Resp. Disease	1.02 (7)	1.07 (9)	1.71 (0.61-4.78)	1.72 (0.63-4.66)
Diseases of Digestive System	0.53 (4)	0.81 (7)	0.57 (0.18-1.78)	0.89 (0.34-2.33)
Cirrhosis of Liver	0.31 (1)	0.57 (2)	0.34 (0.04-2.93)	0.56 (0.11-2.89)
Diseases of Genito-urinary System	0.54 (1)	1.73 (4)	0.42 (0.05-3.72)	1.51 (0.36-6.33)
Chronic & Unspec. Nephritis, Renal Failure	0 (0.88 exp.)	0.92 (1)	--	0.37 (0.04-3.60)
Diseases of Musculo-skeletal System	0 (1.22 exp.)	2.83 (4)	--	6.65 (0.74-59.6)
Symptoms & Ill-Defined Conditions	0.66 (1)	2.91 (5)	1.02 (0.09-11.2)	4.49 (0.85-23.7)
Accidents	1.57† (23)	0.72 (9)	2.90 (1.46-5.76)	1.33 (0.55-3.25)
Transportation Accidents	1.69† (18)	0.82 (7)	2.94† (1.35-6.42)	1.40 (0.51-3.89)
Accidental Falls	2.33 (2)	0 (1.07 exp.)	--	--
Suicide	1.02 (5)	0.24 (1)	1.02 (0.32-3.24)	0.28 (0.03-2.31)
Homicide	0 (1.81 exp.)	2.83 (4)	--	2.29 (0.49-10.6)
Other & Unspecified	1.26 (6)	1.63 (8)	0.90 (0.33-2.44)	1.09 (0.43-2.72)
All Deaths	0.93 (153)	0.94 (176)	Not available	

*SRR not computed if 0 deaths occurred in baseline or comparison category.
†95% CI of SMR excludes 1.00. SRRs are not flagged.

5.4 External Radiation Dose-response Analysis in the Full Cohort

5.4.1 White Males

Slope estimates for standardized rates as a function of on-site dose category are shown in Table 5-12 for causes of death containing five or more cases among badged workers. Plots of the SRR and CI in each dose group (data not shown) as well as examination of the slope values in Table 5-12, indicate that the slope in some instances depended on whether the baseline group consisted of people having just positive doses (up to 1 mSv) or included monitored people with zero dose.

Many causes of death that showed evidence of lowered rates in badged compared to unbadged workers also showed strong negative dose-response trends among those ever-monitored. For example, mortality rates for emphysema and lung cancer, two strongly smoking-related diseases, showed clear evidence of decrease with increasing dose (that is, a highly significant negative dose-response slope). The slope was $-4.07\text{E-}7 \pm$ standard error (SE) $0.455\text{E-}7$ for emphysema, and $-37.1\text{E-}7 \pm$ SE $6.74\text{E-}7$ for lung cancer. Other causes of death with strongly negative standardized rate slopes included cancers of intestine, liver and gall bladder, pancreas, peritoneum, kidney and bladder, and "other and unspecified" sites. The standardized rates for "other respiratory" diseases (non-cancers) showed deficits in some of the higher-dose groups (compared to the 0-0.99 mSv group) and, along with diseases of the digestive system (including cirrhosis of liver), exhibited highly negative slope with increasing dose.

Slope estimates were positive, though not significantly greater than zero, for all lymphatic and hematopoietic cancers combined (Table 5-12). SRR estimates were significantly elevated in the highest dose group (≥ 100 mSv) for this category (data not shown). This elevation appears to have been due primarily to NHL, which was doubled in the ≥ 100 mSv group (Figure 5-1a). Although CIs were wide, the leukemia group SRR was also nearly doubled in the highest dose group (Figure 5-1b). There was no evidence of a higher non-pernicious anemia mortality rate in the higher dose groups (three anemia deaths overall among workers with non-zero dose, two in the lowest dose group and one in the 50-100 mSv category).

Of the four thyroid cancers and two bone cancers observed among the monitored group, all were observed in those receiving less than 10 mSv. The slope estimate for malignant melanoma was not positive; however the SRR was highly elevated in the 10-50 mSv group, when compared to those in the lowest dose category (Figure 5-2a). The SMR was also substantially elevated in this category, compared to the regional population.

Mortality rates of asbestos-related disease (asbestosis and "other respiratory" cancers) and hypertensive heart disease were significantly elevated compared to the state rates but were not dependent on badging practices within the cohort (§5.3). However, among monitored workers the elevation in the asbestosis mortality rate was concentrated in the dose categories ≥ 10 mSv, and for asbestos-related cancer ("other respiratory") was primarily restricted to the lower-dose category.

The inclusion of a lag period into the dose-response analysis had modest to substantial effect on slope estimates. Brain cancer, in particular, showed great change in slope estimate when increasing the lag period to 20 years, with evidence of a positive association with radiation dose at that lag only (Table 5-12, Figure 5-2b). The leukemia slope estimate was also significantly positive at its longest lag (10 years) although the slope estimate itself was lower than for shorter lags, with a lag of 5 years producing the highest absolute slope estimate (Table 5-12).

Table 5-12. Slopes and standard errors (for badged white male workers, using on-site dose only), adjusted for age and calendar year. Slope represents change in standardized rate per mSv-person-year.

Cause of death	Slope, std error x10^{-7}, including zero dose, no lag	Slope, std error x10^{-7}, only pos. dose, no lag	Slope, std error x10^{-7}, only pos. dose, 5-yr lag†	Slope, std error x10^{-7}, only pos. dose, 10-yr lag‡	Slope, std error x10^{-7}, only pos. dose, 20-yr lag§
MN Buccal cavity	-2.75, 3.72 (N=22)	0.600, 3.61 (N=12)	0.270, 3.70 (N=12)	1.65, 4.58 (N=11)	0.234, 1.08 (N=8)
MN Digestive	-15.5, 10.8 (N=375)	-12.5, 9.79 (N=253)	-11.7, 10.2 (N=247)	-19.9*, 7.54 (N=238)	-11.1, 9.12 (N=180)
MN Esophagus	-0.0511, 3.81 (N=35)	1.15, 3.21 (N=22)	1.27, 2.76 (N=20)	1.29, 2.89 (N=20)	1.70, 4.71 (N=15)
MN Stomach	-0.00186, 2.74 (N=51)	1.15, 2.18 (N=33)	1.10, 2.3 (N=32)	0.632, 1.73 (N=31)	2.42, 1.27 (N=23)
MN Intestine	-8.04*, 1.76 (N=134)	-8.13*, 2.60 (N=96)	-6.78*, 2.97 (N=95)	-10.8*, 2.45 (N=93)	-6.69*, 3.30 (N=70)
MN Rectum	-1.20, 4.03 (N=26)	0.558, 2.98 (N=15)	-0.518, 7.36 (N=14)	0.313, 3.28 (N=14)	6.00, 7.46 (N=9)
MN Liver & Gall Bladder	-8.74*, 1.01 (N=26)	-9.90*, 0.727 (N=17)	-10.1*, 0.686 (N=17)	-8.66*, 3.04 (N=16)	-8.64, 7.06 (N=15)
MN Liver Unspecified	0.897, 1.16 (N=8)	2.74*, 0.060 (N=4)	2.71*, 0.0799 (N=4)	2.64*, 0.114 (N=4)	3.77, NA†† (N=3)
MN Pancreas	-4.17*, 1.99 (N=90)	-3.48, 2.06 (N=61)	-2.48, 2.98 (N=60)	-4.12, 2.44 (N=55)	-2.57, 2.66 (N=41)
MN Peritoneum & Other	0.0638, 1.65 (N=5)	-2.63*, 0.918 (N=5)	-2.67*, 0.931 (N=5)	-2.86*, 1.03 (N=5)	-1.79, 1.11 (N=4)
MN Respiratory	-38.5*, 6.62 (N=509)	-37.1*, 6.63 (N=347)	-36.4*, 6.03 (N=338)	-37.2*, 5.64 (N=323)	-37.0*, 8.96 (N=253)
MN Larynx	0.189, 1.12 (N=16)	0.645, 1.09 (N=11)	0.527, 1.03 (N=11)	0.417, 0.846 (N=10)	0.00957, 0.220 (N=6)
MN Trachea, Bronch., Lung	-37.1*, 6.74 (N=478)	-36.4*, 7.29 (N=328)	-35.3*, 6.70 (N=319)	-36.4*, 6.85 (N=307)	-33.8*, 8.96 (N=242)
MN Other Respiratory	-0.691, 2.76 (N=15)	0.800, 1.51 (N=8)	-0.871, 1.53 (N=8)	0.637, 0.631 (N=6)	-0.00676, 2.62 (N=5)
MN Male Genital	-1.53, 2.57 (N=157)	1.76, 6.41 (N=107)	2.56, 6.60 (N=104)	2.73, 6.73 (N=102)	4.94, 9.08 (N=87)
MN Prostate	-1.57, 2.39 (N=156)	1.75, 6.24 (N=106)	2.55, 6.41 (N=103)	2.73, 6.73 (N=102)	4.94, 9.08 (N=87)
MN Urinary Organs	-2.53, 2.24 (N=88)	-3.30*, 1.63 (N=67)	-4.53*, 1.50 (N=66)	-5.47*, 1.32 (N=61)	-7.03*, 0.820 (N=44)
MN Kidney	-1.51, 1.52 (N=47)	-1.69, 1.60 (N=35)	-2.85, 1.14 (N=35)	-1.91, 0.662 (N=30)	-3.01, 0.788 (N=23)
MN Bladder	-1.26, 1.27 (N=41)	-1.75*, 0.823 (N=32)	-1.70*, 0.559 (N=31)	-3.71, 0.743 (N=31)	-2.52, 2.18 (N=21)
MN Other & Unspecified Sites	-8.74*, 1.75 (N=237)	-9.53*, 2.66 (N=169)	-10.2*, 2.39 (N=163)	-6.93, 5.49 (N=145)	-9.75, 7.34 (N=116)
MN Skin Melanoma	-1.28, 2.06 (N=31)	-1.61, 2.19 (N=25)	-1.83, 1.97 (N=24)	-2.62*, 0.711 (N=22)	-1.59, 0.916 (N=15)
MN Brain & Other Nervous System	-3.53, 3.34 (N=53)	-2.33, 3.11 (N=35)	-2.59, 3.15 (N=34)	-0.454, 2.49 (N=28)	3.06*, 0.983 (N=22)
MN Connective Tissue	0.648, 1.22 (N=11)	0.170, 0.995 (N=9)	0.187, 1.01 (N=9)	1.03, 1.33 (N=7)	-0.927, 8.68 (N=6)
MN Other & Unspecified	-6.63*, 2.79 (N=132)	-6.88*, 3.36 (N=94)	-8.19*, 3.39 (N=92)	-6.46, 4.75 (N=84)	-9.15, 5.27 (N=70)
MN Lymphatic & Hematopoietic System	2.24, 9.04 (N=167)	4.75, 9.30 (N=120)	4.49, 7.57 (N=116)	2.30, 7.52 (N=108)	0.566, 9.63 (N=91)
Non-Hodgkin Lymphoma	2.74, 6.16 (N=79)	5.74, 6.42 (N=53)	6.54, 5.31 (N=53)	3.56, 5.51 (N=50)	4.12, 3.27 (N=42)
Hodgkin's Disease	0.224, 4.38 (N=8)	-2.49, 3.24 (N=7)	-4.16, 4.82 (N=6)	1.08, 5.29 (N=4)	-11.7, NA (N=2)
Leukemia	1.70, 2.37 (N=62)	3.80, 3.28 (N=44)	4.24, 3.56 (N=43)	4.66, 3.29 (N=42)	1.26*, 0.507 (N=39)

Table 5-12. Slopes and standard errors (for badged white male workers, using on-site dose only), adjusted for age and calendar year. Slope represents change in standardized rate per mSv-person-year.

Cause of death	Slope, std error x10⁻⁷, including zero dose, no lag	Slope, std error x10⁻⁷, only pos. dose, no lag	Slope, std error x10⁻⁷, only pos. dose, 5-yr lag[†]	Slope, std error x10⁻⁷, only pos. dose, 10-yr lag[‡]	Slope, std error x10⁻⁷, only pos. dose, 20-yr lag[§]
Non-CLL Leukemia	0.123, 3.21 (N=49)	2.14, 3.39 (N=35)	2.69, 3.52 (N=34)	3.33, 3.11 (N=33)	1.39, 1.67 (N=31)
Myeloma	0.233, 0.504 (N=18)	-0.326, 2.80 (N=16)	-0.269, 2.60 (N=15)	-0.318, 2.90 (N=15)	-0.286, 3.24 (N=14)
Benign & Unspecified Neoplasms		0.671, 1.25 (N=14)	0.784, 1.21 (N=13)	0.993, 1.29 (N=12)	1.95*, 0.832 (N=9)
Neoplasms of Unspecified Nature of Nervous System		0.533, 0.572 (N=5)	0.460, 0.606 (N=5)	0.594, 0.462 (N=4)	0.477, NA (N=2)
Diabetes Mellitus	1.73, 4.05 (N=122)	3.90, 4.02 (N=83)	3.31, 4.98 (N=83)	4.94, 5.52 (N=78)	2.70, 7.23 (N=74)
Diseases of Blood		0.186, 0.694 (N=9)	0.182, 0.725 (N=9)	0.208, 0.784 (N=9)	0.289, 0.947 (N=8)
Non-pernicious & Unspecified Anemias		1.55, NA (N=3)	1.40, NA (N=3)	1.34, NA (N=3)	-1.57, NA (N=2)
Alcoholism	-0.566, 0.897 (N=20)	-1.09, 1.51 (N=18)	-0.698, 1.78 (N=17)	-1.03, 1.27 (N=16)	-1.06, 0.875 (N=12)
Other Mental Disorders		0.0803, 3.79 (N=28)	0.171, 0.343 (N=27)	0.070, 3.50 (N=27)	1.27, 2.97 (N=23)
Diseases of Nervous System & Sense Organs	-3.87*, 1.15 (N=109)	-1.91, 2.07 (N=69)	-1.53, 2.54 (N=68)	-0.644, 2.65 (N=64)	-2.71, 2.22 (N=57)
Diseases of Heart		-9.50, 24.7 (N=1218)	-10.6, 30.4 (N=1189)	-8.74, 39.4 (N=1104)	-54.2, 32.8 (N=825)
Ischemic Heart Disease		-5.50, 11.8 (N=1010)	-5.52, 17.2 (N=984)	5.09, 18.4 (N=905)	-56.2, 33.7 (N=659)
Hypertension with Heart Dis.		-0.452, 1.78 (N=29)	-0.345, 1.56 (N=28)	-2.34*, 1.08 (N=27)	-1.26, 1.44 (N=19)
Other Diseases of Circulatory System	-12.2, 11.0 (N=425)	-7.43, 9.43 (N=278)	-6.14, 10.6 (N=273)	-6.98, 11.8 (N=258)	-17.0, 9.10 (N=204)
Hypertension w/o Heart Dis.	-1.07, 0.921 (N=17)	-0.403, 0.614 (N=11)	-0.371, 0.785 (N=11)	-1.43, 2.47 (N=11)	-5.16*, 0.067 (N=10)
Cerebrovascular Disease	-4.38, 9.43 (N=246)	-0.917, 8.43 (N=160)	-0.138, 9.44 (N=156)	0.192, 9.46 (N=145)	-2.82, 8.37 (N=112)
Disease of Respiratory Syst.		-2.28, 6.51 (N=322)	-4.50, 7.04 (N=319)	-2.09, 8.36 (N=311)	-1.59, 9.55 (N=250)
Pneumonia		-2.08, 1.80 (N=75)	-1.77, 1.40 (N=73)	-1.70, 1.13 (N=72)	-0.747, 3.17 (N=61)
Chronic & Unspecified Bronchitis		-0.191, 1.38 (N=11)	-0.212, 1.37 (N=11)	0.209, 0.937 (N=9)	5.54, 8.71 (N=6)
Emphysema		-4.37*, 1.53 (N=52)	-4.01*, 1.86 (N=51)	-4.05, 2.85 (N=51)	-4.19*, 1.70 (N=32)
Asthma		0.459, 0.266 (N=7)	0.397, 0.252 (N=7)	0.450, 0.949 (N=6)	1.47*, 0.473 (N=4)
Asbestosis		-0.891*, 0.261 (N=5)	-0.921*, 0.247 (N=5)	-1.47*, 0.0199 (N=5)	1.55, 1.18 (N=3)
Other Respiratory Disease		0.878, 6.30 (N=166)	-1.97, 5.65 (N=166)	-0.961, 6.64 (N=162)	-2.59, 7.59 (N=140)
Diseases of Digestive System	-10.2*, 5.02 (N=191)	-7.69*, 2.70 (N=123)	-7.72*, 3.44 (N=123)	-10.0, 5.62 (N=117)	-8.12*, 3.71 (N=81)

Table 5-12. Slopes and standard errors (for badged white male workers, using on-site dose only), adjusted for age and calendar year. Slope represents change in standardized rate per mSv-person-year.

Cause of death	Slope, std error x10^{-7}, including zero dose, no lag	Slope, std error x10^{-7}, only pos. dose, no lag	Slope, std error x10^{-7}, only pos. dose, 5-yr lag[†]	Slope, std error x10^{-7}, only pos. dose, 10-yr lag[‡]	Slope, std error x10^{-7}, only pos. dose, 20-yr lag[§]
Cirrhosis of Liver	-6.01, 3.98 (N=76)	-4.99, 3.30 (N=49)	-5.23, 3.65 (N=49)	-12.1, 9.66 (N=46)	-16.6, 11.6 (N=28)
Diseases of Genitourinary System		-3.90, 7.70 (N=37)	-3.89, 7.76 (N=37)	-3.63, 9.63 (N=35)	-3.77, 2.31 (N=28)
Chronic & Unspec. Nephritis & Chronic Renal Failure		-2.53, 3.14 (N=30)	-2.60, 3.16 (N=30)	-1.57, 4.21 (N=29)	-1.09, 3.69 (N=23)
Diseases of Skin	-1.22, 2.29 (N=6)	-1.38, 4.18 (N=5)	-1.35, 4.34 (N=5)	-1.29, 4.55 (N=5)	-1.19, 3.59 (N=4)
Diseases of Musculoskeletal & Connective Tissue		0.280, 0.641 (N=8)	0.0981, 0.776 (N=8)	0.235, 0.354 (N=7)	-0.0132, 1.10 (N=7)
Symptoms & Ill-defined Conditions	3.31, 3.31 (N=56)	1.85, 3.75 (N=44)	2.43, 3.16 (N=43)	0.915, 6.98 (N=43)	0.636, 4.51 (N=35)
Accidents		5.18, 7.51 (N=247)	15.5, 11.4 (N=214)	-11.9, 15.9 (N=188)	-17.4, 14.8 (N=107)
Transportation Accidents		3.06, 6.85 (N=140)	3.30, 6.01 (N=117)	-7.04, 8.30 (N=100)	-4.85*, 2.13 (N=54)
Accidental Falls		-1.92, 1.95 (N=25)	-1.93, 2.06 (N=25)	-2.03, 1.97 (N=25)	10.0, 7.64 (N=17)
Other Accidents		3.75, 2.26 (N=67)	7.93*, 2.04 (N=59)	2.32, 5.60 (N=51)	-1.93, 1.72 (N=30)
Suicide	-0.941, 5.74 (N=167)	0.213, 7.00 (N=118)	-3.70, 11.0 (N=107)	-6.54, 9.08 (N=92)	-4.06, 5.99 (N=56)
Homicide		2.22, 2.55 (N=17)	3.14*, 1.57 (N=14)	-1.25, 2.26 (N=12)	9.18, 6.10 (N=5)
HIV-related	-1.89*, 0.484 (N=8)	-1.96*, 0.070 (N=6)	-1.31, 0.886 (N=5)	-2.73*, 0.904 (N=5)	NA, NA (N=1)
Other & Unspecified		-13.6, 7.39 (N=148)	-14.1*, 6.35 (N=142)	-14.0*, 7.10 (N=129)	-5.97, 8.91 (N=79)

*Slope is significantly different from zero (p<0.05)
[†] 2-year lag for leukemia.
[‡] 5-year lag for leukemia.
[§] 10-year lag for leukemia.
[††] NA: standard error not available because number of deaths in category is too small.

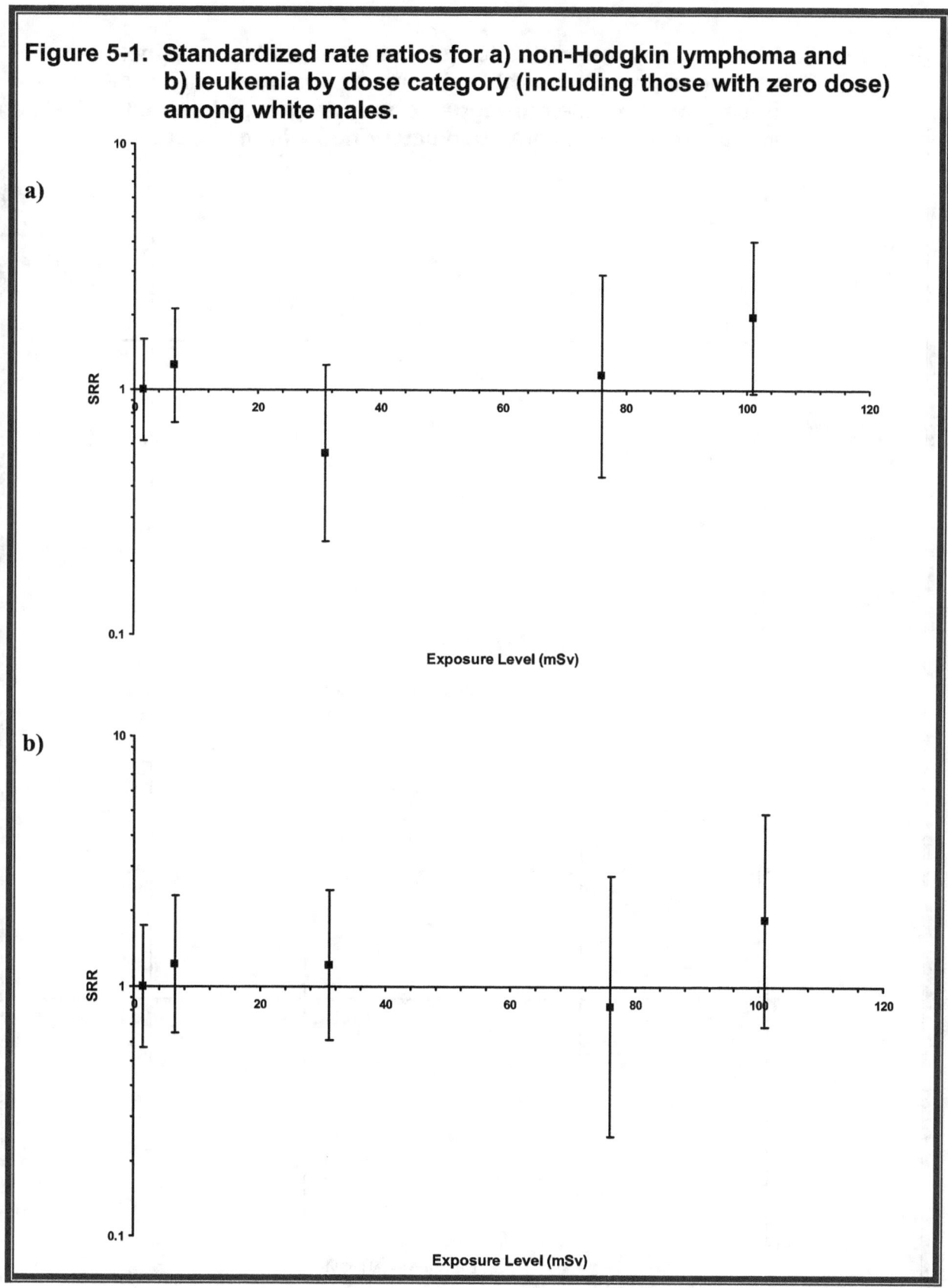

Figure 5-1. Standardized rate ratios for a) non-Hodgkin lymphoma and b) leukemia by dose category (including those with zero dose) among white males.

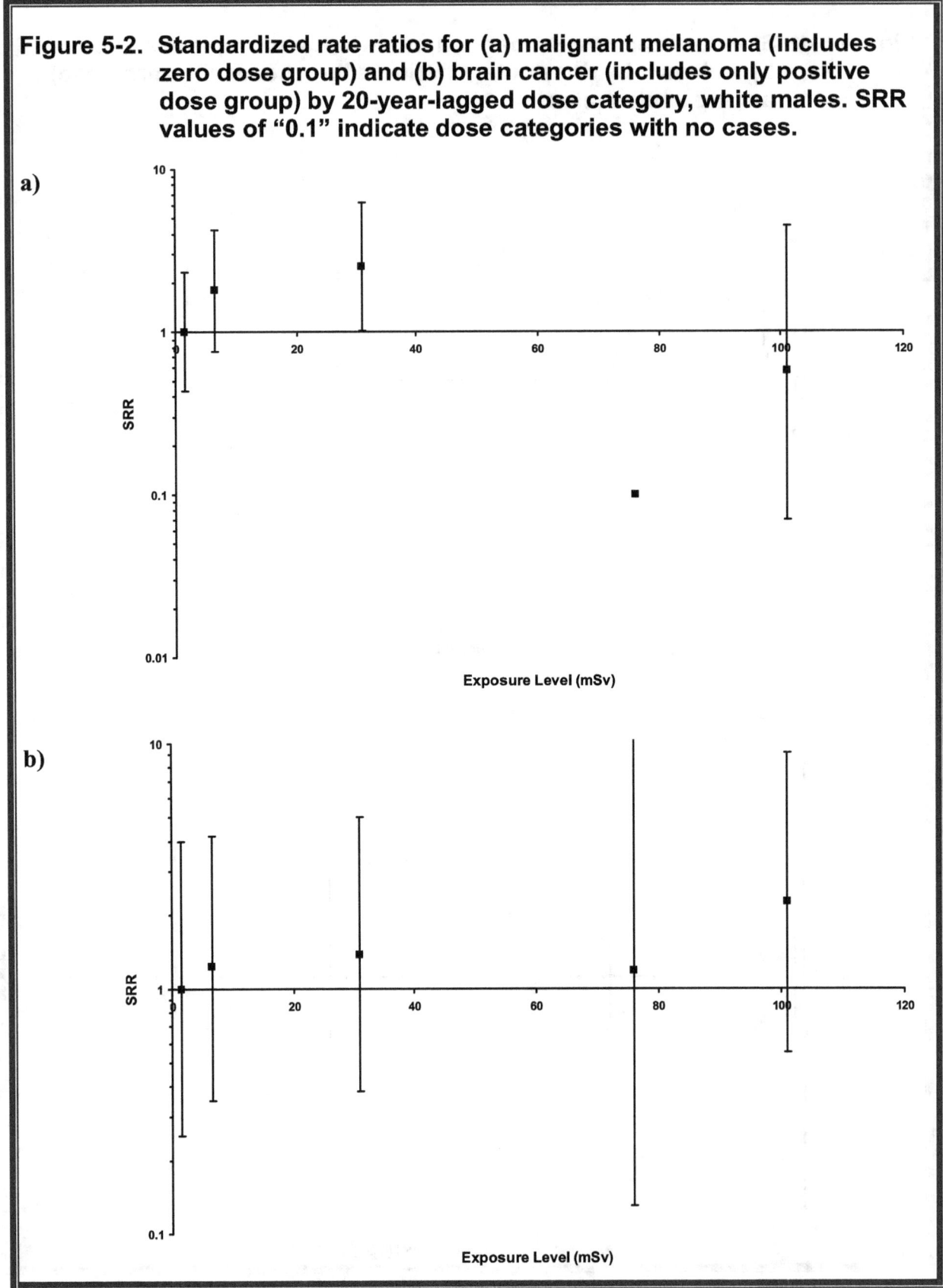

Figure 5-2. Standardized rate ratios for (a) malignant melanoma (includes zero dose group) and (b) brain cancer (includes only positive dose group) by 20-year-lagged dose category, white males. SRR values of "0.1" indicate dose categories with no cases.

5.4.2 White Females

External dose slope results for WF are shown in Table 5-13 for causes of death containing five or more cases among badged workers. Slope results were substantially different from those of WM, and for all disease outcomes, generalization is difficult because of the extremely low number of person-years in dose categories corresponding to 50 mSv or greater. Digestive cancer, particularly pancreatic cancer, showed significantly positive slope, with observed cases in the first three dose categories (Figure 5-3). The breast cancer standardized rate slope was also significantly positive with respect to radiation exposure (Table 5-13), and risks were elevated in the two intermediate dose categories (Figure 5-4). No cases occurred at doses greater than 50 mSv; however, very little person-time was accrued above these doses. All leukemia and NHL deaths among WF occurred in those who received less than 10 mSv cumulative dose (data not shown).

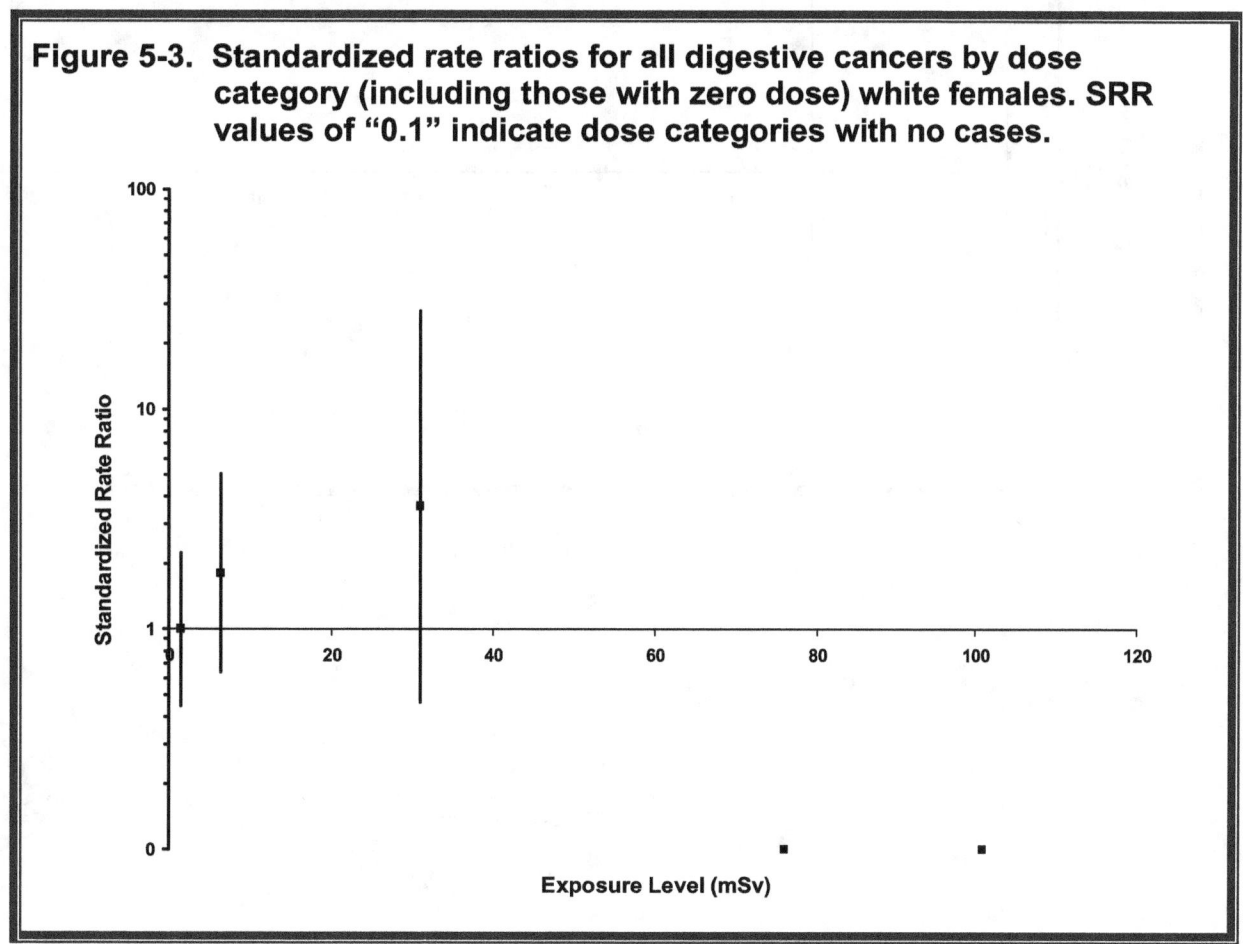

Figure 5-3. Standardized rate ratios for all digestive cancers by dose category (including those with zero dose) white females. SRR values of "0.1" indicate dose categories with no cases.

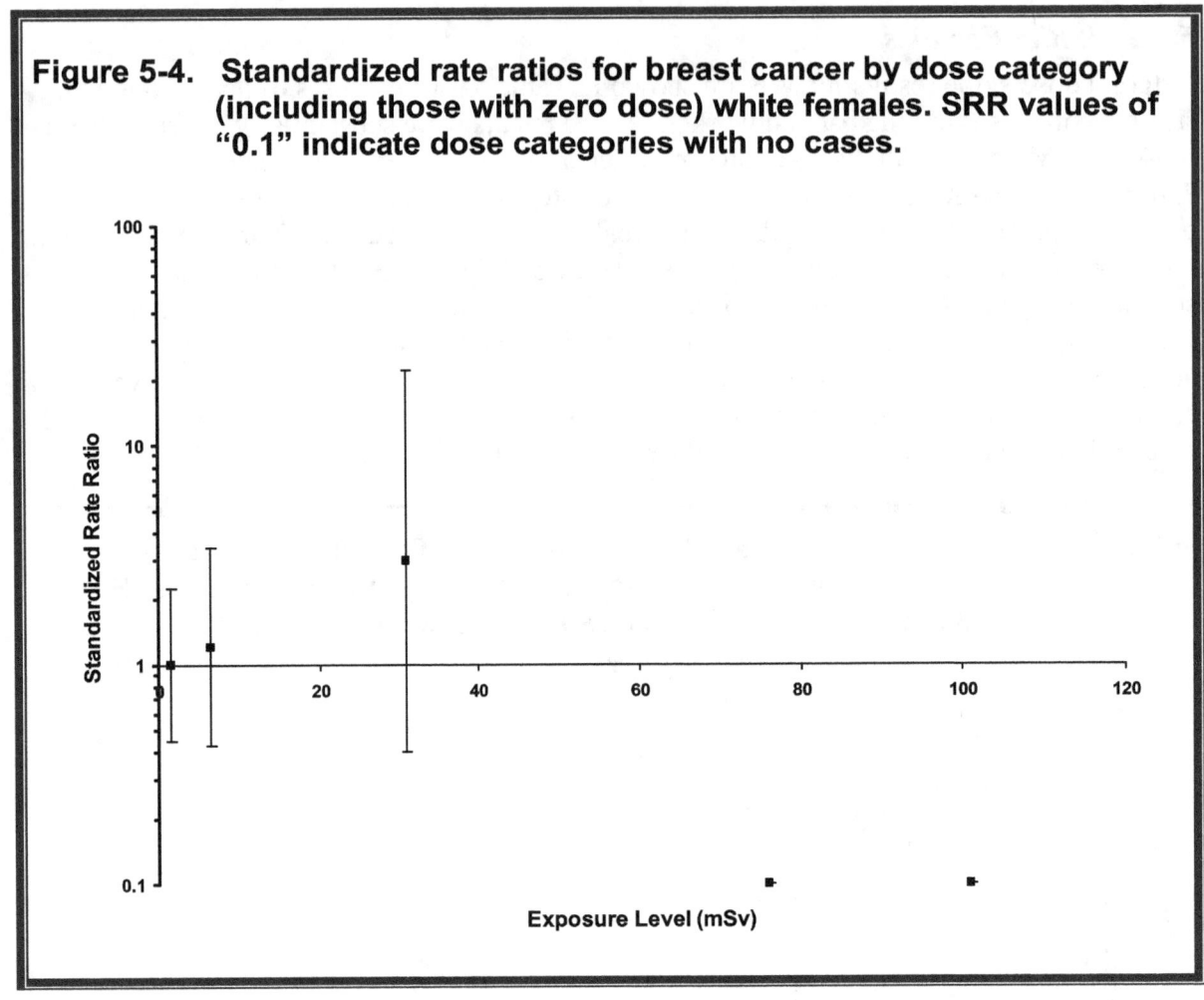

Figure 5-4. Standardized rate ratios for breast cancer by dose category (including those with zero dose) white females. SRR values of "0.1" indicate dose categories with no cases.

Table 5-13. SMR and dose-response trends (for badged white female workers, on-site dose only), adjusted for age and calendar year. Slope represents change in standardized rate per mSv-person-year.

Cause of death	Slope, std error x10⁻⁷, including zero dose, 0 lag	Slope, std error x10⁻⁷, only pos. dose, 0 lag	Slope, std error x10⁻⁷, only positive dose, 5-yr lag[†]	Slope, std error x10⁻⁷, only positive dose, 10-yr lag[‡]	Slope, std error x10⁻⁷, only positive dose, 20-yr lag[§]
MN Digestive	147*, 45 (N=19)	132*, 36.8 (N=13)	76.1*, 34.9 (N=13)	39.8, 113 (N=9)	82.7, 104 (N=6)
MN Intestine	*237, NA[††] (N=9)*	*219, NA (N=8)*	*38.2, NA (N=8)*	*21.4, NA (N=6)*	*-119, NA (N=5)*
MN Pancreas	*113*, 33.4 (N=5)*	*101, 56.3 (N=4)*	*107*, 52.7 (N=4)*	*154, NA (N=3)*	*-109, NA (N=1)*
	-236, NA (N=18)	-458, NA (N=10)	-470, NA, (N=10)	-398, NA (N=9)	-472, NA (N=7)
	-187, NA (16)	*-288, NA (N=8)*	*-299, NA (N=8)*	*-222, NA (N=7)*	*-254, NA (N=5)*
MN Breast	134*, 28.8 (N=27)	157*, 62.7 (N=14)	121*, 44.6 (N=14)	187*, 33.5 (N=13)	-94.5, NA (N=10)
	-75.8, NA (N=12)	-57.2, NA (N=5)	-74.5, NA (N=5)	-122, NA (N=5)	24.6, NA (N=4)
	NA, NA (N=5)	*NA, NA (N=1)*	*NA, NA (N=1)*	*NA, NA (N=1)*	*NA, NA (N=1)*
MN Other & Unspecified Sites	-114, NA (N=13)	44.2, NA (N=4)	44.5, NA (N=4)	35.0, NA (N=4)	132, NA (N=3)
MN Brain & Other Nervous System	*32.5, NA (N=7)*	*-4.02, NA (N=2)*	*159, NA (N=2)*	*44.0, NA (N=2)*	*30.7, NA (N=2)*
	-23.7, NA, (N=15)	-34.9, NA (N=8)	-32.6, NA (N=8)	-27.8, NA (N=8)	194, NA (N=5)
	-9.97, NA (N=5)	8.02, NA (N=2)	11.3, NA (N=2)	21.8, NA (N=2)	45.3, NA (N=1)
	50.8, NA (N=7)	*78.1, NA (N=4)*	*81.5, NA (N=4)*	*82.7, NA (N=4)*	*82.0, NA (N=4)*
Diabetes mellitus	115, NA (N=8)	39.4, NA (N=7)	36.0, NA (N=7)	45.6, NA (N=7)	103, NA (N=6)
	-48.4, NA (N=7)	-66.9, NA (N=4)	-69.1, NA (N=4)	-86.1, NA (N=4)	NA, NA (N=3)
Diseases of Heart	-28.1, 23.7 (N=66)	-24.6, 30.1 (N=33)	-24.9, 30.8 (N=33)	-26.8, 32.2 (N=33)	-25.3, 32.7 (N=30)
Ischemic Heart Disease	*-96.4, NA (N=41)*	*114, NA (N=20)*	*147, NA (N=20)*	*12.3, NA (N=20)*	*-279, NA (N=18)*
	-61.7, NA (N=26)	-372, NA (N=15)	-295, NA (N=14)	-65.3, NA (N=12)	-3290, NA (N=10)
	80.8, NA (N=20)	-254, NA (N=13)	-158, NA (N=12)	-416, NA (N=10)	-362, NA (N=9)
Disease of Respiratory System	25.8, 21.2 (N=22)	9.25, 112 (N=12)	-43.2, 227 (N=11)	-46.0, 246 (N=9)	1070, NA (N=7)
Other Respiratory Disease	*61.7*, 17.8 (N=16)*	*66.9, 37.3 (N=9)*	*15.7, 138 (N=8)*	*32.2, 176 (N=7)*	*1070, NA (N=7)*
	70.9*, 24.2 (N=11)	58.2*, 6.1 (N=7)	69.3*, 2.00 (N=7)	94.3*, 13.8 (N=7)	3.89, 280 (N=6)

Table 5-13. SMR and dose-response trends (for badged white female workers, on-site dose only), adjusted for age and calendar year. Slope represents change in standardized rate per mSv-person-year.

Cause of death	Slope, std error x10^{-7}, including zero dose, 0 lag	Slope, std error x10^{-7}, only pos. dose, 0 lag	Slope, std error x10^{-7}, only positive dose, 5-yr lag†	Slope, std error x10^{-7}, only positive dose, 10-yr lag‡	Slope, std error x10^{-7}, only positive dose, 20-yr lag§
Diseases of Genitourinary System	139*, 4.27 (N=5)	83.4, NA (N=4)	134, NA (N=4)	114, NA (N=4)	-1130, NA (N=3)
	12.5, 75.1 (N=6)	-31.3, 146 (N=5)	-24.7, 151 (N=5)	-29.0, 173 (N=5)	25.8, 146 (N=4)
Accidents	26.8, 60.5 (N=32)	123*, 36.5 (N=9)	169*, 50.5 (N=8)	472*, 97.3 (N=8)	-608, NA (N=7)
Transportation Accidents	*-22.0, 152 (N=25)*	*105*, 0.611, (N=7)*	*142*, 15.8 (N=6)*	*383, NA (N=6)*	*175, NA (N=5)*
	NA, NA (N=6)	NA, NA (N=1)	NA, NA (N=0)	NA, NA (N=0)	NA, NA (N=0)
Other & Unspecified	-230, NA (N=14)	-483, NA (N=8)	-518, NA (N=8)	-569, NA (N=8)	-148, NA (N=3)

*Slope is significantly different from zero (p<0.05).
† 2-year lag for leukemia.
‡ 5-year lag for leukemia.
§ 10-year lag for leukemia.
†† NA: standard error not available because number of deaths in category is too small.

5.5 Internal Exposure Category

5.5.1 White Males

White male workers' person-time was classified into three monitoring categories: unmonitored, monitored but no likely exposure, and monitored with likely positive exposure (Table 5-14). For WM, overall death rates were lower among the monitored but unexposed workers and the exposed workers, compared to the general regional population. SMRs for all causes were 0.89 and 0.83, respectively. The all-cancer mortality rate was substantially elevated among the non-monitored and monitored but unexposed, and was lower among internally exposed workers, compared to the general population.

Estimated SRRs for many cancers, such as oral cavity, digestive tract, brain, connective tissue, NHL and "other and unspecified" sites were elevated among the monitored but unexposed workers and lower among the monitored and exposed workers, compared to unmonitored workers. However, CIs were wide for most of these SRRs. Other diseases, such as respiratory and esophagus cancer, multiple myeloma, non-malignant urinary disease, and lifestyle-related causes of death (e.g., alcoholism, emphysema, cirrhosis of liver) exhibited decreasing rates with increasing likelihood of internal exposure.

"Other respiratory cancers" (including cancers of pleura) were elevated in all groups, and the two monitored categories of workers showed higher rates than the unmonitored group. Testicular cancer appeared elevated (although highly uncertain) among the more-exposed workers. Accounting for the three testicular cancers occurring before 1960, however, greatly reduced this apparent association. The bone cancer death rate was elevated in the most-exposed group, although the SRR estimate (7.33) was based on only two cases and is highly uncertain. Stomach and urinary tract cancers showed slight elevations in the monitored and exposed group, but 95% CIs included one in each instance.

Table 5-14. Standardized rate ratios for internal monitoring compared to unmonitored person-time, white males only. If no cases were observed, the number expected is given, based on state rates.

Cause of death	SMR non-monitored (N)	SRR monitored-not exposed (95% CI)	SRR monitored & exposed (95% CI)
MN Buccal Cavity	**0.77 (28)**	**1.90 (0.76-4.77)**	**0.791 (0.33-1.91)**
MN Pharynx	*0.59 (10)*	*2.99 (0.77-11.6)*	*1.12 (0.31-4.07)*
MN Digestive	**1.07 (457)**	**1.18 (0.89-1.55)**	**0.816 (0.66-1.01)**
MN Esophagus	*1.11 (51)*	*0.949 (0.40-2.24)*	*0.643 (0.32-1.31)*
MN Stomach	*1.02 (59)*	*0.871 (0.37-2.06)*	*1.18 (0.69-2.00)*
MN Intestine	*1.09 (162)*	*1.30 (0.82-2.05)*	*0.622 (0.41-0.94)*
MN Rectum	*0.86 (30)*	*0.491 (0.12-2.07)*	*1.18 (0.58-2.42)*
MN Liver & Gall Bladder	*1.12 (31)*	*1.45 (0.58-3.61)*	*0.830 (0.36-1.89)*
MN Liver Unspecified	*0.89 (9)*	*1.51 (0.19-12.0)*	*1.25 (0.34-4.61)*
MN Pancreas	*1.10 (106)*	*1.32 (0.77-2.26)*	*0.857 (0.55-1.34)*
MN Peritoneum & other	*1.19 (9)*	*1.72 (0.37-8.03)*	*0.779 (0.17-3.61)*
MN Respiratory	**1.20‡ (705)**	**0.990 (0.77-1.27)**	**0.633 (0.52-0.77)**
MN Larynx	*0.98 (18)*	*1.04 (0.24-4.51)*	*0.984 (0.36-2.66)*

Table 5-14. Standardized rate ratios for internal monitoring compared to unmonitored person-time, white males only. If no cases were observed, the number expected is given, based on state rates.

Cause of death	SMR non-monitored (N)	SRR monitored-not exposed (95% CI)	SRR monitored & exposed (95% CI)
MN Trachea, Bronchus & Lung	1.20‡ (672)	0.960 (0.74-1.24)	0.620 (0.51-0.76)
MN Other Respiratory	2.19‡ (15)	2.29 (0.78-6.71)	0.811 (0.17-3.94)
MN Breast	**1.99 (4)**	0 (0.21 exp.)	0.831 (0.09-7.43)
MN Male Genital	**1.12 (210)**	0.862 (0.52-1.42)	0.948 (0.69-1.30)
MN Prostate	1.15† (209)	0.866 (0.52-1.43)	0.938 (0.68-1.28)
MN Testis	0.17† (1)	0 (0.81 exp.)	3.09 (0.19-49.5)
MN Testis, 1940+*	0.37 (3)	2.50 (0.26-24.0)	1.02 (0.11-9.76)
MN Urinary Organs	**1.11 (108)**	0.543 (0.24-1.21)	1.22 (0.82-1.81)
MN Kidney	1.13 (56)	0.671 (0.26-1.72)	1.20 (0.67-2.15)
MN Bladder	1.09 (52)	0.406 (0.09-1.77)	1.23 (0.72-2.12)
MN Other & Unspecified Sites	**1.13† (285)**	1.19 (0.85-1.68)	0.746 (0.55-1.01)
MN Skin Melanoma	1.24 (41)	0.930 (0.34-2.53)	0.768 (0.29-2.03)
MN Brain & Other Nervous System	1.11 (66)	1.33 (0.71-2.50)	0.625 (0.32-1.22)
MN Thyroid	1.70 (5)	0 (0.34 exp.)	0 (0.77 exp.)
MN Bone	0.18 (1)	0 (0.64 exp.)	7.33 (0.66-81.3) N=2
MN Connective Tissue	0.80 (9)	2.94 (0.74-11.8)	0.887 (0.19-4.16)
MN Other & Unspecified	1.19† (155)	1.16 (0.72-1.88)	0.779 (0.53-1.14)
MN Lymphatic & Hematopoietic	**1.11 (211)**	0.838 (0.52-1.35)	0.788 (0.56-1.11)
Non-Hodgkin Lymphoma	1.31† (90)	1.20 (0.64-2.25)	0.825 (0.49-1.38)
Hodgkin's Disease	0.52 (6)	1.23 (0.15-10.3)	2.85 (0.61-13.2)
Leukemia	1.13 (85)	0.495 (0.19-1.30)	0.675 (0.39-1.15)
Myeloma	0.88 (30)	0.641 (0.15-2.70)	0.590 (0.23-1.54)
Benign & Unspec. Nature Neoplasms	**1.19 (29)**	0 (2.80 exp.)	1.01 (0.46-2.21)
Benign of Nervous System	0.81 (2)	0 (0.28 exp.)	0 (0.67 exp.)
Unspecified of Nervous System	0.79 (9)	0 (1.42 exp.)	0.820 (0.18-3.80)
All Cancers, SMRs only (N), 95% CI	**1.13‡ (2008), 1.08-1.18**	1.15† (232), 1.01-1.31	0.89† (409), 0.80-0.98
Diabetes Mellitus	**0.96 (123)**	1.19 (0.72-1.97)	1.30 (0.92-1.85)
Diseases of Blood & Blood-Forming Organs	**0.88 (21)**	0 (2.52 exp.)	0.865 (0.33-2.30)
Non-pernicious Anemia	1.26 (10)	0 (0.81 exp.)	0.360 (0.05-2.81)
Alcoholism	**0.76 (32)**	0.519 (0.10-2.66)	0.321 (0.10-1.05)
Other Mental Disorders	**0.99 (44)**	2.31 (1.15-4.62)	1.02 (0.55-1.90)
Diseases of Nervous System & Sense Organs	**1.20† (151)**	0.554 (0.26-1.16)	0.905 (0.63-1.30)
Diseases of Heart	**0.95† (2391)**	0.822 (0.71-0.96)	0.877 (0.80-0.96)
Ischemic Heart Disease	0.97 (‡1974)	0.800 (0.67-0.95)	0.911 (0.82-1.01)
Hypertens w/Heart Disease	1.78‡ (52)	1.22 (0.51-2.92)	0.840 (0.43-1.63)
Other Diseases of Circ. System	**0.94 (573)**	1.10 (0.83-1.46)	0.886 (0.73-1.07)
Hypertension w/o Heart Disease	1.13 (22)	1.25 (0.37-4.20)	0.690 (0.23-2.02)
Cerebrovascular Disease	0.93 (345)	1.19 (0.83-1.71)	0.842 (0.66-1.08)
Diseases of Respiratory System	**1.06 (725)**	1.10 (0.86-1.41)	0.715 (0.60-0.86)
Pneumonia	0.92 (148)	1.49 (0.90-2.44)	0.859 (0.59-1.25)
Chronic & Unspec. Bronchitis	0.94 (18)	2.26 (0.49-10.3)	0.662 (0.19-2.26)

Table 5-14. Standardized rate ratios for internal monitoring compared to unmonitored person-time, white males only. If no cases were observed, the number expected is given, based on state rates.

Cause of death	SMR non-monitored (N)	SRR monitored-not exposed (95% CI)	SRR monitored & exposed (95% CI)
Emphysema	*1.13 (126)*	*0.931 (0.49-1.76)*	*0.590 (0.36-0.96)*
Asthma	*0.85 (15)*	*0 (1.99 exp.)*	*1.05 (0.35-3.17)*
Asbestosis	*3.58‡ (8)*	*0 (0.24 exp.)*	*0.924 (0.20-4.35)*
Silicosis	*0.39 (1)*	*0 (0.22 exp.)*	*3.93 (0.25-62.8)*
Other Respiratory Disease	*1.13† (403)*	*1.02 (0.73-1.42)*	*0.654 (0.51-0.84)*
Diseases of Digestive System	**1.05 (313)**	**0.682 (0.44-1.06)**	**0.485 (0.35-0.68)**
Cirrhosis of Liver	*1.09 (147)*	*0.582 (0.30-1.13)*	*0.297 (0.16-0.55)*
Diseases of Genito-urin. System	**1.09 (89)**	**0.747 (0.27-2.03)**	**0.422 (0.22-0.82)**
Acute Glomerulonephritis & Acute Renal Failure	*0.98 (8)*	*0 (0.85 exp.)*	*0 (2.22 exp.)*
Chronic & Unspec. Nephritis, Renal Failure	*1.38† (57)*	*0.520 (0.16-1.67)*	*0.530 (0.25-1.12)*
Diseases of Skin	**1.44 (6)**	**1.88 (0.23-15.6)**	**0.565 (0.07-4.69)**
Diseases of Musculoskeletal & Connective Tissue	**0.55 (11)**	**0.671 (0.09-5.20)**	**0.923 (0.26-3.32)**
Symptoms & Ill-Defined Cond.	**0.83 (68)**	**1.19 (0.59-2.43)**	**0.975 (0.58-1.64)**
Accidents	**0.81‡ (563)**	**0.777 (0.58-1.04)**	**0.695 (0.53-0.91)**
Transportation Accidents	*0.79‡ (315)*	*0.878 (0.61-1.27)*	*0.677 (0.48-0.95)*
Accidental Falls	*1.06 (62)*	*0.794 (0.33-1.91)*	*0.634 (0.32-1.25)*
Suicide	**0.78‡ (217)**	**0.793 (0.51-1.23)**	**1.07 (0.69-1.66)**
Homicide	**0.77 (38)**	**0.336 (0.08-1.49)**	**0.412 (0.15-1.16)**
HIV-related	**1.24 (19)**	**0.198 (0.03-1.48)**	**0.191 (0.03-1.43)**
Other & Unspecified	**2.42‡ (309)**	**0.946 (0.63-1.41)**	**0.628 (0.45-0.87)**
All Deaths (SMRs only)	**1.01 (7731)**	**0.89‡ (761)**	**0.83‡ (1603)**

* SMR calculated using U.S. rates for 1940-1999. SRR calculated using cohort rates 1940-1999.
†95% CI of SMR excludes 1.00. SRRs are not flagged.
‡99% CI of SMR excludes 1.00. SRRs are not flagged.

5.5.2 White Females

All-cause and all-cancer mortality rates were very similar to the local regional population among each of the internal exposure categories for WF, although the total number of deaths was very small for the two monitored categories (Table 5-15).

Individual COD categories showed some variability in rates among exposure categories. Cancers of buccal cavity were suggestively elevated, although with a wide CI, among monitored and exposed workers compared to unmonitored workers. Digestive cancer rates were elevated, with relatively high precision, among the monitored but unexposed workers, because of elevations in cancers of intestine, liver and pancreas. The lung cancer mortality rate in exposed white female workers was approximately one-quarter that of unmonitored workers (CI of SRR, 0.05-0.93). Both bladder cancer deaths for WF occurred among the internally exposed group (SMR=8.48; CI, 0.95-36.8), a substantial elevation compared to the regional population. The sole bone cancer death among WF occurred in the non-monitored group.

Ischemic heart disease was more than doubled among internally exposed workers, compared to unexposed workers (CI of SRR, 0.95-8.98). The mortality rate for diseases of the genitourinary system was highly elevated among exposed WF, compared to unmonitored workers (SRR=7.16; CI, 1.43-35.8, N=4). This elevation was due to the non-specific category "other genitourinary diseases" (SRR=4.49; CI, 0.63-31.9, N=2). The mortality rates for transportation accidents were higher among the two internally monitored groups, compared to unmonitored white female workers.

Table 5-15. SRRs for internal monitoring compared to unmonitored person-time, white females. If no cases were observed, the number expected is given, based on state rates.

Cause of death	SMR non-monitored (N)	SRR monitored-not exposed (95% CI)	SRR monitored & exposed (95% CI)
MN Buccal cavity	2.30 (4)	0 (0.28 exp.)	2.70 (0.48-15.0)
MN Digestive	0.71 (19)	2.61* (1.16-5.90)	1.05 (0.37-3.02)
MN Stomach	1.55 (4)	0 (0.41 exp.)	0 (0.50 exp.)
MN Intestine	0.62 (7)	1.97 (0.51-7.69)	1.51 (0.38-5.94)
MN Rectum	0.54 (2)	4.87 (0.30-77.8)	0 (0.37 exp.)
MN Liver & Gall Bladder	0.88 (2)	3.22 (0.29-35.5)	0 (0.41 exp.)
MN Pancreas	0.62 (4)	2.64 (0.48-14.4)	1.81 (0.20-16.2)
MN Respiratory	0.99 (30)	0.97 (0.37-2.50)	0.45 (0.13-1.59)
MN Trachea, Bronchus, Lung	0.98 (29)	1.00 (0.39-2.60)	0.22 (0.05-0.93)
MN Breast	1.16 (41)	0.54 (0.19-1.52)	0.57 (0.20-1.63)
MN Female Genital	0.69 (13)	0.80 (0.18-3.56)	2.39 (0.80-7.11)
MN Cervix Uteri	1.08 (5)	0 (0.78 exp.)	0.85 (0.10-7.28)
MN Ovary	0.48 (5)	2.16 (0.42-11.1)	0.78 (0.09-6.67)
MN Urinary Organs	0.59 (2)	0 (0.53 exp.)	3.81 (0.54-27.1)
MN Kidney	0.90 (2)	0 (0.35 exp.)	0 (0.41 exp.)
MN Bladder	0 (1.18 exp.)	0 (0.18 exp.)	SMR=8.48 (2)
MN Other & Unspecified	0.85 (18)	0.85 (0.25-2.88)	0.75 (0.21-2.68)
MN Skin Melanoma	0.38 (1)	0 (0.45 exp.)	0 (0.36 exp.)
MN Brain & Other Nervous Sys.	1.54 (8)	1.22 (0.26-5.74)	0.45 (0.06-3.59)
MN Other & Unspecified	0.66 (7)	0.77 (0.09-6.27)	1.41 (0.28-7.18)
MN Lymphatic & Hematopoietic	1.29 (18)	0.87 (0.26-2.98)	0.88 (0.19-4.03)
Non-Hodgkin Lymphoma	1.45 (8)	0 (0.90 exp.)	1.24 (0.16-9.95)
Leukemia	1.28 (7)	2.33 (0.60-9.06)	0 (0.90 exp.)
Myeloma	1.37 (3)	0 (0.34 exp.)	0 (0.41 exp.)
Benign & Unspecified Nature Neoplasms	1.73 (4)	0 (0.37 exp.)	0 (0.44 exp.)
Unspecified of Nervous System	3.23 (3)	0 (0.15 exp.)	0 (0.16 exp.)
All Cancers (SMRs only)	0.96 (145)	1.05 (26)	1.01 (26)
Diabetes Mellitus	0.94 (12)	0.55 (0.07-4.21)	1.72 (0.46-6.48)
Other Mental Disorders	1.91 (7)	0 (0.51 exp.)	0.46 (0.06-3.77)
Diseases of Nervous System & Sense Organs	1.23 (15)	0.70 (0.16-3.05)	0 (2.18 exp.)
Diseases of Heart	0.84 (77)	0.85 (0.44-1.67)	2.18 (0.86-5.51)
Ischemic Heart Disease	0.73* (47)	1.13 (0.53-2.43)	2.92 (0.95-8.98)
Hypertension w/Heart Disease	2.45 (5)	0 (0.28 exp.)	1.28 (0.25-6.63)
Other Diseases of Circulatory System	1.00 (43)	0.85 (0.35-2.09)	0.51 (0.19-1.34)
Hypertension w/o Heart Disease	1.39 (2)	0 (0.21 exp.)	0 (0.32 exp.)
Cerebrovascular Disease	1.08 (31)	1.07 (0.40-2.83)	0.70 (0.26-1.89)
Diseases of Respiratory System	0.76 (28)	0.89 (0.31-2.55)	0.48 (0.14-1.73)

Table 5-15. SRRs for internal monitoring compared to unmonitored person-time, white females. If no cases were observed, the number expected is given, based on state rates.

Cause of death	SMR non-monitored (N)	SRR monitored-not exposed (95% CI)	SRR monitored & exposed (95% CI)
Pneumonia	0.56 (5)	0 (1.29 exp.)	0 (1.92 exp.)
Emphysema	0.24 (1)	0 (0.64 exp.)	3.73 (0.23-59.7)
Asthma	0.93 (2)	0 (0.36 exp.)	0 (0.37 exp.)
Other Respiratory Disease	1.01 (20)	0.88 (0.26-2.97)	0.15 (0.02-1.11)
Diseases of Digestive System	0.81 (17)	0.60 (0.14-2.63)	1.09 (0.35-3.40)
Cirrhosis of Liver	0.68 (6)	0.92 (0.11-7.65)	0.64 (0.08-5.31)
Diseases of Genitourinary System	0.96 (5)	0 (0.78 exp.)	7.16* (1.43-35.8)
Chronic & Unspecified Nephritis & Renal Failure	1.21 (3)	0 (0.37 exp.)	1.04 (0.11-9.98)
Diseases of Musculoskeletal & Connective Tissue	0.87 (3)	2.33 (0.24-22.4)	2.07 (0.22-19.9)
Symptoms & Ill-Defined Conditions	1.67 (7)	0 (0.66 exp.)	0.56 (0.07-4.56)
Accidents	0.82 (31)	2.06 (0.96-4.40)	1.95 (0.55-6.85)
Transportation Accidents	0.88 (24)	2.45 (1.08-5.55)	2.11 (0.48-9.19)
Accidental Falls	0.40 (1)	4.75 (0.30-75.9)	0 (0.51 exp.)
Suicide	0.71 (9)	1.43 (0.38-5.32)	0.37 (0.05-2.96)
Homicide	0.87 (4)	1.35 (0.15-12.1)	3.33 (0.61-18.31)
Other & Unspecified	1.40 (18)	2.00 (0.54-7.50)	0.94 (0.25-3.56)
All Deaths (SMRs only)	0.93 (427)	0.98 (71)	0.97 (81)

*95% CI of SMR excludes 1.00. SRRs are not flagged.

5.6 SES Subcohort

5.6.1 White Males

The all-cause mortality rates varied substantially among the SES categories (Table 5-16). Professional, intermediate, skilled non-manual workers and those of unknown SES all showed sharply lower mortality rates. Skilled manual, partly skilled and unskilled workers had substantially higher mortality rates, compared to the general regional population. All cancers as a group showed a much lower mortality rate among professionals, and a much higher mortality rate among skilled manual workers, compared to the regional population.

For many lifestyle-related causes of death, mortality rates were higher among workers of lower SES, compared to professional employees (Table 5-16). The elevation is particularly notable for the SRR for ischemic heart disease, which increased monotonically from professional workers through the intermediate, skilled non-manual, skilled manual, partly skilled and unskilled workers. Mortality rates from cirrhosis of liver and alcoholism showed similar elevations among skilled manual, partly skilled and unskilled workers, compared to professional workers. The death rate from diabetes was approximately doubled for skilled non-manual, skilled manual, partly skilled and unskilled workers, compared to professional workers. Accidental death and homicide rates were substantially elevated among skilled manual, partly skilled and unskilled workers compared to professional workers, although the rates among the professional workers were less than one-third those of the general regional population. Unskilled workers had particularly high rates of death from "other accidents," compared to professional workers. The "other (non-malignant) respiratory disease" mortality rate was highly elevated among these categories in addition to workers of unknown SES.

The rate of pancreatic cancer mortality was nearly doubled among skilled manual and partly skilled workers, compared to professional workers. Lung cancer and emphysema death rates also varied by SES category, with professional workers showing much lower rates than intermediate, skilled manual and partly skilled workers. The emphysema mortality rate was elevated among partly skilled and unskilled workers. The mortality rate from prostate cancer was nearly doubled among intermediate, skilled manual and unskilled workers, compared to professional workers. Skilled manual workers also had elevated rates of kidney and bladder cancer mortality.

The brain cancer mortality rate was elevated, although with wide CIs, among intermediate workers, and the rate of NUN of brain was substantially elevated among partly skilled workers, although based on very small numbers.

Rates of mortality for causes related to asbestos exposure also varied by SES group. No cases of asbestosis and just one case each of "other respiratory" and "peritoneal and other digestive" cancers, occurred among professional WM. Asbestosis rates were highly elevated compared to the general population among workers classified as intermediate and as skilled non-manual workers. Mortality rates for cancers of peritoneum and other digestive organs and of "other respiratory" organs were elevated among skilled manual employees and those of unknown SES, compared to professional workers.

Death rates for cancers of esophagus, stomach, intestine, rectum, liver and gall bladder, breast and skin (melanoma) did not vary substantially by SES category. Very little heterogeneity by SES class was observed in the rates of death from hematopoietic neoplasms, although partly skilled workers showed a low mortality rate from NHL, compared to professional workers.

Table 5-16. SRRs for white males in SES groups Intermediate, Skilled Non-manual, Skilled manual, Partly skilled, Unskilled and unknown, compared to the Professional SES group.

Cause of death	Professional SMR (N)	Intermediate SRR (95% CI)	Skilled Non-man. SRR (95% CI)	Skilled manual SRR (95% CI)	Partly skilled SRR (95% CI)	Unskilled SRR (95% CI)	Unknown SES SRR (95% CI)
MN Buccal cavity	0.77 (7)	0.473 (0.12-1.88)	0.734 (0.09-6.08)	1.48 (0.59-3.71)	1.50 (0.49-4.64)	0.63 (0.12-3.22)	0.14 (0.02-1.13)
MN Pharynx	0.23 (1)	1.14 (0.10-12.6)	0 (0.82 expected)†	2.29 (0.28-18.4)	0.910 (0.06-14.6)	2.31 (0.21-26.1)	0.499 (0.03-7.98)
MN Digestive	1.03 (106)	1.00 (0.75-1.34)	0.881 (0.52-1.49)	1.12 (0.88-1.43)	1.27 (0.92-1.74)	1.30 (0.94-1.81)	1.13 (0.68-1.88)
MN Esophagus	1.00 (12)	0.979 (0.42-2.26)	0.411 (0.05-3.18)	0.944 (0.45-1.97)	1.11 (0.43-2.88)	1.72 (0.69-4.30)	1.39 (0.30-6.46)
MN Stomach	0.89 (12)	1.17 (0.50-2.74)	1.83 (0.57-5.93)	1.47 (0.74-2.95)	1.75 (0.73-4.19)	1.24 (0.45-3.42)	0.906 (0.23-3.61)
MN Intestine	1.24 (44)	0.957 (0.61-1.50)	1.02 (0.47-2.21)	0.805 (0.54-1.20)	0.884 (0.52-1.50)	1.04 (0.60-1.82)	0.863 (0.35-2.13)
MN Rectum	0.49 (4)	1.58 (0.40-6.19)	1.19 (0.12-11.5)	1.34 (0.39-4.65)	1.33 (0.30-5.97)	2.87 (0.72-11.4)	0.830 (0.17-4.15)
MN Liver & Gall Bladder	1.45 (10)	0.545 (0.18-1.61)	0.504 (0.06-3.96)	0.746 (0.33-1.71)	1.04 (0.34-3.17)	0.724 (0.19-2.79)	2.05 (0.51-8.13)
MN Liver Unspec.	1.68 (4)	1.07 (0.24-4.82)	0 (0.47 expected)	0.277 (0.06-1.24)	0 (1.28 expected)	1.59 (0.35-7.13)	0 (0.92 expected)
MN Pancreas	0.82 (19)	1.14 (0.58-2.24)	0.612 (0.14-2.65)	1.94 (1.14-3.30)	2.36 (1.23-4.54)	1.57 (0.77-3.23)	1.30 (0.41-4.00)
MN Peritoneum & Other Digestive				4.06 (0.51-32.4)	1.52 (0.10-24.3)	0 (0.92 expected)	5.60 (0.44-70.9)
MN Respiratory	0.78** (114)	1.29 (1.00-1.67)	1.06 (0.68-1.66)	1.79 (1.44-2.22)	1.48 (1.11-1.97)	1.32 (0.96-1.81)	1.28 (0.85-1.94)
MN Larynx	0.22 (1)	1.04 (0.06-16.6)	0 (0.86 expected)	7.08 (0.92-54.3)	7.34 (0.82-65.8)	7.03 (0.72-68.6)	2.21 (0.20-24.5)
MN Trachea, Bronchus & Lung	0.80* (112)	1.28 (0.99-1.66)	1.05 (0.67-1.65)	1.69 (1.36-2.11)	1.40 (1.05-1.88)	1.26 (0.91-1.75)	1.15 (0.76-1.73)
MN Other Resp.	0.58 (1)	3.14 (0.28-34.7)	4.88 (0.31-78.1)	10.8 (1.39-83.6)	6.80 (0.62-75.2)	2.86 (0.18-45.8)	22.7 (1.67-310)
MN Breast	2.06 (1)	1.28 (0.08-20.5)	0 (0.09 expected)	1.19 (0.11-13.4)	0 (0.27 exp.)	0 (0.23 exp.)	1.27 (0.08-20.3)
MN Male Genital	0.70 (29)	1.87 (1.14-3.06)	1.58 (0.74-3.40)	1.88 (1.23-2.87)	1.42 (0.79-2.54)	1.87 (1.07-3.27)	1.81 (0.85-3.84)
MN Prostate	0.70 (28)	1.92 (1.17-3.16)	1.63 (0.75-3.51)	1.93 (1.26-2.96)	1.46 (0.81-2.62)	1.92 (1.09-3.37)	1.83 (0.85-3.92)
MN Urinary Organs	0.77 (18)	1.59 (0.85-2.97)	1.50 (0.55-4.12)	2.13 (1.24-3.66)	1.43 (0.70-2.94)	0.796 (0.31-2.08)	1.57 (0.46-5.41)
MN Kidney	0.63 (8)	2.28 (0.95-5.49)	1.77 (0.46-6.87)	2.37 (1.05-5.35)	1.41 (0.47-4.22)	0.880 (0.23-3.41)	3.04 (0.70-13.1)
MN Bladder	0.93 (10)	0.999 (0.39-2.58)	1.27 (0.28-5.85)	1.93 (0.94-3.96)	1.45 (0.56-3.75)	0.725 (0.19-2.84)	0.315 (0.07-1.47)
MN Other & Unspecified Sites	0.89 (57)	1.21 (0.84-1.76)	0.761 (0.37-1.56)	1.56 (1.14-2.14)	1.26 (0.82-1.94)	1.28 (0.81-2.01)	1.95 (1.05-3.64)
MN Skin Melanoma	1.22 (11)	0.858 (0.35-2.11)	0.774 (0.17-3.56)	0.815 (0.37-1.80)	0.818 (0.27-2.44)	0.400 (0.08-1.90)	0.758 (0.24-2.35)
MN Brain & Other Nervous System	1.02 (16)	1.62 (0.84-3.14)	0.367 (0.05-2.77)	1.38 (0.74-2.56)	1.01 (0.41-2.48)	1.45 (0.61-3.41)	0.411 (0.15-1.10)
MN Connective Tissue	0.34 (1)	2.64 (0.24-29.2)	5.30 (0.33-84.8)	4.12 (0.49-34.8)	7.44 (0.76-73.1)	2.08 (0.13-33.3)	0 (1.56 expected)
MN Other & Unspecified	0.87 (28)	1.04 (0.60-1.80)	0.566 (0.17-1.87)	1.86 (1.21-2.86)	1.25 (0.68-2.31)	1.61 (0.89-2.94)	3.36 (1.59-7.07)

Epidemiologic Study of Mortality and Radiation-Related Risk of Cancer Among INEEL Workers

Table 5-16. SRRs for white males in SES groups Intermediate, Skilled Non-manual, Skilled manual, Partly skilled, Unskilled and unknown, compared to the Professional SES group.

Cause of death	Professional SMR (N)	Intermediate SRR (95% CI)	Skilled Non-man. SRR (95%CI)	Skilled manual SRR (95% CI)	Partly skilled SRR (95% CI)	Unskilled SRR (95% CI)	Unknown SES SRR (95% CI)
MN Lymphatic & Hematopoietic							1.23 (0.58-2.59)
Non-Hodgkin Lymphoma							1.95 (0.69-5.50)
Hodgkin's Disease	0.68 (2)	0.584 (0.05-6.44)	0 (0.64 expected)	0.770 (0.12-4.83)	2.91 (0.46-18.4)	0 (1.50 expected)	0.687 (0.06-7.59)
Leukemia	1.04 (19)	0.660 (0.32-1.35)	0.775 (0.22-2.72)	0.953 (0.54-1.69)	0.981 (0.45-2.16)	1.46 (0.67-3.17)	0.659 (0.15-2.96)
Myeloma	0.84 (7)	0.587 (0.17-2.03)	0.573 (0.07-4.69)	0.704 (0.26-1.93)	2.35 (0.83-6.66)	0.370 (0.08-1.80)	1.03 (0.32-3.34)
Benign & Unspec. Nature Neoplasms	0.70 (4)	1.17 (0.26-5.32)	2.90 (0.53-15.9)	2.43 (0.78-7.51)	4.84 (1.44-16.2)	2.83 (0.69-11.6)	1.28 (0.22-7.52)
Unspec. Nature Neopl. Nerv. Syst.	0.36 (1)	0 (2.52 expected)	0 (0.56 expected)	4.17 (0.48-36.0)	12.1 (1.35-108)	3.91 (0.24-62.5)	0 (1.26 expected)
All Cancers SMRs only (95% CI), N	0.88 (0.79-0.97), 381	1.02 (0.92-1.12), 402	0.84 (0.66-1.07), 71	1.23 (1.16-1.30), 1134	1.12 (0.99-1.27), 260	1.05 (0.92-1.20), 216	1.09 (0.94-1.26), 185
Diabetes Mellitus	0.71 (22)	1.49 (0.84-2.66)	2.47 (1.15-5.29)	1.80 (1.10-2.95)	1.90 (1.02-3.57)	1.98 (1.04-3.78)	0.570 (0.23-1.42)
Diseases of Blood & Blood-Forming Organs						0 (2.96 expected)	2.69 (0.77-9.46)
Non-pernicious & Unspec. Anemias						0 (1.03 expected)	2.20 (0.14-35.2)
Alcoholism	0.27 (3)	1.30 (0.26-6.46)	1.60 (0.17-15.4)	4.15 (1.21-14.3)	4.03 (1.00-16.2)	4.18 (0.99-17.7)	0.389 (0.04-3.74)
Other Mental Disorders							0.518 (0.11-2.39)
Diseases of Nervous System	0.99 (30)	1.02 (0.60-1.75)	1.27 (0.57-2.84)	1.17 (0.74-1.84)	1.01 (0.55-1.89)	1.64 (0.95-2.83)	0.548 (0.20-1.53)
Diseases of Heart	0.70** (398)	1.16 (1.01-1.34)	1.27 (1.01-1.60)	1.47 (1.31-1.65)	1.50 (1.29-1.75)	1.66 (1.43-1.93)	1.15 (0.89-1.49)
Ischemic Heart	0.72** (330)						1.12 (0.84-1.48)
Hypertension with Heart Disease	1.93* (12)						2.74 (0.75-10.1)
Other Diseases of Circulatory Syst.	0.79* (102)	0.970 (0.73-1.30)	1.08 (0.67-1.76)	1.30 (1.03-1.63)	1.32 (0.97-1.81)	1.61 (1.20-2.16)	0.659 (0.39-1.12)
Hypertension w/o Heart Disease	0.67 (3)	1.49 (0.30-7.45)	0 (0.89 expected)	2.78 (0.80-9.61)	1.64 (0.27-10.2)	3.49 (0.76-16.1)	1.18 (0.12-11.4)

Table 5-16. SRRs for white males in SES groups Intermediate, Skilled non-manual, Skilled manual, Partly skilled, Unskilled and unknown, compared to the Professional SES group.

Cause of death	Professional SMR (N)	Intermediate SRR (95% CI)	Skilled Non-man. SRR (95%CI)	Skilled manual SRR (95% CI)	Partly skilled SRR (95% CI)	Unskilled SRR (95% CI)	Unknown SES SRR (95% CI)
Cerebrovasc. Dis.	0.74* (57)	1.05 (0.72-1.54)	1.27 (0.68-2.36)	1.37 (1.01-1.86)	1.66 (1.12-2.46)	1.68 (1.14-2.47)	0.619 (0.30-1.30)
Dis. of Respir. Sys. Chronic & Unspec.	0.56** (85)	1.36 (1.01-1.83)	1.45 (0.89-2.37)	2.25 (1.76-2.86)	2.95 (2.21-3.93)	2.67 (1.98-3.60)	1.27 (0.75-2.13) 0.235 (0.03-2.11)
Emphysema	0.76 (17)	0.978 (0.47-2.05)	1.32 (0.41-4.25)	1.52 (0.87-2.67)	2.41 (1.25-4.66)	2.50 (1.27-4.90)	0.469 (0.12-1.78)
Asbestosis (SMRs)	0 (0.50 exp.)	6.05* (N=3)	0 (0.10 expected)	4.89** (N=6)	3.68 (N=1)	0 (0.27 expected)	0 (0.27 expected)
Silicosis (SMRs)	0 (0.42 exp.)	0 (0.46 expected)	0 (0.10 expected)	0 (1.70 expected)	3.70 (N=1)	0 (0.36 expected)	9.66 (N=1)
Other Respir. Dis.	0.44** (37)	1.67 (1.08-2.56)	1.02 (0.45-2.34)	3.52 (2.47-5.01)	4.15 (2.75-6.25)	3.66 (2.38-5.64)	2.27 (1.16-4.44)
Dis. of Digestive Sys.	0.64** (46)	0.907 (0.58-1.41)	1.38 (0.75-2.56)	1.75 (1.24-2.47)	1.46 (0.93-2.27)	1.52 (0.96-2.42)	1.63 (0.93-2.85)
Cirrhosis of Liver	0.45** (16)	1.19 (0.60-2.37)	1.64 (0.63-4.26)	2.51 (1.43-4.42)	1.87 (0.93-3.77)	2.14 (1.04-4.42)	3.06 (1.44-6.51)
Diseases of Genito-Urinary System Chronic & Unspec. Nephritis	0.40* (7)		0 (1.88 expected)				1.16 (0.24-5.72)
Symptoms & Ill-Def. Conditions	0.69 (12)	1.31 (0.59-2.92)	0.460 (0.06-3.56)	2.05 (1.06-3.95)	1.23 (0.47-3.22)	1.75 (0.69-4.47)	0.644 (0.08-5.44)
Accidents	0.35** (65)	1.29 (0.73-2.30)	1.04 (0.51-2.14)	2.00 (1.15-3.48)	2.20 (1.23-3.93)	2.00 (1.12-3.60)	0.843 (0.22-3.25)
Transportation Accidents	0.37** (40)	1.55 (1.02-2.36)	1.63 (0.84-3.13)	2.40 (1.66-3.47)	2.78 (1.82-4.25)	2.22 (1.41-3.49)	1.42 (0.76-2.68)
Other Accidents	0.23** (12)	2.35 (1.16-4.73)	0.826 (0.18-3.74)	3.95 (2.13-7.34)	3.08 (1.47-6.46)	5.51 (2.76-11.0)	1.68 (1.03-2.74)
Suicide	0.64** (47)	0.998 (0.65-1.53)	1.52 (0.82-2.81)	1.11 (0.76-1.62)	1.31 (0.82-2.10)	1.99 (1.28-3.10)	2.89 (1.21-6.91)
Homicide	0.22** (3)	3.43 (0.91-13.0)	2.01 (0.21-19.3)	3.93 (1.14-13.6)	4.07 (1.02-16.3)	4.24 (0.95-19.0)	0.655 (0.40-1.07)
Other & Unspecified	1.80** (57)	1.09 (0.74-1.59)	1.38 (0.77-2.46)	1.26 (0.92-1.72)	0.810 (0.50-1.32)	0.874 (0.54-1.40)	2.18 (0.49-9.77)
All Deaths SMRs only (N), 95% CI	0.71** (1280), 0.67-0.75	0.85** (1395), 0.80-0.89	0.85** (306), 0.76-0.96	1.09** (4394), 1.06-1.12	1.10** (1065), 1.03-1.17	1.11** (1033), 1.05-1.18	1.59 (0.88-2.87) 0.90** (624), 0.83-0.98

†SRR not computed if 0 deaths occurred in baseline or comparison category (if the former, SMR based on regional state population is presented in the other categories).
*95% CI of SMR excludes 1.00. SRRs are not flagged.
**99% CI of SMR excludes 1.00. SRRs are not flagged.

5.6.2 White Females

The large majority of female workers were classified as skilled non-manual workers, a category comprising a group of largely clerical workers. Because so few female workers were classified as professional, this is a poor baseline group to evaluate differences in mortality rates among SES groups. In addition, the age distributions were quite different among the groups (Table 5-17); the mean age of the professional group was approximately seven years younger than the skilled non-manual group, and the 75th percentile was approximately twelve years younger. Consequently the ratio between SMRs should not be used to estimate an SRR among SES categories.

Table 5-17. Distribution of age (at DLO) among SES categories for white females.

Age statistics (years)	SES category						
	Professional	Intermediate	Skilled non-manual	Skilled manual	Partly skilled	Unskilled	Unknown
N	735	1421	5678	134	524	315	1865
Mean	44.8	47.9	51.5	49.0	54.3	51.8	45.1
10th %	34.8	33.9	33.0	34.9	38.2	34.2	32.4
25th %	38.1	39.1	42.6	39.6	43.7	40.4	37.2
Median	43.3	45.7	51.8	45.7	51.5	49.2	43.5
75th %	48.8	54.5	60.5	56.7	63.4	63.2	51.5
90th %	57.8	65.2	69.3	68.1	77.0	74.4	60.3

Despite these limitations, several patterns of interest emerge in the SES analysis (Table 5-18). Ischemic heart disease was substantially higher among non-professional women, and rates of death from breast cancer and hypertension with and without heart disease were much lower among non-professional women compared to professionals. These observations are consistent with those among WM. All-cancer and all-cause SMRs showed much less heterogeneity among SES categories for WF than for males.

Table 5-18. SRRs for white females in SES groups Intermediate, Skilled Non-manual, Skilled manual, Partly skilled, Unskilled and unknown, compared to the Professional SES group.

Cause of death	Professional SMR (N)†	Intermediate SRR (95% CI)	Skilled Non-man. SRR (95% CI)	Skilled manual SRR (95% CI)	Partly skilled SRR (95% CI)	Unskilled SRR (95% CI)	Unknown SES SRR (95% CI)
MN Buccal cavity	0 (0.06 exp.)	0 (0.26 exp.)	SMR=2.61 (N=4)	0 (0.03 exp.)	SMR=9.84* (N=2)	0 (0.10 exp.)	0 (0.15 exp.)
MN Digestive		0.554 (0.06-5.07)	0.552 (0.10-3.06)	0 (0.44 exp.)	0.333 (0.04-2.88)	1.17 (0.12-12.2)	1.02 (0.16-6.40)
MN Pancreas		0 (1.01 exp.)	0.068 (0.01-0.75)	0 (0.11 exp.)	0 (0.83 exp.)	1.10 (0.07-17.6)	0.658 (0.07-6.33)
MN Trachea, Bronchus & Lung	0.97 (1)	1.03 (0.11-9.24)	0.591 (0.08-4.44)	2.39 (0.15-38.2)	1.42 (0.16-12.3)	1.19 (0.12-11.7)	1.05 (0.12-9.05)
MN Breast		0.228 (0.03-1.85)	0.237 (0.03-1.62)	0 (0.55 exp.)	0.179 (0.02-1.70)	0.210 (0.02-2.20)	0.081 (0.01-0.85)
MN Female Genital	0 (0.82 exp.)	SMR=0.71 (N=2)	SMR=0.73 (N=12)	SMR=3.32 (N=1)	SMR=1.92 (N=4)	SMR=0.98 (N=1)	0 (1.62 exp.)
MN Urinary Organs		0.441 (0.04-4.87)	0.035 (0.00-0.57)	0 (0.06 exp.)	0 (0.43 exp.)	0 (0.22 exp.)	0 (0.26 exp.)
MN Other & Unspecified Sites	1.02 (1)	0.966 (0.09-10.9)	1.47 (0.19-11.1)	0 (0.33 exp.)	0 (2.21 exp.)	3.20 (0.27-37.2)	2.12 (0.23-19.2)
MN Brain & Other Nervous System	0 (0.26 exp.)	SMR=1.32 (N=1)	SMR=1.33 (N=6)	0 (0.08 exp.)	0 (0.48 exp.)	SMR=4.19 (N=1)	SMR=5.78* (N=3)
MN Other & Unspec.	2.27 (1)	0.378 (0.02-6.04)	0.793 (0.10-6.35)	0 (0.17 exp.)	0 (1.21 exp.)	0 (0.62 exp.)	0 (0.91 exp.)
MN Lymphatic & Hematopoietic		SMR=1.44 (N=3) 95% CI 0.29-4.73	SMR=1.32 (N=16) 95% CI 0.75-2.17	SMR=4.62 (N=1) 95% CI 0.06-37.7	SMR=0.64 (N=1) 95% CI 0.01-5.23	SMR=1.27 (N=1) 95% CI 0.02-10.3	SMR=0.82 (N=1) 95% CI 0.01-6.66
Benign & Unspecified Neoplasms	0 (0.10 exp.)	0 (0.35 exp.)	SMR=1.47 (N=3)	0 (0.04 exp.)	SMR=3.71 (N=1)	0 (0.13 exp.)	0 (0.18 exp.)
All Cancer, SMRs only (N), 95% CI		0.89 (20), 0.54-1.38	0.91 (120), 0.75-1.08	1.25 (3), 0.25-4.10	1.02 (17), 0.59-1.65	1.43 (12), 0.74-2.55	1.35 (18), 0.80-2.16
Diabetes Mellitus	0 (0.55 exp.)	SMR=1.02 (N=2)	SMR=0.74 (N=8)	0 (0.20 exp.)	SMR=0.64 (N=1)	SMR=2.49 (N=2)	SMR=2.76 (N=3)
Diseases of Nervous Sys. & Sense Org.		SMR=0.53 (N=1)	SMR=1.25 (N=13)	SMR=5.40 (N=1)	SMR=0.69 (N=1)	0 (0.76 exp.)	SMR=0.99 (N=1)
Diseases of Heart	1.15 (4)	0.968 (0.29-3.27)	1.03 (0.34-3.12)	1.91 (0.31-11.7)	1.81 (0.54-6.02)	0.575 (0.14-2.39)	1.05 (0.21-5.32)
Ischemic Heart Disease	0.42 (1)	3.66 (0.45-29.8)	4.31 (0.59-31.4)	13.0 (1.13-149)	10.6 (1.37-82.0)	2.33 (0.23-23.2)	6.38 (0.59-68.7)
Hypertension with Heart Disease	25.0** (2)	0.155 (0.01-1.77)	0.084 (0.01-0.63)	0 (0.03 exp.)	0.146 (0.01-1.66)	0.183 (0.02-2.09)	0 (0.10 exp.)

129

Table 5-18. SRRs for white females in SES groups Intermediate, Skilled non-manual, Skilled manual, Partly skilled, Unskilled and unknown, compared to the Professional SES group.

Cause of death	Professional SMR (N)	Intermediate SRR (95% CI)	Skilled Non-man. SRR (95% CI)	Skilled manual SRR (95% CI)	Partly skilled SRR (95% CI)	Unskilled SRR (95% CI)	Unknown SES SRR (95% CI)
Other Diseases of Circulatory Syst.	1.15 (2)	1.46 (0.25-8.39)	1.28 (0.25-6.47)	2.90 (0.23-36.0)	1.18 (0.17-8.26)	2.60 (0.36-18.8)	3.25 (0.45-23.5)
Hypertension w/o Heart Disease	19.0 (1)	0 (0.23 exp.)	0.0403 (0.00-0.64)	0 (0.02 exp.)	0 (0.22 exp.)	0 (0.12 exp.)	0 (0.08 exp.)
Cerebrovascular Disease		SMR=1.27 (N=6)	SMR=0.99 (N=24)	SMR=2.23 (N=1)	SMR=0.66 (N=3)	SMR=1.30 (N=3)	SMR=2.24 (N=4)
Disease of Respiratory System	0.72 (1)	1.14 (0.13-9.86)	0.928 (0.13-6.85)	0 (0.58 exp.)	0.475 (0.5-4.63)	0 (2.59 exp.)	0.141 (0.01-2.25)
Diseases of Digestive System		0 (3.19 exp.)	2.01 (0.27-15.0)	0 (0.33 exp.)	1.07 (0.07-17.1)	0 (1.25 exp.)	0.934 (0.08-11.0)
Diseases of Genito-urinary System	0 (0.21 exp.)	SMR=3.67 (N=3)	SMR=1.09 (N=5)	0 (0.08 exp.)	SMR=1.35 (N=1)	0 (0.38 exp.)	0 (0.31 exp.)
Symptoms & Ill-Defined Conditions			=	=	=	=	0 (0.34 exp.)
Accidents	1.23 (3)	0.983 (0.18-5.42)	0.816 (0.18-3.65)	1.36 (0.12-15.6)	0.395 (0.05-3.04)	0 (1.52 exp.)	0.456 (0.07-2.94)
Suicide		1.29 (0.08-20.6)	1.31 (0.16-10.7)	0 (0.18 exp.)	0 (0.92 exp.)	0 (0.45 exp.)	3.36 (0.36-31.0)
Homicide	2.86 (1)	1.03 (0.06-16.5)	0.444 (0.05-4.29)	0 (0.06 exp.)	0 (0.30 exp.)	0 (0.15 exp.)	4.42 (0.31-62.3)
All Deaths SMRs only (N), 95% CI	0.60-1.5	0.86 (61), 0.66-1.11	0.91 (358), 0.81-1.01	1.40 (10), 0.67-2.65	0.95 (54), 0.71-1.24	0.94 (27), 0.62-1.37	1.29 (49), 0.96-1.72

†SRR not computed if 0 deaths occurred in baseline or comparison category (if the former, SMR based on regional state population is presented in the other categories)
*95% CI of SMR excludes 1.00. SRRs are not flagged.
**99% CI of SMR excludes 1.00. SRRs are not flagged.

5.6.3 Non-white Males

Very small numbers of deaths caused difficulty in stratifying on SES with adequate precision, and insufficient numbers of workers in any particular category precluded use of SRRs. Among causes of death with more than five cases, SMRs are shown in Table 5-19.

The all-cause mortality rate was low compared to the regional population, for each category except skilled non-manual and unskilled workers (although SMRs for these groups were estimated at well below one). All-cancer SMRs did not differ substantially from one for any category and were relatively uniform given the very low number of deaths in each SES group. Most individual cancer SMRs did not vary meaningfully among SES categories; although age structure differences among SES groups precluded direct comparison of SMRs, the SRRs among the groups were very similar, and infrequently differed from unity (data not shown).

The leukemia mortality rate was highly elevated among skilled manual workers, but with a very small number of cases (SMR=13.9, N=2). One unspecified neoplasm of brain was observed among unskilled workers, with 0.0085 expected, resulting in a greatly elevated SMR. No malignant brain cancers were observed.

Table 5-19. SMR results for non-white males, for SES groups Professional, Intermediate, Skilled non-manual, Skilled manual, Partly skilled, and Unskilled.

Cause of death	SMR (number observed)					
	Professional	Intermediate	Skilled Non-manual	Skilled manual	Partly skilled	Unskilled
MN Digestive Organs	0.95 (2)	1.30 (2)	6.87 (1)	0 (1.66 expected)	0 (0.65 expected)	0 (1.01 expected)
MN Trachea, Bronchus, Lung	0.43 (1)	0.61 (1)	0 (0.13 expected)	1.58 (3)	1.46 (1)	1.73 (2)
Ischemic Heart Disease	0.25* (2)	0.69 (4)	0 (0.59 expected)	0.62 (4)	0.79 (2)	1.24 (5)
Cerebrovascular Disease	0.58 (1)	1.70 (2)	0 (0.12 expected)	0.69 (1)	0 (0.62 expected)	0 (0.89 expected)
Transportation Accidents	0 (5.97 expected)	0.15* (1)	0 (1.86 expected)	0.67 (3)	0.64 (2)	0.81 (3)
All Cancers	0.89 (6)	0.82 (4)	2.23 (1)	1.28 (7)	0.47 (1)	1.50 (5)
All Deaths	0.24** (12)	0.47** (20)	0.53 (4)	0.55** (21)	0.32** (6)	0.77 (20)

*95% CI of SMR excludes 1.00.
**99% CI of SMR excludes 1.00.

5.7 Local vs. Migrant Subcohorts

There were large differences in mortality patterns between local (i.e., SSN of issue was ID, MT, WY or UT, N=31,010) and migrant workers (N=21,425). The elevation in the mortality rate from hypertension with heart disease existed primarily among the migrants to the region, for both men and women (Table 5-20). For women, the pattern was particularly striking: there was a near five-fold elevation in the mortality rate (SRR=4.96; CI, 0.95-26.0) from this cause among migrants compared to local workers and only a very slight elevation among the local workers compared to the regional population.

Many cancer rates tended to be higher among migrants than among the local workers, for both sexes, although CIs were generally wide. For individual cancers the patterns differed, to some extent, by sex. Cancers of oral cavity, most of digestive tract (esophagus, stomach, intestine, rectum and liver), lung and bladder, as well as benign and unspecified neoplasms, were elevated among male migrants (compared to local workers). This pattern suggests smoking-related cancer mortality rates were higher among the migrant subcohort than among the local subcohort. Mortality rates for cancers of pancreas, prostate, NHL and leukemia were substantially lower among migrants compared to local workers. White male migrants experienced lower rates of death from diabetes, alcoholism and other mental disorders, non-cardiac diseases of circulatory system, non-malignant respiratory disease, and chronic and unspecified nephritis than local males. White male migrants showed higher mortality rates from non-pernicious anemia and hypertensive heart disease, compared to local workers. Both groups had similar elevations in asbestosis, compared to each other and to the general population

Among WF, the migrants were younger as a population than the local workers. Generalizing findings among females is difficult because of very small numbers of deaths in each category. However, female migrants showed quite different cancer mortality patterns from male migrants (Table 5-20). Women migrants had elevations in death rates for cancers of pharynx, stomach, pancreas, larynx, lung, breast, kidney, brain, leukemia and multiple myeloma, compared to local women, although CIs were quite wide for all these cancers. Cancers that were substantially lower (although imprecisely estimated) among women migrants include intestinal and genital cancers (particularly ovarian). Non-cancer mortality rates that showed elevation among female migrants include hypertensive heart disease, which exhibited a near five-fold elevation among female migrants, cerebrovascular disease, and chronic and unspecified nephritis. Migrant female workers experienced reduced mortality rates from other mental disorders and cirrhosis of liver, although CIs were very wide for these causes of death.

Table 5-20. Standardized rate ratios for migrants compared to local workers, for white males and white females. 95% CIs shown for selected causes of death.

Cause of death	White males		White females	
	SMR local (N)	SRR migrant (95% CI)	SMR local (N)	SRR migrant (95% CI)
MN Buccal cavity	**0.46 (10)**	**1.78 (0.84-3.77)**	**2.69 (4), 0.72-7.47**	**0.795 (0.15-4.34)**
MN Pharynx	0.40 (4)	1.87 (0.58-6.08)	0† (0.68 expected)	SMR=5.29 (N=2)
MN Respiratory	**0.94 (240)**	**1.15 (0.97-1.35)**	**0.83 (19)**	**1.06 (0.50-2.24)**
MN Larynx	0.84 (23)	1.28 (0.76-2.16)	1.08 (1)	1.44 (0.09-23.1)
MN Trachea, Bronchus & Lung	0.82 (28)	1.48 (0.93-2.35)	0.45 (1)	4.50 (0.40-50.6)
MN Other Respiratory	0.84 (74)	1.40 (1.05-1.86)	0.83 (8)	0.615 (0.16-2.34)
	0.63 (13)	1.46 (0.74-2.88)	1.25 (2)	0 (0.88 expected)
	0.91 (15)	1.28 (0.67-2.44)	1.03 (2)	0.955 (0.09-10.6)
	0.67 (4)	1.63 (0.49-5.44)	1.73 (1)	0 (0.31 expected)
	1.32* (75)	0.688 (0.49-0.96)	0.55 (3)	2.46 (0.55-11.1)
	1.78 (8)	0.506 (0.17-1.55)	1.74 (1)	0 (0.31 expected)
MN Respiratory	**0.88* (305)**	**1.41 (1.22-1.62)**	**0.82 (21)**	**1.34 (0.69-2.60)**
MN Larynx	0.83 (9)	1.16 (0.50-2.72)	2.20 (1)	1.83 (0.11-29.3)
MN Trachea, Bronchus & Lung	0.86** (284)	1.44 (1.24-1.66)	0.80 (20)	1.31 (0.66-2.61)
MN Other Respiratory	2.87** (12)	0.759 (0.33-1.77)	0 (0.32 expected)	0 (0.17 expected)
	3.40 (4)	0.220 (0.02-1.97)	0.87 (26)	1.35 (0.75-2.45)
MN Genital	**1.19 (129)**	**0.809 (0.63-1.04)**	**0.93 (15)**	**0.467 (0.15-1.44)**
MN Prostate	1.24* (129)	0.796 (0.62-1.02)	--	--
MN Cervix	--	--	1.00 (4)	0.759 (0.14-4.15)
MN Ovary	--	--	0.67 (6)	0.267 (0.03-2.22)
	1.06 (61)	1.08 (0.78-1.51)	1.03 (3)	0.600 (0.06-5.77)
	1.14 (34)	0.979 (0.62-1.55)	0.53 (1)	1.79 (0.11-28.6)
	0.98 (27)	1.21 (0.74-1.96)	1.99 (2)	0 (0.56 expected)
MN Other & Unspecified Sites	**1.10 (168)**	**0.944 (0.77-1.16)**	**0.72 (13)**	**1.50 (0.67-3.35)**
MN Skin Melanoma	0.96 (20)	1.22 (0.68-2.16)	0 (2.22 expected)	SMR=0.83 (N=1)
MN Brain & Other Nervous System	1.11 (41)	0.969 (0.63-1.48)	1.35 (6)	1.44 (0.44-4.71)

Table 5-20. Standardized rate ratios for migrants compared to local workers, for white males and white females. 95% CIs shown for selected causes of death.

Cause of death	White males		White females	
	SMR local (N)	SRR migrant (95% CI)	SMR local (N)	SRR migrant (95% CI)
MN Bone	0.58 (2)	0.496 (0.04-5.53)	3.03 (1)	0 (0.15 expected)
MN Connective Tissue	1.00 (7)	0.804 (0.28-2.30)	0 (1.31 expected)	SMR=1.44 (N=1)
MN Other & Unspecified	1.16 (90)	0.917 (0.69-1.22)	0.67 (6)	1.24 (0.34-4.43)
	1.20* (138)	0.750 (0.59-0.96)	1.00 (12)	1.45 (0.62-3.37)
	1.50** (63)	0.708 (0.49-1.02)	1.06 (5)	1.12 (0.27-4.71)
	0.65 (5)	1.02 (0.29-3.58)	1.41 (1)	0 (0.30 expected)
	1.17 (53)	0.746 (0.50-1.10)	1.06 (5)	1.63 (0.47-5.68)
	0.84 (17)	0.852 (0.44-1.66)	0.54 (1)	3.51 (0.32-38.8)
Benign & Unspec. Neoplasms	0.88 (13)	1.41 (0.70-2.82)	1.96 (4)	0 (1.03 expected)
	1.00 (1055)	1.10** (1414)	0.87 (113)	1.04 (73)
Diabetes mellitus	1.19 (92)	0.735 (0.55-0.99)	1.01 (11)	0.619 (0.20-1.95)
	0.43* (6)	2.50 (1.00-6.27)	1.27 (2)	0 (0.82 expected)
	0.65 (3)	1.79 (0.46-6.94)	0 (0.55 expected)	0 (0.28 expected)
Alcoholism	0.92 (24)	0.377 (0.18-0.77)	0.76 (1)	0 (0.71 expected)
	1.44* (38)	0.558 (0.34-0.92)	1.87 (6)	0.387 (0.05-3.23)
Diseases of Nervous System & Sense Organs	1.14 (87)	0.919 (0.69-1.23)	1.04 (11)	0.714 (0.22-2.29)
	0.88** (1298)	1.01 (0.94-1.09)	0.84 (67)	0.992 (0.66-1.50)
	0.91** (1084)	0.997 (0.92-1.08)	0.80 (44)	1.01 (0.60-1.68)
	1.35 (23)	1.46 (0.88-2.43)	1.12 (2)	4.96 (0.95-26.0)
Other Diseases of Circulatory System	0.99 (351)	0.846 (0.73-0.98)	0.84 (31)	1.24 (0.71-2.18)
Hypertension w/o Heart Disease	1.47 (17)	0.471 (0.22-1.03)	0 (1.27 expected)	SMR=3.01 (N=2)
Cerebrovascular Disease	0.97 (209)	0.867 (0.72-1.05)	0.88 (22)	1.50 (0.80-2.80)
	1.03 (407)	0.833 (0.73-0.95)	0.69 (22)	1.00 (0.49-2.04)
	1.01 (11)	0.661 (0.27-1.60)	1.47 (1)	0 (0.37 expected)

Table 5-20. Standardized rate ratios for migrants compared to local workers, for white males and white females. 95% CIs shown for selected causes of death.

Cause of death	White males SMR local (N)	White males SRR migrant (95% CI)	White females SMR local (N)	White females SRR migrant (95% CI)
Emphysema	1.07 (68)	0.828 (0.59-1.16)	0.56 (2)	0 (1.96 expected)
Asthma	0.85 (9)	0.846 (0.34-2.14)	0 (1.87 expected)	SMR=2.03 (N=2)
Asbestosis	3.05 (4)	0.976 (0.26-3.64)	0 (0.003 expected)	0 (0.002 expected)
Silicosis	0.70 (1)	0.766 (0.05-12.2)	0 (0.0011 expected)	0 (0.0005 expected)
Other Respiratory Disease	1.11 (229)	0.796 (0.66-0.96)	0.94 (16)	0.944 (0.40-2.22)
Diseases of Digestive System	0.85* (153)	1.03 (0.83-1.27)	1.04 (19)	0.258 (0.08-0.87)
Cirrhosis of Liver	0.81 (67)	1.07 (0.78-1.48)	0.92 (7)	0.251 (0.03-2.04)
Diseases of Genitourinary System	1.08 (52)	0.669 (0.45-1.01)	1.07 (5)	1.54 (0.41-5.76)
Acute Glomerulonephritis & Acute Renal Failure	0.62 (3)	1.34 (0.32-5.63)	0 (0.37 expected)	0 (0.18 expected)
Chronic & Unspec. Nephritis, Renal Failure & Other Renal Sclerosis	1.42* (35)	0.617 (0.37-1.03)	0.45 (1)	6.16 (0.64-59.6) N=3
Symptoms & Ill-Defined Conditions	0.93 (45)	0.873 (0.58-1.32)	1.36 (5)	0.664 (0.13-3.43)
Accidents	0.76** (362)	0.838 (0.72-0.98)	0.95 (32)	0.722 (0.37-1.41)
Suicide	0.82* (151)	0.836 (0.66-1.06)	0.73 (8)	0.916 (0.28-3.05)
Homicide	0.62* (21)	1.08 (0.59-2.00)	1.00 (4)	0.817 (0.15-4.48)
HIV-related	1.29 (14)	0.494 (0.19-1.30)	0 (0.25 exp.)	0 (0.15 exp.)
Other and unspecified	1.57** (125)	1.57 (1.26-1.95)	1.77* (20)	0.44 (0.16-1.19)
All deaths SMRs only (N), 95% CI	0.94** (4306), 0.91-0.97	0.94** (5038), 0.91-0.97	0.91 (363), 0.82-1.01	0.88 (185), 0.76-1.01

†SRR not computed if 0 deaths occurred in baseline or comparison category.
*95% CI of SMR excludes 1.00. SRRs are not flagged.
**99% CI of SMR excludes 1.00. SRRs are not flagged.

5.8 Construction and Service Worker Subcohort

5.8.1 White Males

A slight minority of the total study cohort (N=29,631, or 46.6%) were ever construction or service workers. However, their age distribution was older, contributing to a higher mortality expectation and more than twice the observed number of deaths for all causes combined compared to non-construction workers (Table 5-21). Construction and service workers also have shown higher cumulative external radiation doses, on average, than non-construction workers, particularly since the early 1960s (Figure 3-12).

The mortality experience of construction and service workers was quite different from non-construction workers. The all-cause mortality rate was higher for construction workers and was lower for other workers (SMRs of 1.08 and 0.77, respectively), compared to the regional population rate. All-cancer mortality exhibited the same pattern (SMRs of 1.19 and 0.90, respectively, for construction and non-construction workers). No precisely estimated COD exhibited lower mortality rate among construction workers, although the estimated rate ratio for hypertensive heart disease mortality was reduced.

Among WM, mortality rates for many lifestyle-related diseases (respiratory and digestive cancers, alcoholism, and cirrhosis of liver) were elevated among the construction and service workers, compared to non-construction workers. Pancreatic, lung and kidney cancers showed particularly high rate ratios, with CIs excluding unity, compared to non-construction workers (SRRs of 1.90, 1.61 and 2.05, respectively). The estimated rates of death from these cancers were also elevated with respect to the regional population (data not shown).

Asbestos-related disease mortality rates varied substantially between construction and non-construction workers. The mortality rate from peritoneal and other digestive cancers for construction and service workers was particularly elevated compared to non-construction workers (SRR of 8.76; CI, 1.14-67.6). The mortality rate for "other respiratory" neoplasms, which includes pleura cancers, was also higher among construction workers (but was elevated in non-construction workers as well, compared to the regional population). All cases of asbestosis occurred among those classified as construction workers, with an SMR of 4.92 (CI, 2.35-9.26) for this group. The mortality rate for the "other and unspecified" cancer subcategory, which included mesotheliomas of unspecified site, was also elevated (with CIs that exclude one) in construction workers. A special rate file, comprising all pleura and peritoneal cancers, was developed to analyze these cancers together as likely mesotheliomas (as described in §2.6.2.2). The SMR for non-construction workers was 0.87 (CI, 0.10-3.77) and for construction workers was 2.98 (CI, 1.54-5.31), both compared to the general population. For construction workers compared to non-construction workers, the SRR for the combined pleura and peritoneum cancers was 4.54 (CI, 1.01-20.4, N=12).

The anemia mortality rate was much higher among construction and service workers than among other workers (SRR of 8.55) although the leukemia SRR was not elevated. The rate ratio of accidents was also much higher among construction workers than non-construction workers (SRR=1.80, 1.52-2.13) although still lower than expected compared to the general public (SMR=0.88). The highest SRR for accidental COD was from falls (SRR=2.59, 1.38-4.87) and this cause was also elevated compared to the general public.

Construction and service workers exhibited higher rates of "other and unspecified" causes of death than non-construction workers (though both groups were elevated compared to the general population). This finding was likely due to the lower rates of death certificate retrieval for decedents in this category. Such deaths were classified in the "other and unspecified" category, and death certificates could not be obtained for a greater number of deceased construction workers (179, or 2.4%) than non-construction workers (65, or 2.0%).

5.8.2 White Females

Among white female construction workers, the combined cancer mortality rate was slightly higher than in the regional population, with an SMR of 1.13, yet the SMR of non-construction workers was slightly lower than expected (Table 5-21). CIs were, however, wide for both subcohorts. For all deaths combined, SMRs were higher among construction workers than among non-construction workers and were lower than expected among non-construction workers, compared to the general regional population. Specific cancers showing elevated mortality rates among female construction workers compared to other workers included cancers of oral cavity, digestive tract, bladder, respiratory system and brain (including NUN of nervous system) although all were based on small numbers and CIs on SRRs were wide.

Lymphatic and hematopoietic cancer SRRs were elevated among female construction workers. This elevation was driven primarily by NHL (SRR=4.07; CI, 1.08-15.3) and by multiple myeloma (SRR=7.07; CI, 0.64-78.0). For the latter cancer, a substantial deficit was observed among the baseline group (SMR=0.45).

For non-cancer causes of death, the ischemic heart disease mortality rate was substantially elevated among female construction workers compared to non-construction workers. Some non-malignant diseases of respiratory system, including emphysema and asthma, also showed elevated rates among female construction workers, although CIs were wide. Death rates from cirrhosis of liver, symptoms and ill-defined conditions, transportation accidents, homicide and suicide were also higher among female construction workers compared to other women workers, although all the 95% CIs overlapped unity. Mortality rates for no precisely estimated causes of death were lower among female construction than non-construction workers.

Table 5-21. SRR results for construction workers compared to non-construction workers, for white males and white females. Where no cases were observed among non-construction workers, SMRs are reported for construction workers.

Cause of death	White males		White females	
	SMR non-construction (N)	SRR construction (95% CI)	SMR non-construction (# obs)	SRR construction (95% CI)
MN Buccal cavity	0.55 (10)	1.83 (0.87-3.82)	2.24 (4)	1.77 (0.32-9.66)
MN Pharynx	0.24* (2)	3.00 (0.64-14.0)	2.44 (2)	0† (0.25 exp.)
MN Digestive	0.94 (191)	1.29 (1.08-1.54)	0.87 (24)	1.30 (0.60-2.82)
MN Esophagus	1.08 (26)	0.94 (0.57-1.56)	1.78 (2)	0 (0.34 exp.)

Table 5-21. SRR results for construction workers compared to non-construction workers, for white males and white females. Where no cases were observed among non-construction workers, SMRs are reported for construction workers.

Cause of death	White males SMR non-construction (N)	White males SRR construction (95% CI)	White females SMR non-construction (# obs)	White females SRR construction (95% CI)
MN Stomach	0.87 (23)	1.40 (0.84-2.32)	0.75 (2)	3.16 (0.43-23.1)
MN Intestine	1.10 (77)	1.02 (0.76-1.37)	0.78 (9)	1.55 (0.47-5.11)
MN Rectum	0.68 (11)	1.39 (0.67-2.88)	0.52 (1)	3.37 (0.21-53.8)
MN Liver & Gall Bladder	1.02 (14)	1.26 (0.66-2.41)	0.86 (2)	1.93 (0.17-21.3)
MN Liver unspecified	0.85 (4)	1.25 (0.38-4.12)	1.45 (1)	0 (0.21 exp.)
MN Pancreas	0.77 (35)	1.90 (1.28-2.82)	0.91 (6)	0.64 (0.08-5.33)
MN Peritoneum & other	0.29 (1)	8.76 (1.14-67.6)	1.47 (1)	0 (0.21 exp.)
MN Respiratory	0.80** (230)	1.63 (1.40-1.90)	0.90 (28)	1.18 (0.57-2.45)
MN Larynx	0.68 (6)	2.15 (0.85-5.44)	1.84 (1)	2.89 (0.18-46.2)
MN Trachea, Bronchus & Lung	0.80** (218)	1.61 (1.37-1.89)	0.90 (27)	1.12 (0.52-2.39)
MN Other Respiratory	1.73 (6)	2.16 (0.84-5.56)	0 (0.39 exp.)	0 (0.12 exp.)
MN Breast	1.07 (1)	2.77 (0.31-24.8)	1.02 (37)	0.99 (0.50-1.96)
MN Genital	0.87 (70)	1.39 (1.05-1.84)	0.83 (16)	0.92 (0.30-2.75)
MN Prostate	0.90 (69)	1.39 (1.05-1.85)	--	--
MN Cervix	--	--	1.05 (5)	0.780 (0.09-6.68)
MN Ovary	--	--	0.47 (5)	2.16 (0.51-9.11)
MN Urinary Organs	0.94 (43)	1.34 (0.93-1.95)	0.86 (3)	1.02 (0.11-9.80)
MN Kidney	0.71 (18)	2.05 (1.19-3.52)	0.88 (2)	0 (0.69 exp.)
MN Bladder	1.21 (25)	0.919 (0.55-1.53)	0.82 (1)	2.97 (0.19-47.5)
MN Other & Unspecified Sites	0.96 (124)	1.31 (1.05-1.64)	0.79 (17)	1.39 (0.57-3.37)
MN Skin Melanoma	1.13 (21)	1.01 (0.57-1.78)	0.37 (1)	0 (0.75 exp.)
MN Brain & Other NS	1.06 (34)	1.13 (0.72-1.78)	1.32 (7)	2.05 (0.60-7.03)
MN Thyroid	0.68 (1)	3.30 (0.36-29.9)	0 (0.35 exp.)	0 (0.11 exp.)
MN Bone	0.37 (1)	1.53 (0.14-17.1)	0 (0.38 exp.)	SMR=9.95 (N=1)
MN Connective Tissue	0.99 (6)	0.89 (0.30-2.61)	0.64 (1)	0 (0.44 exp.)
MN Other & Unspec.	0.94 (60)	1.43 (1.05-1.95)	0.74 (8)	0.691 (0.15-3.26)
MN Lymphatic & Hematopoietic	1.13 (107)	0.920 (0.71-1.19)	1.05 (15)	1.83 (0.77-4.35)
Non-Hodgkin Lymphoma	1.27 (45)	1.06 (0.72-1.56)	0.71 (4)	4.07 (1.08-15.3)
Hodgkin's Disease	0.99 (6)	0.467 (0.13-1.71)	1.25 (1)	0 (0.22 exp.)
Leukemia	1.11 (41)	0.867 (0.57-1.31)	1.61 (9)	0.378 (0.05-2.99)
Myeloma	0.91 (15)	0.838 (0.42-1.66)	0.45 (1)	7.07 (0.64-78.0)
Benign & Unspecified Neoplasms	0.95 (11)	1.71 (0.84-3.50)	0.84 (2)	3.59 (0.50-25.6)
Benign neoplasms of brain & nervous syst.	0.85 (1)	0.708 (0.04-11.3)	0 (0.42 exp.)	0 (0.13 exp.)
Neoplasms of nervous system, unspec. nature	0.35 (2)	3.76 (0.81-17.5)	1.04 (1)	7.39 (0.67-81.7)
All Cancers SMRs only (N), 95% CI	0.90** (776), 0.84-0.97	1.19** (1862), 1.13-1.24	0.93 (144), 0.78-1.09	1.13 (52), 0.84-1.48
Diabetes mellitus	0.93 (58)	1.26 (0.91-1.74)	0.77 (10)	2.13 (0.77-5.91)
Diseases of Blood & Blood-Forming Organs	0.84 (9)	1.17 (0.51-2.69)	0.54 (1)	3.31 (0.21-53.0)
Anemias of other & unspecified type	0.30 (1)	8.55 (1.09-66.9)	0 (0.64 exp.)	0 (0.19 exp.)
Alcoholism	0.35** (8)	2.30 (1.02-5.22)	0.63 (1)	0 (0.45 exp.)

Table 5-21. SRR results for construction workers compared to non-construction workers, for white males and white females. Where no cases were observed among non-construction workers, SMRs are reported for construction workers.

Cause of death	White males		White females	
	SMR non-construction (N)	SRR construction (95% CI)	SMR non-construction (# obs)	SRR construction (95% CI)
Other Mental Disorders	0.75 (15)	1.72 (0.94-3.13)	1.31 (5)	2.58 (0.58-11.6)
Diseases of Nervous Syst.	0.91 (55)	1.21 (0.87-1.70)	1.12 (14)	0.65 (0.19-2.28)
Diseases of Heart	0.77** (856)	1.31 (1.21-1.43)	0.78* (74)	1.48 (0.98-2.23)
Ischemic Heart Disease	0.78** (703)	1.32 (1.21-1.45)	0.68** (45)	1.65 (0.99-2.74)
Hypertens. w/Heart Dis.	2.29** (28)	0.72 (0.43-1.19)	2.35 (5)	1.22 (0.23-6.38)
Other Diseases of Circulatory System	0.71** (179)	1.48 (1.24-1.77)	0.88 (39)	1.13 (0.61-2.09)
Hypertension w/o Heart Disease	0.68 (6)	1.62 (0.60-4.40)	0.67 (1)	3.37 (0.21-53.8)
Cerebrovascular Dis.	0.74** (110)	1.43 (1.14-1.79)	0.97 (29)	1.29 (0.66-2.56)
Disease of Respiratory System	0.65** (191)	1.86 (1.57-2.20)	0.71 (27)	0.956 (0.43-2.11)
Chronic & Unspecified Bronchitis	0.82 (6)	1.10 (0.42-2.87)	1.24 (1)	0 (0.26 exp.)
Emphysema	0.74 (32)	1.72 (1.14-2.61)	0.24 (1)	3.06 (0.19-48.9)
Asthma	0.37 (3)	3.67 (1.06-12.7)	0.45 (1)	3.91 (0.24-62.5)
Asbestosis (SMRs only)	0 (1.02 exp.)	4.92** (95% CI: 2.35-9.26, N=10)	0 (0.004 exp.)	0 (0.001 exp.)
Silicosis	1.24 (1)	0.72 (0.05-11.6)	0 (0.001 exp.)	0 (0.001 exp.)
Other Respiratory Dis.	0.63** (103)	2.10 (1.68-2.64)	0.93 (19)	0.851 (0.32-2.29)
Diseases of Digestive System	0.68** (98)	1.53 (1.20-1.96)	0.78 (17)	1.16 (0.46-2.96)
Cirrhosis of Liver	0.59** (42)	1.82 (1.26-2.62)	0.55 (5)	2.06 (0.49-8.65)
Diseases of Genito-Urinary System	0.58* (20)	2.15 (1.30-3.57)	1.28 (7)	1.03 (0.21-4.98)
Acute Glomeruloneph. & Acute Renal Failure	0 (3.72 exp.)	SMR=0.94 (7)	0 (0.43 exp.)	0 (0.13 exp.)
Chronic & Unspecified Nephritis, Renal Fail.	0.89 (16)	1.69 (0.94-3.02)	1.55 (4)	0 (0.78 exp.)
Symptoms & Ill-Defined Conditions	0.79 (27)	1.22 (0.76-1.95)	0.92 (4)	3.16 (0.79-12.7)
Accidents	0.52** (204)	1.80 (1.52-2.13)	0.88 (34)	1.22 (0.62-2.42)
Transport. accidents	0.55** (129)	1.64 (1.32-2.03)	0.94 (26)	1.32 (0.62-2.83)
Accidental falls	0.46** (13)	2.59 (1.38-4.87)	0.78 (2)	0 (0.78 exp.)
Suicide	0.64** (102)	1.36 (1.06-1.75)	0.62 (8)	2.39 (0.78-7.33)
Homicide	0.40** (12)	2.23 (1.14-4.37)	0.86 (4)	2.66 (0.59-12.0)
Other and unspecified	1.63** (106)	1.31 (1.03-1.66)	1.58 (21)	0.63 (0.21-1.85)
All deaths SMRs only (N), 95% CI	0.77** (2747), 0.74-0.80	1.08 (7294), 1.05-1.10	0.88 (416), 0.80-0.97	1.13 (159), 0.96-1.32

†SRR not computed if 0 deaths occurred in baseline or comparison category.
*95% CI of SMR excludes 1.00. SRRs are not flagged.
**99% CI of SMR excludes 1.00. SRRs are not flagged.

5.9 Asbestos Worker Subcohort

Asbestos workers consisted of a small (N=2741), somewhat older subcohort (median age at DLO was 61, about five years older than non-asbestos workers). Only one death occurred among white female asbestos workers. Among WM, rates of all-cause and all-cancer mortality were higher than in the general regional population. Substantial elevations in mortality rates were observed among specific outcomes as well. Lung cancer death rates were greater than in non-asbestos workers (SRR=1.53; CI, 1.22-1.93). Accidental death rates were also higher among asbestos workers, due primarily to accidental falls, the rate of which was more than doubled among asbestos workers compared to other workers (Table 5-22).

The rate of mortality from asbestosis was 25 times that of non-asbestos workers (Table 5-22). Rates of "other respiratory diseases" and cancers of "other respiratory" and "peritoneum and other digestive" were slightly elevated (SRR of 1.70 and 1.13, respectively, for the latter two causes) but had very wide CIs.

Given the very high elevation of asbestosis, it seemed counterintuitive that rates of cancer categories containing mesotheliomas were not much elevated among asbestos workers. Mortality rates for "other and unspecified" cancer, a category that includes mesotheliomas of unspecified site in ICD-9, were also elevated among the asbestos workers (SRR=1.63; CI, 1.03-2.57). An evaluation of cohort deaths occurring in 1999 (which included both ICD-10 and -9 coding) showed that most mesotheliomas (71%) were coded in the "unspecified cancer" cause-of-death category (code 199) in ICD-9. Furthermore, inspection of ICD codes for deceased cohort members showed that, among the 22 "other respiratory" cancers occurring among WM, just 9 were pleura cancers (Table 4-4). All the non-pleura cancers in the "other respiratory" cancer grouping occurred within the non-asbestos exposed cohort. Therefore, the SRR would be expected to increase to about 3.5 if just pleura cancers were evaluated. Similarly, among the 13 peritoneal and other unspecified digestive cancers occurring among WM, just 5 were peritoneal (and all non-peritoneal occurred among non-asbestos workers). It was estimated that the SRR would increase to about 3.4 for the exposed group for just peritoneal cancers, if just these deaths were evaluated.

Therefore, a separate analysis of pleura and peritoneum cancer rates combined was conducted, through the creation of a special rate file described in §2.6.2.2 (Table 2-14). Analysis using this rate file indicated that 14 pleura and peritoneal cancer deaths occurred in the cohort (all among WM), three of which occurred in asbestos workers (Table 5-23). The resulting SRR was 4.28 (CI, 1.19-15.5) for asbestos workers compared to other workers.

Table 5-22. SRR results for asbestos workers compared to non-asbestos workers, for white males.

Cause of death	SMR non-asbestos workers (N)	SRR asbestos workers (95% CI), N
MN Buccal cavity	0.81 (37)	0.60 (0.18-2.03), 3
MN Pharynx	0.65 (14)	0.87 (0.20-3.83), 2
MN Digestive	1.06 (566)	1.16 (0.85-1.58), 53
MN Esophagus	1.03 (60)	1.49 (0.59-3.73), 6
MN Stomach	1.05 (76)	0.94 (0.39-2.24), 7
MN Intestine	1.04 (193)	1.13 (0.68-1.89), 20

Table 5-22. SRR results for asbestos workers compared to non-asbestos workers, for white males.

Cause of death	SMR non-asbestos workers (N)	SRR asbestos workers (95% CI), N
MN Rectum	0.95 (41)	0.30 (0.04-2.15), 1
MN Liver & Gall Bladder	1.18 (41)	0.76 (0.15-3.92), 2
MN Liver Unspecified	0.95 (12)	0.49 (0.06-3.75), 1
MN Pancreas	1.09 (131)	1.65 (0.94-2.88), 15
MN Peritoneum & Other Respiratory	1.28 (12)	1.13 (0.15-8.72), 1
MN Respiratory	**1.07 (793)**	**1.54 (1.23-1.92), 101**
MN Larynx	0.91 (21)	1.67 (0.54-5.13), 4
MN Trachea, Bronchus & Lung	1.06 (752)	1.53 (1.22-1.93), 95
MN Other Respiratory	2.31** (20)	1.70 (0.39-7.31), 2
MN Breast	**1.99 (5)**	**0† (0.23 expected)**
MN Genital	**1.08 (249)**	**1.25 (0.81-1.92), 30**
MN Prostate	1.11 (247)	1.26 (0.82-1.94), 30
MN Urinary Organs	**1.11 (135)**	**1.39 (0.75-2.57), 14**
MN Kidney	1.10 (69)	1.80 (0.81-4.03), 8
MN Bladder	1.12 (66)	0.96 (0.38-2.42), 6
MN Other & Unspecified Sites	**1.08 (342)**	**1.38 (0.96-2.00), 36**
MN Skin Melanoma	1.14 (48)	1.33 (0.44-3.97), 4
MN Brain & Other Nervous System	1.08 (81)	0.95 (0.38-2.37), 5
MN Thyroid	1.34 (5)	0 (0.31 expected)
MN Bone	0.44 (3)	0 (0.59 expected)
MN Connective Tissue	0.84 (12)	2.20 (0.49-9.91), 2
MN Other & Unspecified Sites	1.12 (184)	1.63 (1.03-2.57), 24
MN Lymphatic & Hematopoietic	**1.06 (252)**	**1.11 (0.70-1.76), 22**
Non-Hodgkin Lymphoma	1.28* (111)	1.15 (0.57-2.31), 10
Leukemia	1.02 (96)	1.30 (0.67-2.55), 10
Myeloma	0.81 (35)	0.79 (0.18-3.45), 2
Benign & Unspecified Nature Neoplasms	**1.09 (33)**	**1.21 (0.39-3.78), 4**
All Cancers, SMRs only (95% CI), N	**1.07** (1.02-1.11), 2379**	**1.32** (1.17-1.50), 259**
Diabetes Mellitus	**1.05 (168)**	**1.39 (0.81-2.40), 16**
Diseases of Blood	**0.78 (23)**	**1.54 (0.33-7.26), 2**
Alcoholism	**0.64** (34)**	**1.62 (0.47-5.62), 3**
Other Mental Disorders	**1.15 (63)**	**0.76 (0.27-2.15), 4**
Diseases of Nervous System & Sense Organs	**1.14 (179)**	**1.20 (0.69-2.09), 16**
Diseases of Heart	**0.91** (2817)**	**1.19 (1.05-1.35), 316**
Ischemic Heart Disease	0.92** (2323)	1.28 (1.11-1.47), 276
Hypertension with Heart Disease	1.74** (62)	0.79 (0.29-2.17), 5
Other Diseases of Circulatory System	**0.92* (686)**	**1.09 (0.83-1.42), 75**
Hypertension w/o Heart Disease	1.12 (27)	0.59 (0.13-2.71), 2
Cerebrovascular Disease	0.92 (413)	1.03 (0.73-1.44), 45
Disease of Respiratory System	**0.97 (820)**	**1.55 (1.26-1.91), 119**
Chronic & Unspecified Bronchitis	0.86 (20)	2.09 (0.54-8.08), 3
Emphysema	0.99 (133)	1.54 (0.95-2.51), 22
Asthma	0.82 (18)	0.96 (0.13-7.21), 1
Asbestosis	1.07 (3)	25.6 (6.25-104.8), 7
Silicosis	0.67 (2)	0 (0.42 expected)
Other Respiratory Disease	1.03 (459)	1.55 (1.16-2.08), 59
Diseases of Digestive System	**0.90* (335)**	**1.53 (1.07-2.19), 40**
Cirrhosis of Liver	0.86 (147)	2.11 (1.28-3.48), 20
Diseases of Genitourinary System	**0.90 (90)**	**1.65 (0.88-3.11), 13**

Table 5-22. SRR results for asbestos workers compared to non-asbestos workers, for white males.

Cause of death	SMR non-asbestos workers (N)	SRR asbestos workers (95% CI), N
Acute Glomerulonephritis & Acute Renal Failure	0.59 (6)	2.03 (0.24-16.9), 1
Chronic & Unspecified Nephritis & Renal Failure	1.17 (60)	1.60 (0.71-3.60), 8
Symptoms & Ill-Defined Conditions	0.80* (80)	1.76 (0.95-3.28), 14
Accidents	0.73** (639)	1.31 (0.98-1.76), 58
Transportation Accidents	0.72** (364)	1.35 (0.91-2.00), 32
Accidental Falls	0.94 (68)	2.16 (1.05-4.45), 10
Other Accidents	0.65** (160)	1.12 (0.62-2.02), 14
Suicide	0.77** (271)	1.01 (0.59-1.74), 16
Homicide	0.61** (38)	2.52 (0.93-6.79), 5
Other & Unspecified	2.10** (336)	1.57 (1.12-2.19), 45
All Deaths, SMRs only, (95% CI), N	0.95** (0.93-0.97), 9034	1.15** (1.08-1.22), 1007

†SRR not computed if 0 deaths occurred in baseline or comparison category.
*95% CI of SMR excludes 1.00. SRRs are not flagged.
**99% CI of SMR excludes 1.00. SRRs are not flagged.

Table 5-23. SMRs (compared to combined ID, MT, UT and WY) and SRRs for combined pleura and peritoneal cancers, for asbestos workers and other workers.

	White males	Total
Asbestos workers SMR (N, 95% CI)	6.02* (3, 1.21-19.73)	6.00* (3, 1.21-19.7)
Other workers SMR (N, 95% CI)	1.89 (11, 0.94-3.45)	1.74 (11, 0.87-3.18)
All workers SMR (N, 95% CI)	2.21† (14, 1.21-3.77)	2.05* (14, 1.12-3.49)
Asbestos workers SRR (95% CI)	4.28 (1.19-15.5)	4.28 (1.19-15.5)

*95% CI of SMR excludes 1.00. SRRs are not flagged.
†99% CI of SMR excludes 1.00. SRRs are not flagged.

However, as stated above, the categories containing cancers of pleura and peritoneum miss many mesotheliomas (i.e., those of unspecified site) that are included in the "other and unspecified cancer" category in ICD-9. Although comparing mesothelioma mortality rates to the general population before ICD-10 was not possible, an additional analysis was done to evaluate risks for possible mesotheliomas among asbestos workers compared to non-asbestos workers in the INEEL cohort. Specific causes of death were reviewed for mention of mesothelioma among "Other and unspecified" cancer deaths occurring before 1979 (i.e., the period during which death certificates were collected) and in 1999 (when mesothelioma was added as a separate ICD code). Among all workers who died of "Other Respiratory," "Other Digestive," or "Unknown primary site" cancers during these time periods (i.e., before 1979 and in 1999) the odds of being an asbestos worker was ten times as high (odds ratio 95% CI: 1.55-109) for those with likely mesothelioma than for those without. Thus, although the LTAS groups within which mesotheliomas were classified had just slight elevations in asbestos workers, when restricting analysis to those causes that were most likely to have been mesothelioma (i.e., cancers of pleura, peritoneum and mesothelioma as stated on the death certificate) a much higher proportion were asbestos workers. Lung cancer rates were elevated among these workers, as well.

5.10 Chemical Worker Subcohort

Chemical workers (N=5332), a fairly large subcohort within INEEL, had a younger age distribution than other workers (median attained age was 53, about four years younger than non-chemical workers). Chemical workers had lower overall cancer mortality rates compared to the general population (all-cancer SMR=0.71; CI, 0.58-0.86), yet other workers showed elevated cancer rates compared to the general regional population (SMR=1.11; CI, 1.07-1.15). For smoking-related cancers, such as lung, chemical workers had much lower mortality rates than non-chemical workers, although CIs were wide for several of these causes of death such as bladder and oral cavity (Table 5-24). By measurement of most other lifestyle-associated disease (cancer of pancreas, ischemic heart disease, cirrhosis of liver, alcoholism and emphysema), those classified as chemical workers have exhibited far lower mortality rates than the rest of the cohort.

White male chemical workers showed elevations in cancers of stomach (SRR=1.58) and benign and unspecified neoplasms (SRR=3.40, Table 5-24). The last elevation was due to brain and other nervous system NUN, which was highly elevated compared to non-chemical workers (SRR=9.74, 1.59-59.7, N=2). The brain cancer mortality rate was slightly elevated with very wide CIs. This result was evaluated further using a combined "brain neoplasms—malignant, benign and unspecified" rate file created for this study, to ensure this elevation was not due to regional differentials in diagnosing brain malignancies. The combined brain tumor SRR was 2.12 (CI, 0.82-5.49, N=8) among white male chemical workers.

Female chemical workers had highly elevated digestive tract (particularly esophagus, intestine, rectum and pancreas) cancer mortality rates, compared to non-chemical workers (Table 5-24). Mortality rates from cancers of larynx and lung as well as leukemia were also elevated among female chemical workers, although CIs for these SRRs included one.

Table 5-24. SRR results for chemical workers compared to non-chemical workers, for white males and white females.

Cause of death	White males		White females	
	SMR non-chem (N)	SRR chemical workers (95% CI)	SMR non-chem (N)	SRR chemical workers (95% CI)
MN Buccal cavity	0.83 (39)	0.292 (1)	2.79* (6)	0† (0.18 expected)
MN Pharynx	0.73 (16)	0 (1.45 expected)	2.02 (2)	0 (0.08 expected)
MN Digestive	1.08 (594)	0.857 (0.55-1.33)	0.72 (24)	4.03 (1.86-8.74)
MN Esophagus	1.05 (62)	1.25 (0.41-3.77)	0.74 (1)	8.45 (0.53-135)
MN Stomach	1.04 (78)	1.58 (0.55-4.55)	1.24 (4)	0 (0.26 expected)
MN Intestine	1.07 (204)	0.876 (0.42-1.85)	0.64 (9)	4.60 (1.39-15.3)
MN Rectum	0.91 (41)	0.297 (0.04-2.16)	0.43 (1)	11.5 (0.72-183)

Table 5-24. SRR results for chemical workers compared to non-chemical workers, for white males and white females.

Cause of death	White males SMR non-chem (N)	White males SRR chemical workers (95% CI)	White females SMR non-chem (N)	White females SRR chemical workers (95% CI)
MN Liver & Gall Bladder	1.24 (44)	0 (2.32 expected)	1.07 (3)	0 (0.23 expected)
MN Liver Unspecified	0.92 (12)	1.17 (0.15-8.98)	1.19 (1)	0 (0.07 expected)
MN Pancreas	1.14 (141)	0.582 (0.23-1.48)	050 (4)	8.87 (1.98-39.7)
MN Peritoneum & Other	1.23 (12)	1.77 (0.23-13.7)	1.21 (1)	0 (0.06 expected)
MN Respiratory	**1.15** (875)	**0.453 (0.28-0.72)**	**0.86 (32)**	**2.05 (0.85-4.94)**
MN Larynx	1.05 (25)	0 (1.52 expected)	1.52 (1)	11.1 (0.69-177)
MN Trachea, Bronch., Lung	1.14** (829)	0.462 (0.29-0.74)	0.86 (31)	1.76 (0.68-4.56)
MN Other Respiratory	2.39** (21)	0.649 (0.09-4.82)	0	0 (0.04 expected)
MN Breast	**1.92 (5)**	**0 (0.16 expected)**	**1.09 (47)**	**0.441 (0.11-1.82)**
MN Genital	**1.09 (265)**	**0.882 (0.49-1.57)**	**0.78 (18)**	**1.31 (0.30-5.67)**
MN Prostate	1.12 (263)	0.888 (0.50-1.58)	--	--
MN Testis	0.27* (2)	0 (0.65 expected)	--	--
MN Cervix	--	--	1.07 (6)	0 (0.54 expected)
MN Ovary	--	--	0.54 (7)	1.58 (0.19-12.83)
MN Urinary Organs	**1.16 (146)**	**0.403 (0.11-1.44)**	**0.95 (4)**	**0 (0.35 expected)**
MN Kidney	1.17 (75)	0.598 (0.13-2.82)	0.73 (2)	0
MN Bladder	1.15 (71)	0.196 (0.03-1.41)	1.36 (2)	0
MN Other & Unspecified Sites	**1.11* (359)**	**0.855 (0.53-1.38)**	**0.90 (23)**	**0.512 (0.07-3.79)**
MN Skin Melanoma	1.17 (49)	0.633 (0.19-2.06)	0.32 (1)	0 (0.31 expected)
MN Brain & Other NS	1.06 (80)	1.15 (0.49-2.71)	1.76 (11)	0 (0.60 expected)
MN Thyroid	1.31 (5)	0 (0.25 expected)	0	0 (0.04 expected)
MN Bone	0.43 (3)	0 (0.45 expected)	2.25 (1)	0 (0.04 expected)
MN Connective Tissue	0.90 (13)	0.937 (0.12-7.16)	0.54 (1)	0 (0.18 expected)
MN Other & Unspecified	1.19* (199)	0.860 (0.44-1.70)	0.69 (9)	1.31 (0.17-10.4)
MN Lymphatic & Hematopoietic	**1.07 (260)**	**0.766 (0.44-1.34)**	**1.17 (20)**	**1.76 (0.52-5.93)**
Non-Hodgkin Lymphoma	1.27* (112)	1.15 (0.57-2.33)	1.19 (8)	1.48 (0.18-11.8)
Hodgkin's Disease	0.67 (10)	0 (1.10 expected)	1.06 (1)	0 (0.08 expected)
Leukemia	1.06 (102)	0.483 (0.17-1.34)	1.20 (8)	2.88 (0.61-13.7)
Myeloma	0.81 (36)	0.586 (0.08-4.27)	1.10 (3)	0 (0.22 expected)
Benign & Unspec. Nature Neoplasms	**1.05 (33)**	**3.40 (0.83-13.9)**	**1.39 (4)**	**0 (0.25 expected)**
Neoplasms of Nervous Syst. of Unspec. Nature	0.62 (9)	9.74 (1.59-59.7)	2.62 (3)	0
All Cancers, SMRs only (N), 95% CI	**1.11* (2543), 1.07-1.15**	**0.71** (103), 0.58-0.86**	**0.94 (174), 0.80-1.09**	**1.42 (23), 0.90-2.15**
Diabetes Mellitus	**1.06 (174)**	**1.04 (0.54-1.99)**	**0.96 (15)**	**1.02 (0.14-7.74)**
Diseases of Blood & Blood-Forming Organs	**0.78 (24)**	**1.74 (0.36-8.27)**	**0.89 (2)**	**0 (0.19 expected)**
Other & Unspec. Anemias	1.08 (11)	0 (0.53 expected)	--	--
Alcoholism	**0.67* (36)**	**0.331 (0.05-2.42)**	**0.54 (1)**	**0 (0.18 expected)**
Other Mental Disorders	**1.17 (67)**	**0.452 (0.06-3.26)**	**1.74 (8)**	**0 (0.37 expected)**
Diseases of Nervous System & Sense Organs	**1.18* (191)**	**0.485 (0.19-1.21)**	**0.94 (14)**	**2.46 (0.71-8.60)**
Diseases of Heart	**0.93** (3024)**	**0.672 (0.54-0.83)**	**0.90 (105)**	**0.524 (0.19-1.46)**
Ischemic Heart Disease	0.95** (2503)	0.688 (0.55-0.86)	0.84 (68)	0.159 (0.02-1.14)
Hypertension w/Heart Disease	1.78** (67)	0.385 (0.09-1.58)	1.90 (5)	6.80 (1.31-35.2)

Table 5-24. SRR results for chemical workers compared to non-chemical workers, for white males and white females.

Cause of death	White males		White females	
	SMR non-chem (N)	SRR chemical workers (95% CI)	SMR non-chem (N)	SRR chemical workers (95% CI)
Other Diseases of Circulatory System	0.94 (741)	0.797 (0.48-1.33)	0.98 (53)	0.236 (0.03-1.71)
Hypertension w/o Heart Disease	1.15 (29)	0 (1.44 expected)	1.09 (2)	0 (0.14 expected)
Cerebrovascular. Disease	0.92 (443)	0.807 (0.43-1.53)	1.13 (41)	0 (2.82 expected)
Disease of Respir. System	1.03 (913)	0.492 (0.33-0.74)	0.70* (32)	1.14 (0.32-4.03)
Chronic & Unspecified Bronchitis	0.88 (22)	0.643 (0.09-4.77)	1.01 (1)	0 (0.08 expected)
Emphysema	1.05 (151)	0.541 (0.21-1.37)	0.39 (2)	0 (0.42 expected)
Asthma	0.79 (18)	0.657 (0.09-4.92)	0.76 (2)	0 (0.24 expected)
Asbestosis	3.10** (9)	1.61 (0.20-12.7)	0 (expected)	0 (0.00 expected)
Other Respiratory Disease	1.09* (505)	0.408 (0.23-0.73)	0.89 (22)	0.8320 (0.18-3.75)
Diseases of Digestive System	0.94 (363)	0.732 (0.39-1.38)	0.84 (22)	0.381 (0.05-2.83)
Cirrhosis of Liver	0.96 (165)	0.474 (0.11-2.09)	0.74 (8)	0 (0.97 expected)
Diseases of Genitourinary System	0.99 (104)	0 (5.54 expected)	0.91 (6)	5.16 (1.28-20.9)
Acute Glomerulonephritis & Acute Renal Failure	0.75 (8)	0 (0.62 expected)	0 (0.53 expected)	0 (0.04 expected)
Chronic & Unspecified Nephritis, Renal Failure	1.27 (68)	0 (2.94 expected)	0.96 (3)	3.93 (0.41-37.8)
Symptoms & Ill-Defined Conditions	0.85 (90)	0.525 (0.19-1.43)	1.54 (8)	0 (0.45 expected)
Accidents	0.76** (665)	0.596 (0.42-0.84)	0.83 (37)	1.48 (0.64-3.42)
Suicide	0.79** (274)	0.505 (0.29-0.89)	0.88 (13)	0 (1.64 expected)
Homicide	0.73* (44)	0 (5.61 expected)	1.33 (7)	0 (0.64 expected)
All deaths, SMRs only (N), 95% CI	0.99 (9707), 0.97-1.01	0.63** (383), 0.57-0.70	0.93* (529), 0.85-1.02	1.00 (49), 0.74-1.32

†SRR not computed if 0 deaths occurred in baseline or comparison category.
*95% CI of SMR excludes 1.00. SRRs are not flagged.
**99% CI of SMR excludes 1.00. SRRs are not flagged.

5.11 Drivers

Approximately 20% of the full cohort was of unknown driver status, as job titles were not available. The subcohort of drivers (N=1947) was predominantly white and male. White male drivers did not exhibit a healthy worker effect overall (Table 5-25) as the SMR compared to the regional population was 1.18 (CI, 1.09-1.28). Cancers showing particularly elevated mortality rate ratios compared to other workers included digestive cancers, connective tissue cancers, and benign and unspecified neoplasms. The lung cancer mortality rate was not elevated among the drivers, although rates of emphysema and the miscellaneous class of "other" (non-malignant, non-pneumoconioses) respiratory deaths were substantially elevated. The mortality rate from acute glomerulonephritis and acute renal failure was also quite elevated among drivers. One asbestosis case occurred among the small cohort of drivers, leading to an elevated rate ratio with very wide CIs.

The death rate from transportation accidents was also elevated among drivers (SRR, 1.63; CI, 1.07-2.48). For non-drivers, the SMR increased with increasing time since hire; however, for drivers no discernable pattern existed (Table 5-26).

Table 5-25. SRR results for drivers compared to non-drivers, for white males only.

Cause of death	SMR non-drivers (N)	SRR drivers (95% CI)
MN Buccal cavity	0.84 (37)	0.967 (0.23-4.01)
MN Digestive	1.03 (536)	1.34 (0.97-1.85)
MN Respiratory	1.12** (797)	0.900 (0.65-1.24)
MN Breast	1.62 (4)	0 (0.13 expected)†
MN Genital	1.09 (250)	0.799 (0.43-1.48)
MN Urinary Organs	1.14 (135)	0.850 (0.39-1.86)
MN Other & Unspecified Sites	1.11 (334)	0.923 (0.55-1.56)
MN Connective Tissue	*0.83 (11)*	*6.54 (1.66-25.7)*
MN Lymphatic & Hematopoietic	1.07 (243)	0.765 (0.42-1.41)
Benign & Unspecified Nature Neoplasms	1.02 (30)	2.93 (1.13-7.61)
All Cancers SMRs only (N), 95% CI	1.09** (2336), 1.04-1.13	1.06 (128), 0.88-1.26
Diabetes Mellitus	1.11 (170)	0.592 (0.24-1.45)
Diseases of Blood & Blood-Forming Organs	0.65 (19)	0.926 (0.12-6.92)
Alcoholism	0.62** (31)	2.51 (0.97-6.47)
Other Mental Disorders	1.15 (62)	1.03 (0.31-3.41)
Diseases of Nervous System & Sense Organs	1.17* (178)	1.21 (0.65-2.22)
Diseases of Heart	0.92** (2814)	1.19 (1.02-1.39)
Other Diseases of Circulatory System	0.92* (688)	1.43 (1.03-1.99)
Disease of Respiratory System	0.98 (823)	1.82 (1.44-2.30)
Emphysema	*1.00 (137)*	*1.94 (1.13-3.31)*
Asbestosis	*3.30** (9)*	*2.42 (0.31-19.1), N=1*
"Other" Respiratory Disease	*1.03 (450)*	*1.77 (1.29-2.44)*
Diseases of Digestive System	0.90 (325)	1.08 (0.68-1.70)
Diseases of Genitourinary System	0.92 (92)	2.02 (1.04-3.92)
Acute Glomerulonephritis & Renal Failure	*0.50 (5)*	*8.74 (1.69-45.1)*
Chronic Glomerulonephritis & Renal Failure	*1.20 (61)*	*1.69 (0.71-4.00)*
Symptoms & Ill-Defined Conditions	0.83 (83)	1.43 (0.62-3.30)
Accidents	0.71** (575)	1.52 (1.11-2.09)
Transportation Accidents	*0.69** (320)*	*1.63 (1.07-2.48)*
Accidental Falls	*0.99 (70)*	*1.36 (0.55-3.39)*
Suicide	0.74** (234)	1.46 (0.88-2.41)
Homicide	0.61** (34)	2.28 (0.77-6.76)
Other and Unspecified Causes	2.27** (343)	0.983 (0.61-1.57)
All Deaths, SMRs only (N), 95% CI	0.96** (8878), 0.94-0.98	1.18** (595), 1.09-1.28

†SRR not computed if 0 deaths occurred in baseline or comparison category.
*95% CI of SMR excludes 1.00. SRRs are not flagged.
**99% CI of SMR excludes 1.00. SRRs are not flagged.

Table 5-26. Standardized mortality ratios by time since hire for transportation accidents among white male drivers and non-drivers at INEEL. Comparison population was white males in ID, MT, WY combined.

Time since hire	SMR non-drivers (95% CI), N	SMR drivers (95% CI), N
0-5 years	0.40* (0.28-0.54), 41	1.11 (0.36-2.76), 5
5-10 years	0.60* (0.46-0.78), 58	0.82 (0.22-2.28), 4
10-15 years	0.51* (0.37-0.69), 43	1.10 (0.35-2.73), 5
15-20 years	0.90 (0.68-1.17), 58	1.65 (0.60-3.77), 6
20-25 years	1.04 (0.77-1.39), 47	0.38 (0.00-3.09), 1
25-30 years	0.76 (0.48-1.15), 23	1.09 (0.12-4.74), 2
30 years and over	1.26 (0.94-1.67), 50	1.65 (0.44-4.59), 4
Total	0.69* (0.62-0.77), 320	1.10 (0.73-1.62), 27

*99% CI of SMR excludes 1.00.

5.12 Reactor Workers

The small subcohort of reactor workers (N=1440) was overwhelmingly white and male, consisting of 95% of total subcohort person-time. The age distribution also was quite different from the rest of the cohort; that is with a median attained age of 52.6, reactor workers were seven years younger on average than other workers. In general, mortality rates among this subcohort were far lower than the non-reactor worker cohort (Table 5-27). The all-cause mortality SMR was 0.98 for non-reactor workers and 0.57 for reactor workers (the latter with a CI excluding one). The multiple myeloma and benign and unspecified neoplasm mortality rates were elevated, although the CIs included one. The reactor worker subcohort exhibited a greatly reduced mortality rate for non-malignant respiratory disease (SRR=0.16; CI, 0.04-0.61) compared to the rest of the cohort. The suicide and homicide SRRs were similarly low among reactor workers compared to other workers.

One notable exception was death from diseases of nervous system and sense organs (SRR=7.03, 1.25-39.5, N=4). No deaths from multiple sclerosis occurred in the reactor workers; the excess was observed in other diseases of the nervous system, a category that includes Alzheimer's disease, motor neuron disease, and various other neuropathies.

5.13 Painter Subcohort

The small subcohort of identifiable painters at INEEL (N=690) was predominantly white and male (approximately 98% of person-time). Very little power was available to detect departures from the number of expected deaths in this subcohort. For most causes of death, mortality rates among painters were similar to other workers at INEEL, although the all-cause and all-cancer SMRs were elevated in the painters compared to the regional population (Table 5-28).

Death rates from NHL and diabetes were each more than doubled among painters compared to the other workers, although numbers were small (and CIs for the former overlap one). Painters exhibited nearly a 60% elevation in death rates from non-malignant respiratory disease compared to the rest of the cohort. This elevation was due to a doubling of the rate for "other non-malignant respiratory diseases," 90% of which were chronic airways obstruction, not elsewhere classified. The SRR for suicide was also elevated among painters.

Table 5-27. SRR results for reactor workers compared to non-reactor workers, for white males only.

Cause of death	SMR non-reactor workers (N)	SRR reactor workers (95% CI)
MN Buccal Cavity	0.86 (39)	0 (0.77 expected)†
MN Digestive	1.06 (569)	0.737 (0.34-1.59)
MN Respiratory	1.12** (832)	0.340 (0.13-0.89)
MN Breast	1.55 (4)	0 (0.04 expected)
MN Genital	1.08 (260)	0.282 (0.04-2.01)
MN Urinary Organs	1.13 (139)	0.967 (0.30-3.07)
MN Other & Unspecified Sites	1.10 (343)	1.33 (0.43-4.14)
Other & Unspecified Sites	*1.13 (185)*	*2.37 (0.73-7.75)*
MN Lymphatic & Hematopoietic	1.07 (252)	0.483 (0.11-2.09)
Multiple Myeloma	*0.70 (30)*	*4.06 (0.90-18.28)*
Benign & Unspecified Neoplasms	1.11 (34)	7.08 (0.97-51.7) N=1
All Cancers, SMRs only (N), 95% CI	1.09** (2438), 1.05-1.13	0.75 (26), 0.49-1.11
Diabetes Mellitus	1.08 (172)	1.04 (0.32-3.39)
Diseases of Blood & Blood-Forming Organs	0.66 (20)	0 (0.39 expected)
Alcoholism	0.70* (36)	0 (1.21 expected)
Other Mental Disorders	1.14 (64)	3.71 (0.51-26.7)
Diseases of Nervous System & Sense Organs	1.17* (185)	7.03 (1.25-39.5) N=4
Multiple Sclerosis	*0.69 (10)*	*0 (0.365 expected)*
Other Diseases of Nervous System	*1.22* (175)*	*7.43 (1.32-41.8)*
Diseases of Heart	0.93** (2979)	0.713 (0.38-1.33)
Other Diseases of Circulatory System	0.94 (731)	1.54 (0.36-6.60)
Disease of Respiratory System	1.03 (899)	0.155 (0.04-0.61)
Emphysema	*1.06 (152)*	*0 (1.27 expected)*
Asbestosis (SMRs)	*3.52** (10)*	*0 (0.03 expected)*
Diseases of Digestive System	0.90 (340)	0.624 (0.25-1.57)
Diseases of Genitourinary System	0.98 (102)	0 (1.16 expected)
Symptoms & Ill-Defined Conditions	0.83 (87)	1.31 (0.31-5.61)
Accidents	0.74** (613)	0.792 (0.29-2.20)
Suicide	0.78** (250)	0.131 (0.03-0.53)
Homicide	0.66** (37)	0.482 (0.07-3.51)
Other and Unspecified Causes	2.30** (358)	0.793 (0.19-3.24)
All deaths, SMRs only (N), 95% CI	0.98 (9386), 0.96-1.00	0.57** (87), 0.46-0.70

†SRR not computed if 0 deaths occurred in baseline or comparison category.
*95% CI of SMR excludes 1.00. SRRs are not flagged.
**99% CI of SMR excludes 1.00. SRRs are not flagged.

Table 5-28. SRR results for painters compared to non-painters, for white males only. SMR for non-painters based on comparison to regional rates (ID, MT, WY).

Cause of death	SMR non-painters (N)	SRR painters (95% CI)
MN Buccal Cavity	0.82 (40)	0 (0.80 expected)†
MN Digestive	1.05 (605)	1.40 (0.80-2.43)
MN Respiratory	1.11** (878)	1.22 (0.74-2.02)
MN Breast	1.48 (4)	15.9 (1.78-143) N=1
MN Genital	1.10 (274)	1.02 (0.41-2.53)
MN Urinary Organs	1.12 (146)	1.23 (0.37-4.07)
MN Other & Unspecified Sites	1.09 (368)	1.67 (0.86-3.23)
MN Lymphatic & Hematopoietic System	1.06 (269)	1.31 (0.53-3.22) N=5
Non-Hodgkin Lymphoma	*1.26* (117)*	*2.46 (0.89-6.80) N=4*
Benign & Unspecified Neoplasms	1.14 (37)	0 (0.56 expected)
All Cancers SMRs only (N), 95% CI	1.08** (2584), 1.04-1.12	1.35* (54), 1.01-1.76
Diabetes Mellitus	1.04 (178)	2.49 (1.09-5.71) N=6
Diseases of Blood & Blood-Forming Organs	0.79 (25)	0 (0.56 expected)
Alcoholism	0.62** (35)	2.62 (0.62-11.0)
Other Mental Disorders	1.11 (66)	1.22 (0.17-8.79)
Diseases of Nervous System & Sense Organs	1.13 (191)	1.35 (0.50-3.69)
Diseases of Heart	0.92** (3082)	0.970 (0.73-1.30)
Other Diseases of Circulatory System	0.92* (741)	1.41 (0.89-2.25)
Disease of Respiratory System	1.00 (914)	1.58 (1.05-2.38)
Emphysema	*1.04 (153)*	*0.61 (0.15-2.46)*
Asbestosis (SMRs)	*3.33** (10)*	*0 (0.05 expected)*
Other Respiratory Diseases	*1.04 (499)*	*2.29 (1.44-3.64)*
Diseases of Digestive System	0.92 (367)	1.36 (0.67-2.79)
Diseases of Genitourinary System	0.94 (102)	0.508 (0.07-3.64)
Symptoms & Ill-Defined Conditions	0.85 (93)	0.347 (0.05-2.49)
Accidents	0.74** (682)	1.29 (0.76-2.19)
Suicide	0.76** (278)	2.07 (1.03-4.16)
Homicide	0.65** (42)	1.59 (0.22-11.5)
Other and Unspecified Causes	2.18** (371)	1.52 (0.79-2.93)
All deaths, SMRs only (N), 95% CI	0.97** (9833), 0.95-0.98	1.18* (208), 1.03-1.35

†SRR not computed if 0 deaths occurred in baseline or comparison category,
*95% CI of SMR excludes 1.00. SRRs are not flagged.
**99% CI of SMR excludes 1.00. SRRs are not flagged.

5.14 Security Worker Subcohort

Only white male deaths were evaluated in this category (158 total deaths) as there were just 2 deaths among white female security workers and no deaths among non-white workers. In general, the mortality experience reflected the patterns of the remainder of the cohort (Table 5-29). Respiratory cancer death rates were slightly higher among security workers than among other workers, an observation caused by a nearly five-fold elevation in the death rate from cancer of the larynx (based on only 2 deaths).

Death rates from malignant neoplasms of nervous system, and especially from nervous system NUN, were elevated among the subcohort of security workers (Table 5-29). The estimated SRR for security workers for all brain neoplasms combined was 2.29 (CI, 0.78-6.71).

In contrast to expectation, security workers exhibited lower mortality rates from ischemic heart disease, compared to both other workers and to the general population. However, the point estimates for the SRRs of mortality from hypertensive heart disease and stroke were elevated, compared to the other workers in the cohort.

Lifestyle-related deaths (e.g., cirrhosis of liver and emphysema) were much lower than expected compared to other workers in the cohort (Table 5-29). The accidental death rate was about three-fold higher among security workers compared to other workers. This excess was due to deaths from accidental falls and "other accidents." Within the latter category, two (33%) were accidental firearm deaths among security workers, which occurred prior to their separation from INEEL. It was not clear whether these deaths occurred while on the job; however, among non-security workers, just 6 of 151 "other accidental" deaths (4%) were from firearm accidents that occurred prior to retirement from INEEL.

Table 5-29. SRR results for security workers compared to non-security workers, for white males. Causes for which there were 2 or more observed or expected deaths are presented. If no cases were observed, the number expected is given, based on state rates.

Cause of death	SMR non-security workers (N)	SRR security workers (95% CI)
MN Digestive	1.06 (567)	0.819 (0.43-1.56)
MN Intestine	1.05 (194)	1.02 (0.37-2.80)
MN Respiratory	1.11** (817)	1.04 (0.65-1.67)
MN Larynx	0.91 (21)	4.92 (1.12-21.7) N=2
MN Trachea, Bronchus & Lung	1.10** (777)	0.965 (0.59-1.58)
MN Genital	1.07 (254)	1.13 (0.52-2.47)
MN Prostate	1.10 (253)	1.14 (0.52-2.48)
MN Urinary Organs	1.16 (142)	0 (2.59 expected)†
MN Other & Unspecified Sites	1.10 (342)	1.07 (0.50-2.31)
MN Brain & Other Nervous System	1.11 (80)	1.26 (0.28-5.72)
MN Other & Unspecified	1.14 (185)	1.43 (0.58-3.50)
MN Lymphatic & Hematopoietic	1.06 (249)	0.621 (0.25-1.52)
Non-Hodgkin Lymphoma	1.29 (109)	0.526 (0.13-2.13)
Leukemia	1.06 (99)	0.983 (0.31-3.15)
Benign & Unspecified Neoplasms	1.02 (31)	5.28 (1.77-15.8)
Neoplasms of Nervous System, Unspec. Nature	0.64 (9)	10.5 (2.09-52.6) N=2
All-Cancer, SMRs only (N), 95% CI	1.09** (2413), 1.04-1.13	1.04 (51), 0.77-1.37
Diabetes Mellitus	1.06 (168)	1.24 (0.58-2.64)
Diseases of Nervous System & Sense Organs	1.19* (187), 1.02-1.37	0.382 (0.09-1.57)
Diseases of Heart	0.93** (2955)	0.660 (0.48-0.90)
Ischemic Heart Disease	0.95* (2452)	0.634 (0.45-0.90)
Hypertension with Heart Disease	1.66** (61)	2.04 (0.58-7.14)
Other Diseases of Circulatory System	0.94 (723)	1.06 (0.58-1.94)
Cerebrovascular Disease	0.93 (436)	1.35 (0.65-2.77)
Disease of Respiratory System	1.03 (892)	0.623 (0.29-1.34)
Emphysema	1.06 (151)	0.191 (0.03-1.36)
Other Respiratory Disease	1.08 (485)	1.00 (0.43-2.35)
Diseases of Digestive System	0.91 (341)	0.487 (0.17-1.36)
Cirrhosis of Liver	0.89 (148)	0 (3.86 expected)
Symptoms & Ill-Defined Conditions	0.86 (89)	0 (2.04 expected)
Accidents	0.72** (606)	3.26 (0.81-13.2), N=15
Transportation Accidents	0.72** (340)	0.618 (0.27-1.41)
Accidental Falls	1.02 (74)	18.9 (2.62-136) N=1
Other Accidents	0.63** (151)	2.34 (0.90-6.06) N=6
Suicide	0.77** (249)	0.455 (0.14-1.43)
Homicide	0.67* (38)	0 (1.46 expected)
Other & Unspecified	2.32** (361)	0.114 (0.02-0.81)
All deaths, SMRs only (N), 95% CI	0.98* (9315), 0.96-1.00	0.79** (158), 0.67-0.92

†SRR not computed if 0 deaths occurred in baseline or comparison category.
*95% CI of SMR excludes 1.00. SRRs are not flagged.
**99% CI of SMR excludes 1.00. SRRs are not flagged.

5.15 Employer Type Subcohorts

The person-time distribution in the INEEL cohort was predominantly contributed by subcontractors (38.4%), those who worked for multiple contractor types (i.e., both prime and subcontractors; 38.5%), and those who worked for prime contractors (21.6%). Those whose employer status was unknown contributed very little person-time to the cohort (1.5%).

White male subcontractor employees exhibited generally higher mortality rates than "prime" contractor employees (defined as employees of DOE, ANL-W, NRF, or any of the prime contractors operating the INEEL facility) and workers who were employed by several types of employers at the facility (Table 5-30). SRRs for those who worked for unknown contractors were highly variable because of the very low person-time accrual by this group. Rates of all-cause and all-cancer mortality were markedly elevated among the subcontractors and were substantially lower among the prime and multiple contractors, in comparison to the regional population. There was evidence of substantial lifestyle-related differences in mortality rates among employer types. Ischemic heart disease, lung cancer and emphysema death rates were lower among the prime and multiple contractor groups than among the subcontractor employees.

In contrast, death rates from asbestosis did not differ substantially among employer types (Table 5-30). The sole COD whose rates appeared meaningfully elevated among workers of unknown contractor type (compared to subcontractor employees) was non-malignant "other respiratory diseases." Accidental death rates for prime and multiple contractor workers were approximately half that of subcontractor employees. Death rates from suicide and, particularly, homicide were also lower among prime and multiple contractor employees, compared to subcontractor employees. Mortality rates for "Other and unspecified" causes were similarly elevated among all groups except the multiple contractor employees, who showed a lower standardized rate than the subcontractor employees.

Employer type differences in mortality rates for WF showed similar patterns, although CIs were much wider than for WM (Table 5-31). Respiratory and digestive cancer death rates were higher among white female subcontractor employees than among prime contractor employees and women employed by multiple types of contractors. The breast cancer mortality rate, in contrast, was more than doubled in prime and multiple contractor employees, compared to subcontractor employees. The brain cancer mortality rate was elevated among WF who worked for multiple contractors, compared to the regional population, although based on small numbers. The leukemia death rates were somewhat elevated among the white female employees of prime and multiple contractors, but the CIs were very wide because of the small numbers in the baseline category. Only five deaths occurred overall among white female employees of unknown contractor type.

Table 5-30. SRRs for white males by employer types (prime contractor, multiple contractors, unknown contractor) compared to subcontractor employee rates. Subcontractor employees were compared to ID, WY and MT combined rates, 1960-1999.

Cause of death	SMR Subcontractor (N)	SRR Prime contractor (95% CI)	SRR Multiple contractors (95% CI)	SRR Unknown contractor (95% CI)†
MN Buccal cavity	1.03 (23)	0.297 (0.09-0.99)	0.747 (0.38-1.46)	0 (0.56 expected)
MN Pharynx	1.05 (11)	0 (4.10 expected)	0.643 (0.22-1.86)	0 (0.26 expected)
MN Respiratory	1.10 (299)	0.897 (0.71-1.13)	0.862 (0.72-1.03)	1.68 (0.65-4.30)
	1.20 (33)	0.867 (0.43-1.74)	0.727 (0.42-1.26)	0.587 (0.08-4.30)
	1.01 (38)	1.11 (0.60-2.07)	1.00 (0.61-1.62)	0 (0.75 expected)
	1.04 (98)	0.964 (0.65-1.44)	0.928 (0.68-1.26)	2.98 (0.68-13.0)
	0.89 (20)	0.684 (0.26-1.83)	1.14 (0.59-2.20)	0 (0.47 expected)
	1.28 (22)	0.929 (0.41-2.12)	0.670 (0.33-1.36)	2.36 (0.53-10.6)
	1.07 (7)	0.703 (0.14-3.50)	0.748 (0.21-2.65)	0 (0.14 expected)
	1.16 (71)	0.853 (0.52-1.40)	0.860 (0.60-1.24)	1.60 (0.42-6.05)
	2.07* (10)	0.298 (0.04-2.34)	0.117 (0.01-0.92)	2.04 (0.26-16.0)
MN Respiratory	1.31** (479)	0.584 (0.47-0.72)	0.768 (0.66-0.89)	0.754 (0.43-1.33)
MN Larynx	1.38 (16)	0.310 (0.07-1.36)	0.517 (0.21-1.26)	0 (0.27 expected)
MN Trachea, Bronchus, Lung	1.30** (455)	0.592 (0.47-0.74)	0.766 (0.66-0.89)	0.751 (0.42-1.36)
MN Other Respiratory	1.93 (8)	0.699 (0.18-2.72)	1.35 (0.52-3.49)	2.38 (0.30-19.2)
	3.11 (4)	0 (0.45 expected)	0.258 (0.03-2.31)	0 (0.02 expected)
MN Male Genital	1.22* (154)	0.877 (0.60-1.28)	0.760 (0.58-0.99)	0.276 (0.04-1.97)
MN Prostate	1.24* (153)	0.884 (0.61-1.29)	0.758 (0.58-0.99)	0.278 (0.04-1.99)
	1.08 (67)	1.09 (0.68-1.73)	0.982 (0.68-1.41)	0.893 (0.22-3.65)
	1.12 (34)	0.778 (0.39-1.56)	1.08 (0.66-1.77)	0.829 (0.11-6.06)
	1.03 (33)	1.42 (0.77-2.64)	0.873 (0.51-1.51)	0.963 (0.13-7.05)
MN Other & Unspecified Site	1.17 (177)	0.702 (0.51-0.96)	0.978 (0.78-1.22)	0.782 (0.27-2.30)
MN Skin Melanoma	1.34 (25)	0.655 (0.28-1.55)	0.857 (0.47-1.57)	0.535 (0.07-3.95)
MN Brain & Other Nervous Syst.	0.97 (33)	0.787 (0.41-1.53)	1.29 (0.81-2.06)	0.853 (0.20-3.63)
MN Thyroid	1.65 (3)	0 (0.71 expected)	0.880 (0.15-5.34)	0 (0.05 expected)
MN Bone	0.60 (2)	0 (1.34 expected)	0.615 (0.05-6.94)	0 (0.07 expected)
MN Connective Tissue	0.76 (5)	0.861 (0.17-4.49)	1.66 (0.53-5.27)	0 (0.20 expected)

Table 5-30. SRRs for white males by employer types (prime contractor, multiple contractors, unknown contractor) compared to subcontractor employee rates. Subcontractor employees were compared to ID, WY and MT combined rates, 1960-1999.

Cause of death	SMR Subcontractor (N)	SRR Prime contractor (95% CI)	SRR Multiple contractors (95% CI)	SRR Unknown contractor (95% CI)
MN Other & Unspecified	1.27* (103)	0.745 (0.49-1.13)	0.878 (0.65-1.19)	0.936 (0.21-4.15)
	1.04 (121)	1.11 (0.80-1.55)	0.953 (0.73-1.25)	2.18 (0.31-15.6)
	1.15 (48)	1.23 (0.75-2.02)	1.14 (0.76-1.71)	5.40 (0.74-39.1)
	0.29 (2)	4.79 (0.92-24.8)	0.911 (0.12-6.64)	0 (0.17 expected)
	1.11 (52)	0.977 (0.56-1.71)	0.82 (0.53-1.27)	0 (1.09 expected)
	0.88 (19)	0.639 (0.23-1.78)	0.825 (0.41-1.68)	0 (0.50 expected)
Benign & Unspecified Neoplasms	1.11 (17)	0.878 (0.33-2.36)	0.905 (0.44-1.85)	0 (0.33 expected)
	1.18** (1324), 1.12-1.25	0.92 (383), 0.83-1.02	1.03 (908), 0.96-1.10	1.33 (34), 0.92-1.86
Diabetes Mellitus	0.87 (70)	1.02 (0.65-1.61)	1.48 (1.08-2.05)	0.770 (0.24-2.45)
	0.90 (14)	0.714 (0.23-2.23)	0.562 (0.23-1.40)	5.91 (0.78-45.0)
	1.50 (8)	0 (1.65 expected)	0.436 (0.12-1.65)	0 (0.09 expected)
Alcoholism	0.70 (17)	0.688 (0.24-1.94)	1.04 (0.52-2.09)	0 (0.73 expected)
	1.14 (33)	0.678 (0.30-1.55)	1.13 (0.67-1.88)	0 (0.52 expected)
Diseases of Nervous System & Sense Organs	1.32** (104)	0.756 (0.49-1.16)	0.795 (0.58-1.09)	0.659 (0.09-4.72)
	1.03 (1697)	0.698 (0.62-0.78)	0.800 (0.74-0.87)	1.26 (0.61-2.60)
	1.05 (1417)	0.673 (0.59-0.76)	0.798 (0.73-0.87)	1.17 (0.49-2.84)
	1.48* (30)	1.40 (0.73-2.67)	1.00 (0.58-1.74)	7.14 (0.97-52.4)
Other Diseases of Circulatory System	1.00 (421)	0.714 (0.57-0.90)	0.835 (0.71-0.98)	1.66 (0.26-10.6)
Hypertension w/o Heart Disease	1.32 (17)	0.931 (0.36-2.41)	0.575 (0.22-1.53)	0 (0.24 expected)
Cerebrovascular Disease	1.02 (264)	0.638 (0.47-0.86)	0.761 (0.62-0.94)	2.53 (0.35-18.0)
	1.20** (551)	0.532 (0.42-0.67)	0.717 (0.62-0.83)	3.39 (0.52-22.1)
	1.03 (14)	0.516 (0.12-2.31)	0.767 (0.31-1.92)	0 (0.18 expected)

Table 5-30. SRRs for white males by employer types (prime contractor, multiple contractors, unknown contractor) compared to subcontractor employee rates. Subcontractor employees were compared to ID, WY and MT combined rates, 1960-1999.

Cause of death	SMR Subcontractor (N)	SRR Prime contractor (95% CI)	SRR Multiple contractors (95% CI)	SRR Unknown contractor (95% CI)
	1.22 (95)	0.466 (0.26-0.83)	0.737 (0.52-1.05)	0 (1.12 expected)
	0.89 (10)	0.904 (0.28-2.92)	0.740 (0.25-2.22)	0 (0.24 expected)
	3.43* (5)	0.467 (0.05-4.02)	1.30 (0.34-4.95)	0 (0.03 expected)
	1.03 (2)	0 (0.40 expected)	0 (1.06 expected)	0 (0.02 expected)
	1.33** (306)	0.457 (0.34-0.62)	0.665 (0.55-0.81)	5.92 (0.91-38.6)
Diseases of Digestive System	1.02 (192)	0.960 (0.72-1.28)	0.727 (0.57-0.92)	0.957 (0.31-2.97)
Cirrhosis of Liver	1.09 (86)	0.877 (0.57-1.36)	0.655 (0.46-0.93)	1.64 (0.42-6.42)
	1.08 (60)	0.402 (0.19-0.83)	0.763 (0.50-1.17)	1.33 (0.18-9.62)
	1.32 (7)	0 (1.81 expected)	0.164 (0.02-1.34)	0 (0.10 expected)
	1.31 (36)	0.479 (0.21-1.10)	0.852 (0.50-1.44)	2.19 (0.30-16.0)
Symptoms & Ill-Defined Cond.	0.70* (39)	0.934 (0.47-1.84)	1.49 (0.95-2.32)	0.246 (0.03-1.79)
	1.01 (380)	0.511 (0.41-0.64)	0.555 (0.47-0.66)	1.11 (0.51-2.41)
	1.00 (206)	0.550 (0.41-0.74)	0.584 (0.47-0.73)	0.740 (0.34-1.60)
	1.22 (45)	0.453 (0.22-0.94)	0.654 (0.39-1.10)	0.448 (0.06-3.26)
	0.92 (99)	0.479 (0.31-0.75)	0.481 (0.34-0.68)	2.44 (0.71-8.40)
Suicide	0.85 (128)	0.812 (0.58-1.13)	0.845 (0.65-1.10)	0.504 (0.18-1.38)
	0.98 (24)	0.421 (0.17-1.04)	0.549 (0.28-1.07)	0 (0.891 expected)
Other & Unspecified	2.77** (213)	1.03 (0.78-1.35)	0.556 (0.43-0.71)	1.44 (0.60-3.42)
	1.09*** (5309), 1.07-1.12	0.80** (1391), 0.76-0.84	0.88** (3296), 0.85-0.92	1.00 (101), 0.81-1.21

†SRR not computed if 0 deaths occurred in baseline or comparison category.
*95% CI of SMR excludes 1.00. SRRs are not flagged.
**99% CI of SMR excludes 1.00. SRRs are not flagged.

Table 5-31. SRRs for white females by prime contractor and multiple contractor employer types, compared to subcontractor employer type rates, for causes of death with more than 5 expected or observed.

Cause of death	SMR Subcontractor (N)	SRR Prime contractor (95% CI)	SRR Multiple contractors (95% CI)
MN Buccal Cavity & Pharynx	0 (0.44 expected)†	SMR=3.60, N=3	SMR=2.88, N=3
MN Digestive	1.88 (13)	0.416 (0.19-0.93)	0.265 (0.11-0.63)
MN Intestine	*1.38 (4)*	*0.624 (0.17-2.33)*	*0.393 (0.10-1.58)*
MN Pancreas	*2.39 (4)*	*0.340 (0.08-1.54)*	*0 (3.97 expected)*
MN Respiratory	1.60 (12)	0.437 (0.19-1.01)	0.544 (0.26-1.15)
MN Trachea, Bronchus & Lung	*1.65 (12)*	*0.437 (0.19-1.01)*	*0.480 (0.22-1.04)*
MN Breast	0.46 (4)	2.61 (0.90-7.58)	2.34 (0.80-6.81)
MN Female Genital	0.64 (3)	0.861 (0.20-3.62)	1.73 (0.49-6.14)
MN Other & Unspecified Sites	0.58 (3)	1.68 (0.46-6.11)	1.66 (0.46-5.99)
MN Brain & Other Nervous Syst.	*0 (1.24 expected)*	*SMR=1.53, N=4*	*SMR=2.38*, N=7*
MN Other & Unspecified	0.38 (1)	2.56 (0.30-22.0)	1.57 (0.17-14.1)
MN Lymphatic & Hematopoietic	2.01 (7)	0.575 (0.21-1.59)	0.481 (0.17-1.34)
Non-Hodgkin Lymphoma	*3.59* (5)*	*0.196 (0.04-1.02)*	*0.176 (0.03-0.92)*
Leukemia	*0.75 (1)*	*2.58 (0.29-23.1)*	*2.59 (0.30-22.4)*
All Cancers, SMRs only (N), 95% CI	1.11 (42), 0.80-1.51	0.97 (71), 0.76-1.22	0.94 (84), 0.75-1.16
Diabetes Mellitus	0.61 (2)	1.34 (0.26-7.00)	1.80 (0.39-8.38)
Other Mental Disorders (not Alcoholism)	1.92 (2)	0.469 (0.04-5.21)	1.14 (0.22-5.97)
Diseases of Nervous System & Sense Organs	0.97 (3)	1.53 (0.40-5.92)	0.948 (0.23-3.88)
Diseases of Heart	1.32 (33)	0.435 (0.25-0.77)	0.624 (0.39-1.00)
Ischemic Heart Disease	*1.09 (19)*	*0.391 (0.19-0.81)*	*0.734 (0.40-1.36)*
Hypertension with Heart Disease	*3.42 (2)*	*0.636 (0.09-4.74)*	*0.610 (0.10-3.69)*
Other Diseases of Circulatory System	1.23 (14)	0.582 (0.28-1.22)	0.740 (0.38-1.44)
Cerebrovascular Disease	*1.16 (9)*	*0.710 (0.29-1.73)*	*0.990 (0.45-2.20)*
Disease of Respiratory System	0.62 (6)	1.69 (0.64-4.44)	0.978 (0.36-2.63)
Diseases of Digestive System	1.13 (6)	0.375 (0.11-1.23)	0.809 (0.30-2.16)
Cirrhosis of Liver	*0.94 (2)*	*0.207 (0.02-2.31)*	*1.07 (0.20-5.58)*
Symptoms & Ill-Defined Cond.	2.78 (3)	0.574 (0.11-2.90)	0.322 (0.05-1.98)
Accidents	1.16 (10)	0.569 (0.25-1.29)	0.819 (0.37-1.79)
Transportation Accidents	*1.32 (8)*	*0.637 (0.26-1.57)*	*0.747 (0.30-1.83)*
Suicide	1.75 (5)	0.328 (0.09-1.23)	0.391 (0.10-1.47)
Homicide	0.98 (1)	1.15 (0.10-12.8)	2.11 (0.24-19.0)
Other & Unspecified	1.27 (4)	1.84 (0.56-6.01)	0.970 (0.28-3.34)
All deaths SMRs only (N), 95% CI	1.14 (133), 0.95-1.35	0.88 (190), 0.76-1.02	0.90 (252), 0.80-1.02

†SRR not computed if 0 deaths occurred in baseline or comparison category.
*95% CI of SMR excludes 1.00. SRRs are not flagged.

6 Multivariable Modeling Results

6.1 All Solid Cancers

6.1.1 Development of Baseline Model

This section describes the establishment of a baseline model to determine important potential confounders of the external radiation association with all-solid-cancer risk. Known confounders (that is, based on the findings described in Chapter 5) included sex, age, calendar year and duration of employment (dichotomized as less than ten years, or greater than or equal to ten years). These factors, therefore, were included as stratification variables in the Poisson regression analysis. Other potential confounding variables, such as SES, migrant status (i.e., state of origin was Idaho, Utah, Montana or Wyoming), and internal exposure monitoring status, were evaluated separately and are described here.

The following four paragraphs refer to results for all monitored workers with DOB, date of first monitoring and monitoring status available (i.e., the second-to-last column of Table 6-1). Before evaluating any confounding variables, all solid cancers were significantly negatively associated with radiation exposure, showing a monotonic decrease with increasing dose category (Models 1 and 2, Table 6-1). Cancer risk was particularly low in the highest dose category, after adjusting just for age, sex, calendar year and duration of employment.

In the evaluation of SES, it was decided initially, based on the results of the SRR analysis described in §5.6 and on information from other studies, to combine the SES professional and intermediate groups, and the partly skilled and unskilled workers, to reduce the level of stratification required for the Poisson regression analysis. For all solid cancers combined, these SES groupings were associated with risk (Model 3, Table 6-1). The all-cancer mortality risk was greater for partly skilled and unskilled workers, for those of unknown SES, and especially skilled manual workers than for professional and intermediate workers, as reported in §5.6 above.

Migrants showed a slightly higher solid cancer mortality risk than local workers (Model 4, Table 6-1), which was exacerbated by adjustment for SES (Model 6, Table 6-1). Two expressions are shown for these interaction tests: the SES·migrant interaction is expressed as migrant worker RR compared to local workers, at each level of SES (the first Model 7 in Table 6-1), and as each SES group's RR compared to the combined groups of Professional & Intermediate workers, for migrants and locals separately (the second Model 7 in Table 6-1). The p-values associated with these two expressions of Model 7 in the table (among "all workers") indicate that the migrant compared to local RR pattern depends on the level of SES (e.g., the RR for migrants is much higher for those of unknown SES than for the combined group of partly skilled and unskilled workers). Similarly, using the second expression of Model 7 for "all workers," the RR for partly skilled and unskilled workers compared to professional and intermediate workers is significantly different for locals (1.523) and migrants (0.858) with a p=0.003.

Migrants were much more likely to have died from cancer than locals among the SES professional/intermediate and unknown groups and showed about the same risk of death among skilled non-manual and skilled manual workers. The migrant partly skilled and unskilled workers, however, showed lower risk of death from cancer compared to local workers in that employment category.

The removal of short-term workers (who were employed less than one year) as in the IARC analysis that included the INEEL cohort, had very little effect on these findings (last column, Table 6-1). One factor that did change was the strength of the interaction between migrant status and SES. The variability in RR of migrants did not vary significantly by SES category, which was likely due to the removal of a large number of short-term workers of unknown SES.

A regrouping of all monitored workers (with DOB, date first monitored and migrant status available) into the six original SES categories (plus unknown) showed different interactions between migrant status and SES for the professional/intermediate grouping and the partly skilled/unskilled grouping (Model 7, Table 6-2). The skilled non-manual grouping, in contrast, was very similar to the professional grouping in both baseline risk and in the interaction between SES and migrant status, after adjusting for sex and the other stratifiers. Therefore, it was decided to combine the professional and skilled non-manual groups, and to keep the others separate, to reduce the total number of strata. This strategy also had the effect of creating a baseline stratum that contained both men and women of lower cancer risk, as the professional group consisted primarily of males (92%) and skilled non-manual of females (77%). Race showed very little independent association with disease (data not shown) and was removed as a covariate, to reduce stratification. Data for all races were included, however, in the analyses.

The final baseline model for all solid cancer included only workers with known DOB, date first monitored and migrant status. The model stratified on attained age and calendar time (in 5-year intervals), sex, and duration of employment (<10 years and ≥10 years). The model also adjusted for SES (in 6 categories including unknown), migrant status, SES·migrant interaction, and internal monitoring status (Table 6-3).

6.1.2 Dose-response Analysis with External Ionizing Radiation

When stratifying on the main factors (sex, calendar year, age and duration), dose was strongly and negatively associated with risk, using a 10-year lag (Model 1, Tables 6-1 and 6-2). The addition of risk factors such as SES, internal dose, migrant status, and the interaction of migrant status and SES did not meaningfully change this association (Models 8-11, Tables 6-1 and 6-2). The strength of the negative association of all cancers combined with dose was attenuated only very slightly with the addition of these factors. Recategorizing SES into the groupings of the final baseline model described above further attenuated this negative dose-response coefficient; however, all solid cancers were still strongly and negatively associated with dose (Model 11, Table 6-3).

Table 6-1. Comparison of all solid cancer (excluding lymphomas, myeloma and leukemias, and with 10-year lag) maximum-likelihood risk estimates and likelihood-based CIs produced with (N=34,916) and without (N=29,585) monitored employees who worked < 1 year, with DOB, date first monitored and migrant status available.*

Model number	Term type	Factor	Risk estimate (RR or ERR[†]); 95% CI All workers	Excluding <1 year duration
1	loglinear	dose coefficient (at 1 mSv)	0.9978; 0.9964, 0.9991	0.9979; 0.9965, 0.9991
2	loglinear	dose category 0-<1 mSv	Baseline	Baseline
		dose category 1-<10 mSv	0.9219; 0.8054, 1.053	0.9945; 0.8592, 1.149
		dose category 10-<50 mSv	0.8758; 0.7447, 1.026	0.8963; 0.7531, 1.063
		dose category 50-<100 mSv	0.8550; 0.6534, 1.100	0.9011; 0.6865, 1.163
		dose category ≥100 mSv	0.5557; 0.4027, 0.7484	0.5778; 0.4178, 0.7803
3	loglinear	SES professional & intermediate	Baseline	Baseline
		SES skilled non-manual	0.9287; 0.7075, 1.205	0.9432; 0.7015, 1.251
		SES skilled manual	1.368; 1.203, 1.556	1.393; 1.207, 1.608
		SES partly skilled & unskilled	1.158; 0.9872, 1.355	1.187; 0.9981, 1.408
		SES unknown	1.152; 0.9124, 1.439	0.9922; 0.7264, 1.327
4	loglinear	Local (SSN=ID, UT, MT, WY)	Baseline	Baseline
		Migrant	1.034; 0.9276, 1.154	1.001; 0.8883, 1.129
5	loglinear	Internal—not monitored	Baseline	Baseline
		Internal—monitored, unexposed	1.129; 0.9791, 1.300	1.263; 1.073, 1.483
		Internal—likely exposed	0.8948; 0.7868, 1.017	0.9682; 0.8421, 1.114
6	loglinear	SES professional & intermediate	Baseline	Baseline
		SES skilled non-manual	0.9500; 0.7229, 1.235	0.9612; 0.7136, 1.277
		SES skilled manual	1.398; 1.226, 1.595	1.419; 1.225, 1.645
		SES partly skilled & unskilled	1.206; 1.020, 1.422	1.225; 1.021, 1.465
		SES unknown	1.145; 0.9066, 1.430	0.9870; 0.7225, 1.320
		Local (SSN=ID, UT, MT, WY)	Baseline	Baseline
		Migrant	1.097; 0.9774, 1.231	1.075; 0.9474, 1.221
7	loglinear	Migrant vs local, SES=prof/int	1.306	1.289
		Migrant vs local, SES=sknman	1.141	1.144
		Migrant vs local, SES=skman	1.055	1.013
		Migrant vs local, SES=ptsk/unsk	0.7353	0.7907
		Migrant vs local, SES=unknown	2.129 (p=0.0093)	1.671 (p=0.1121)
7	loglinear	SES sknm vs prof/int, local	1.062	1.066
		SES sknm vs prof/int, migrant	0.9281 (p>0.5)	0.9460 (p>0.5)
		SES skman vs prof/int, local	1.631	1.670
		SES skman vs prof/int, migrant	1.318 (p=0.128)	1.312 (p=0.114)
		SES ptsk/unsk vs prof/int, local	1.523	1.511
		SES ptsk/unsk vs prof/int, migrant	0.8577 (p=0.003)	0.9266 (p=0.018)
		SES unkn vs prof/int, local	0.7420	0.7756
		SES unkn vs prof/int, migrant	1.187 (p=0.205)	1.005 (p>0.5)
8	loglinear	SES in 5 classes	--	--
	linear	Dose (ERR and SE per mSv[‡])	-0.001728; 0.0003931	-0.001717; 0.0003944
9	loglinear	SES in 5 classes, Migrant	--	--
	linear	Dose (ERR and SE per mSv[‡])	-0.001701; 0.0003989	-0.001698; 0.0003988
10	loglinear	SES in 5 classes, Migrant, Migrant·SES	--	--
	linear	Dose (ERR and SE per mSv[‡])	-0.001686; 0.0004005	-0.001678; 0.0004013
11	loglinear	SES in 5 classes, Migrant, Migrant·SES, Internal	--	--
	linear	**Dose (ERR and SE per mSv[‡])**	**-0.001599; 0.0004300**	**-0.001651; 0.0004221**

* All models stratified on age group (14 strata) calendar time (8 strata) duration of employment (<10 years and ≥10 years) and sex. Factors within each model number were adjusted for all other factors in that model. Migrant status refers to those whose SSN was issued outside of ID, MT, WY or UT.
† RR: Relative risk; ERR: Excess relative risk; CI: CI
‡ Standard errors (SE) are reported for the ERR when likelihood-based CIs could not be computed.

Table 6-2. Comparison of all-solid-cancer (excluding lymphomas, myeloma and leukemias, and with 10-year lag) maximum-likelihood risk estimates and likelihood-based CIs produced with (N=34,916) monitored employees with DOB, date first monitored and migrant status available.*

Model number	Term type	Factor	Risk estimate (RR or ERR[†]); 95% CI
1	loglinear	dose coefficient (at 1 mSv)	0.9979; 0.9965, 0.9991
2	loglinear	dose category 0-<1 mSv	Baseline
		dose category 1-<10 mSv	0.9219; 0.8054, 1.053
		dose category 10-<50 mSv	0.8758; 0.7447, 1.026
		dose category 50-<100 mSv	0.8550; 0.6534, 1.100
		dose category ≥100 mSv	0.5557; 0.4027, 0.7484
3	loglinear	SES professional	Baseline
		SES intermediate	1.093; 0.9142, 1.307
		SES skilled non-manual	0.9723; 0.7296, 1.282
		SES skilled manual	1.427; 1.224, 1.667
		SES partly skilled	1.206; 0.9696, 1.492
		SES unskilled	1.211; 0.9583, 1.520
		SES unknown	1.200; 0.9373, 1.523
4	loglinear	Local (SSN=ID, UT, MT, WY)	Baseline
		Migrant	1.034; 0.9276, 1.154
5	loglinear	Internal—not monitored	Baseline
		Internal—monitored, unexposed	1.129; 0.9791, 1.300
		Internal—likely exposed	0.8948; 0.7868, 1.017
6	loglinear	Race White	Baseline
		Race Other	1.141; 0.6884, 1.765
7	loglinear	SES professional	Baseline
		SES intermediate	1.103; 0.9222, 1.319
		SES skilled non-manual	1.000; 0.7492, 1.322
		SES skilled manual	1.466; 1.253, 1.719
		SES partly skilled	1.258; 1.006, 1.566
		SES unskilled	1.274; 1.000, 1.612
		SES unknown	1.197; 0.9349, 1.519
		Local (SSN=ID, UT, MT, WY)	Baseline
		Migrant	1.101; 0.9811, 1.236
8	loglinear	Migrant vs local, SES=professional	1.043
		Migrant vs local, SES=intermediate	1.659
		Migrant vs local, SES=sk manual	1.142
		Migrant vs local, SES=sk non-man	1.056
		Migrant vs local, SES=pt-skilled	0.6499
		Migrant vs local, SES=unskilled	0.8627
		Migrant vs local, SES=unknown	2.131 (p=0.0039)

Table 6-2. Comparison of all-solid-cancer (excluding lymphomas, myeloma and leukemias, and with 10-year lag) maximum-likelihood risk estimates and likelihood-based CIs produced with (N=34,916) monitored employees with DOB, date first monitored and migrant status available.*

Model number	Term type	Factor	Risk estimate (RR or ERR[†]); 95% CI
8	loglinear	SES int vs professional, local	0.7820
		SES int vs professional, migrant	1.244 (p=0.034)
		SES sknm vs professional, local	0.9343
		SES sknm vs professional, migrant	1.023 (p>0.5)
		SES skman vs professional, local	1.436
		SES skman vs professional, migrant	1.454 (p>0.5)
		SES ptskill vs professional, local	1.382
		SES ptskill vs professional, migrant	0.8611 (p=0.075)
		SES unsk vs professional, local	1.294
		SES unsk vs professional, migrant	1.070 (p=>0.5)
		SES unkn vs professional, local	0.6394
		SES unkn vs professional, migrant	1.306 (p=0.074)
9	loglinear	SES in 7 classes	--
	linear	Dose (ERR and SE per mSv[‡])	-0.001666; 0.0003958
10	loglinear	SES in 7 classes, Migrant	--
	linear	Dose (ERR and SE per mSv[‡])	-0.001638; 0.0004017
11	loglinear	SES in 7 classes, Migrant, Migrant·SES	--
	linear	Dose (ERR and SE per mSv[‡])	-0.001594; 0.0004088
12	loglinear	SES in 7 classes, Migrant, Migrant·SES, Internal	--
	linear	Dose (ERR and SE per mSv[‡])	-0.001504; 0.0004397
13	loglinear	SES in 7 classes, Migrant, Migrant·SES, Internal, Race	--
	linear	Dose (ERR and SE per mSv[‡])	-0.001503; 0.0004400

* All models stratified on age group (14 strata) calendar time (8 strata) duration of employment (<10 years and ≥10 years) and sex. Factors within each model number were adjusted for all other factors in that model. Migrant status refers to those whose SSN was issued outside of ID, MT, WY or UT.
† RR: Relative risk; ERR: Excess relative risk; CI: Confidence interval
‡ Standard errors (SE) are reported for the ERR when likelihood-based CIs could not be computed.

Table 6-3. Final baseline model: comparison of all-solid-cancer (excluding lymphomas, myeloma and leukemias, and with 10-year lag) maximum-likelihood risk estimates and likelihood-based CIs produced with (N=34,916) monitored employees with DOB, date first monitored and migrant status available.*

Model number	Term type	Factor	Risk estimate (RR or ERR[†]); 95% CI
1	loglinear	dose coefficient (at 1 mSv)	0.9979; 0.9965, 0.9992
2	loglinear	dose category 0-<1 mSv	Baseline
		dose category 1-<10 mSv	0.9219; 0.8054, 1.053
		dose category 10-<50 mSv	0.8758; 0.7447, 1.026
		dose category 50-<100 mSv	0.8550; 0.6534, 1.100
		dose category ≥100 mSv	0.5557; 0.4027, 0.7484
3	loglinear	SES professional & skilled non-man	Baseline
		SES intermediate	1.100; 0.9286, 1.301
		SES skilled manual	1.434; 1.242, 1.659
		SES partly skilled	1.214; 0.9851, 1.486
		SES unskilled	1.218; 0.9708, 1.517
		SES unknown	1.207; 0.9483, 1.521
4	loglinear	Local (SSN=ID, UT, MT, WY)	Baseline
		Migrant	1.034; 0.9276, 1.154
5	loglinear	Internal—not monitored	Baseline
		Internal—monitored, unexposed	1.129; 0.9791, 1.300
		Internal—likely exposed	0.8948; 0.7868, 1.017
6	loglinear	SES professional & skilled non-man	Baseline
		SES intermediate	1.103; 0.9315, 1.305
		SES skilled manual	1.466; 1.266, 1.700
		SES partly skilled	1.258; 1.017, 1.548
		SES unskilled	1.274; 1.009, 1.596
		SES unknown	1.197; 0.9405, 1.509
		Local (SSN=ID, UT, MT, WY)	Baseline
		Migrant	1.101; 0.9820, 1.236
7	loglinear	Migrant vs local, SES=prof & sknm	1.081
		Migrant vs local, SES=intermediate	1.658
		Migrant vs local, SES=sk manual	1.056
		Migrant vs local, SES=pt-skilled	0.6496
		Migrant vs local, SES=unskilled	0.8627
		Migrant vs local, SES=unknown	2.130 (p=0.002 for no interaction)

Table 6-3. Final baseline model: comparison of all-solid-cancer (excluding lymphomas, myeloma and leukemias, and with 10-year lag) maximum-likelihood risk estimates and likelihood-based CIs produced with (N=34,916) monitored employees with DOB, date first monitored and migrant status available.*

Model number	Term type	Factor	Risk estimate (RR or ERR[†]); 95% CI
7	loglinear	SES int vs prof/sknm, local	0.8073
		SES int vs prof/sknm, migrant	1.239 (p=0.029) for no interaction
		SES skman vs prof/sknm, local	1.481
		SES skman vs prof/sknm, migrant	1.447 (p>0.5) for no interaction
		SES ptskill vs prof/sknm, local	1.427
		SES ptskill vs prof/sknm, migrant	0.8576 (p=0.04) for no interaction
		SES unsk vs prof/sknm, local	1.335
		SES unsk vs prof/sknm, migrant	1.066 (p=0.395) for no interaction
		SES unkn vs prof/sknm, local	0.6601
		SES unkn vs prof/sknm, migrant	1.313 (p=0.081) for no interaction
8	loglinear	SES in 6 classes	--
	linear	Dose (ERR and SE per mSv[‡])	-0.001662; 0.0003965
9	loglinear	SES in 6 classes	--
		Migrant	--
	linear	Dose (ERR and SE per mSv[‡])	-0.001634; 0.0004024
10	loglinear	SES in 6 classes	--
		Migrant	--
		Migrant·SES	--
	linear	Dose (ERR and SE per mSv[‡])	-0.001588; 0.0004096
11	loglinear	SES in 6 classes	--
		Migrant	--
		Migrant·SES	--
		Internal	--
	linear	Dose (ERR and SE per mSv[‡])	-0.001498; 0.0004405

* All models stratified on age group (14 strata) calendar time (8 strata) duration of employment (<10 years and ≥10 years) and sex. Factors within each model number were adjusted for all other factors in that model. Migrant status refers to those whose SSN was issued outside of ID, MT, WY or UT.
† RR: Relative risk; ERR: Excess relative risk; CI: Confidence interval
‡ Standard errors (SE) are reported for the ERR when likelihood-based CIs could not be computed.

6.2 Smoking-related and Non-smoking-related Cancers

Solid cancers grouped into smoking-related and non-smoking related types showed important differences in risk coefficients. The analysis by dose category for smoking-related cancers (i.e., trachea, bronchus, lung, stomach, esophagus, bladder, oral and nasal cavities, pharynx, larynx, pancreas, kidney and ureter, liver and cervix) showed a very strong negative trend with increasing dose, and the categorical analysis suggested a monotonic decline in the RR with each increasing external radiation dose category (Figure 6-1).

By contrast, for non-smoking-related cancers (i.e., small intestine, colon, rectum, gall bladder, peritoneum, thymus, heart, mediastinum, bone, connective tissue, skin, breast, uterus, other female genital, male genital, eye, brain, thyroid and other endocrine) only the highest dose category showed a lower RR compared to the baseline group (Table 6-4) although CIs overlap unity. After adjusting for SES, migrant status, their interaction, and internal dose, the decreased RR with increasing dose was greatly attenuated for non-smoking-related, but not smoking-related, cancers (Figure 6-1).

The ERR/mSv estimate for non-smoking-related cancers (using the model that adjusted for SES, migrant status and their interaction, and internal dose), although negative, was about half the magnitude of the estimate for smoking-related cancers, and its CI included zero (Table 6-4).

Table 6-4. Results of risk estimation for smoking-related* (N=811) and non-smoking-related[†] (N=483) cancer mortality. Non-specific and ill-defined cancers (N=146), lymphomas, multiple myeloma and leukemias were not included in either definition[‡].

Model number	Term type	Factor	Risk estimate (RR or ERR[§]); 95% CI	
			Smoking-related cancers	Non-smoking-related cancers
1	loglinear	dose coefficient (at 1 mSv)	0.9971; 0.9952, 0.9989	0.9989; 0.9967, 1.001
2	loglinear	dose category 0-<1 mSv	Baseline	Baseline
		dose category 1-<10 mSv	0.9020; 0.7542, 1.075	1.008; 0.7965, 1.268
		dose category 10-<50 mSv	0.8418; 0.6792, 1.036	0.9771; 0.7338, 1.286
		dose category 50-<100 mSv	0.6842; 0.4625, 0.975	1.019; 0.6385, 1.549
		dose category ≥100 mSv	0.5090; 0.3272, 0.757	0.6805; 0.3927, 1.105
3	loglinear	SES professional & sk-nman	Baseline	Baseline
		SES intermediate	1.038; 0.8249, 1.301	1.175; 0.8900, 1.544
		SES skilled manual	1.439; 1.190, 1.744	1.158; 0.8979, 1.497
		SES partly skilled	1.335; 1.019, 1.732	1.065; 0.7354, 1.509
		SES unskilled	1.139; 0.8325, 1.534	1.303; 0.8982, 1.854
		SES unknown	1.200; 0.8702, 1.627	0.9948; 0.6357, 1.502
4	loglinear	Local (SSN=ID, UT, MT, WY)	Baseline	Baseline
		Migrant	1.111; 0.9602, 1.287	0.9818; 0.8143, 1.185
5	loglinear	Internal—not monitored	Baseline	Baseline
		Internal—monitored, unexp.	1.146; 0.9469, 1.380	1.072; 0.8335, 1.367
		Internal—likely exposed	0.8921; 0.7524, 1.057	0.9118; 0.7285, 1.139
6	loglinear	Race—White	Baseline	Baseline
		—Other	1.346; 0.6943, 2.326	0.8220; 0.2926, 1.790

Table 6-4. Results of risk estimation for smoking-related* (N=811) and non-smoking-related† (N=483) cancer mortality. Non-specific and ill-defined cancers (N=146), lymphomas, multiple myeloma and leukemias were not included in either definition‡.

Model number	Term type	Factor	Risk estimate (RR or ERR§); 95% CI	
			Smoking-related cancers	Non-smoking-related cancers
7	loglinear	SES professional & sk-nman	Baseline	Baseline
		SES intermediate	1.045; 0.8304, 1.310	1.175; 0.8903, 1.545
		SES skilled manual	1.501; 1.238, 1.826	1.162; 0.8977, 1.508
		SES partly skilled	1.433; 1.088, 1.872	1.071; 0.7350, 1.528
		SES unskilled	1.244; 0.9024, 1.690	1.312; 0.8956, 1.888
		SES unknown	1.184; 0.8588, 1.606	0.9930; 0.6343, 1.501
		Local (SSN=ID, UT, MT, WY)	Baseline	Baseline
		Migrant	1.206; 1.034, 1.408	1.016; 0.8345, 1.238
8	loglinear	Migrant vs local, SES=prof/sk-nman	1.143	1.074
			2.441	1.275
		Migrant vs local, SES=sk-nman	1.151	0.9478
		Migrant vs local, SES=sk-man	0.4613	1.051
		Migrant vs local, SES=ptsk	1.241	0.4374
		Migrant vs local, SES=unsk	3.233 (p=0.0001)	1.694 (p=0.3231)
		Migrant vs local, SES=unknown		
8	loglinear	SES int vs prof/sknm, local	0.5775	1.048
		SES int vs prof/sknm, migrant	1.234 (p=0.01)	1.243 (p>0.5)
		SES skman vs prof/sknm, local	1.481	1.255
		SES skman vs prof/sknm, migrnt	1.491 (p>0.5)	1.108 (p>0.5)
		SES ptsk vs prof/sknm, local	1.746	1.102
		SES ptsk vs prof/sknm, migrant	0.7048 (p=0.009)	1.078 (p>0.5)
		SES unsk vs prof/sknm, local	1.186	1.588
		SES unsk vs prof/sknm, migrant	1.288 (p>0.5)	0.6466 (p=0.081)
		SES unkn vs prof/int, local	0.4651	0.6653
		SES unkn vs prof/int, migrant	1.316 (p=0.092)	1.049 (p=0.477)
9	linear	Dose (ERR; SE or 95% CI per mSv)	-0.001884; SE=0.0005029	-0.0009445; 95% CI= NB**, 0.0009627
10	loglinear linear	SES in 6 classes, Migrant, Migrant·SES, Internal	--	--
			--	--
		Dose (ERR and SE or 95% CI per mSv)	-0.001825; SE=0.0005345	-0.0008553; 95% CI= NB**, 0.001268

* Cancers defined as smoking-related included trachea, bronchus, lung, stomach, esophagus, bladder, oral and nasal cavities, pharynx, larynx, pancreas, kidney and ureter, liver and cervix.

† Cancers defined as non-smoking-related included small intestine, colon, rectum, gall bladder, peritoneum, thymus, heart, mediastinum, bone, connective tissue, skin, breast, uterus, other female genital, male genital, eye, brain, thyroid and other endocrine.

‡ All models stratified on age group (14 strata) calendar time (8 strata) duration of employment (<10 years and ≥10 years) and sex. Factors within each model number were adjusted for all other factors in that model. Migrant status refers to those whose SSN was issued outside of ID, MT, WY or UT.

§ RR: Relative risk; ERR: Excess relative risk; CI: Confidence interval

** NB: lower confidence limit could not be computed, as it was below the boundary (Preston et al. 1993).

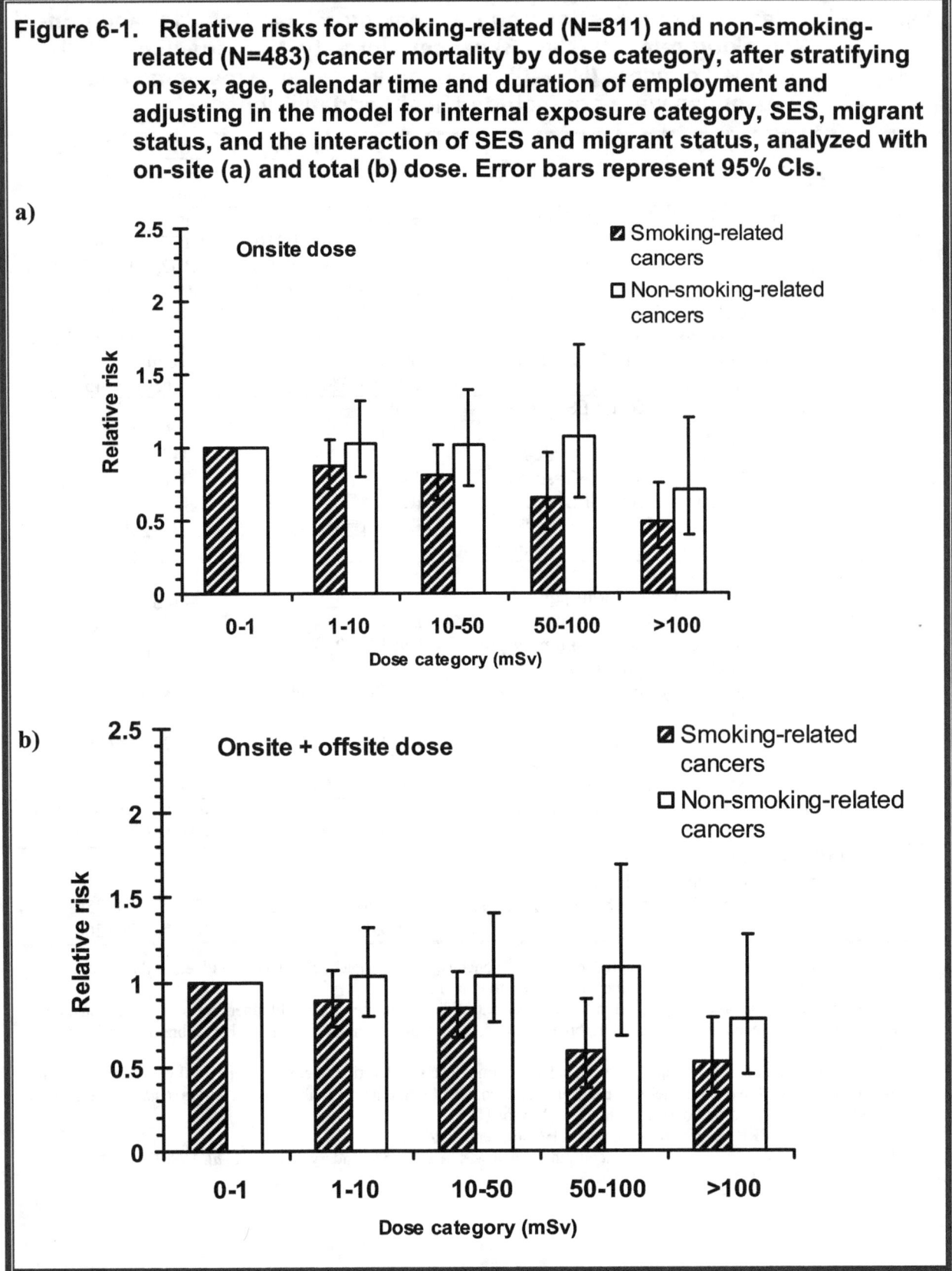

Figure 6-1. Relative risks for smoking-related (N=811) and non-smoking-related (N=483) cancer mortality by dose category, after stratifying on sex, age, calendar time and duration of employment and adjusting in the model for internal exposure category, SES, migrant status, and the interaction of SES and migrant status, analyzed with on-site (a) and total (b) dose. Error bars represent 95% CIs.

6.3 Inclusion of Off-site Dose: Smoking-related and Non-smoking-Related Cancers

Off-site dose (i.e., dose received before beginning work at the INEEL facility) was added separately to models for both smoking-related and non-smoking-related cancers. The effect of this addition was to redistribute cases and person-years from the unexposed and lower-dose into the higher-dose categories (Table 6-5 and 6-6). The non-smoking related cancers showed a three-fold increase in the ERR estimate (although the point estimate was still negative). Smoking-related cancers showed only a 25% increase in the ERR estimate after including off-site dose (and the estimated ERR per mSv was still negative) after adjusting for calendar time, age, sex, duration, SES, migrant status (and their interaction), and internal dose. The effect of adding off-site dose on the RR estimates by dose category, with adjustment for all factors above, was slightly greater for non-smoking-related than for smoking-related cancers (Figure 6-1b). All analyses below describe results including off-site dose.

6.4 "Radiogenic" Non-smoking-related Cancers

Solid cancers defined as "radiogenic" (adapted from Boice *et al.* 1996) were also analyzed separately. Because the radiation-related risk appeared different for cancers related to smoking (i.e., there was apparent negative confounding by smoking), this analysis was restricted to radiogenic non-smoking-related cancers and included cancers of thyroid, breast, colon, ovary, skin, bone, and connective tissue. Several of these cancers had low case-fatality rates; therefore, analyses were conducted both excluding and including non-underlying cancers. Analyses included off-site dose.

Risk estimates per unit of dose were very similar to the complete group of non-smoking-related cancers (Table 6-5). However, CIs were much wider because of the reduced number of cases. The addition of non-underlying radiogenic cancers increased the total number by 20 cases. The categorical RRs in most categories increased compared to the baseline; however, ERR per unit dose changed little (Table 6-5) and was still negative with a CI that included zero.

After adjustment for SES, migrant status, their interaction and internal dose, the point estimates for most dose categories were higher than for the total group of non-smoking cancers, as observed by a comparison of Figure 6-2 to Figure 6-1b.

Table 6-5. Non-smoking-related cancer risk coefficients (N=34,916 monitored employees with DOB, date of first monitoring and migrant status available) with inclusion of off-site dose.*

Group	Term type	Factor	Number of cases	Number of person-years	Risk estimate (RR or ERR[†]); 95% CI
All non-smoking-related underlying cancer (on-site dose only)	loglinear	dose coefficient (at 1 mSv)	--	--	0.9989; 0.9967, 1.001
	loglinear	dose category 0-<1 mSv	260	589,930	Baseline
		dose category 1-<10 mSv	112	124,825	1.008; 0.7965, 1.268
		dose category 10-<50 mSv	71	77,430	0.9771; 0.7338, 1.286
		dose category 50-<100 mSv	23	24,641	1.019; 0.6385, 1.549
		dose category ≥100 mSv	17	21,860	0.6805; 0.3927, 1.105
	linear	dose (ERR and 95% CI per mSv)	--	--	-0.0009445; NB[‡], 0.0009627
	loglinear	SES in 6 classes, migrant, SES·mig Internal	--	--	--
	linear	**dose (ERR and 95% CI per mSv)**	--	--	**-0.0008553; NB[‡], 0.001272**
All non-smoking related underlying cancer (on-site + off-site dose)	loglinear	dose coefficient (at 1 mSv)	--	--	0.9992; 0.9959, 1.002
	loglinear	dose category 0-<1 mSv	250	569,953	Baseline
		dose category 1-<10 mSv	113	131,579	1.004; 0.7930, 1.264
		dose category 10-<50 mSv	75	85,299	0.9909; 0.7482, 1.298
		dose category 50-<100 mSv	25	27,497	1.030; 0.6569, 1.547
		dose category ≥100 mSv	20	24,345	0.7384; 0.4438, 1.165
	linear	dose (ERR and 95% CI per mSv)	--	--	-0.0004111; NB[‡], 0.001455
	loglinear	SES in 6 classes, migrant, SES·mig Internal	--	--	--
	linear	**dose (ERR and 95% CI per mSv)**	--	--	**-0.0002877; NB[‡], 0.001806**
Radiogenic underlying cancer (on-site +off-site dose)	loglinear	dose coefficient (at 1 mSv)	--	--	0.9986; 0.9949, 1.002
	loglinear	dose category 0-<1 mSv	119	569,953	Baseline
		dose category 1-<10 mSv	49	131,579	0.9956; 0.6972, 1.402
		dose category 10-<50 mSv	35	85,299	1.134; 0.7470, 1.683
		dose category 50-<100 mSv	10	27,497	1.018; 0.4900, 1.895
		dose category ≥100 mSv	7	24,345	0.6760; 0.2775, 1.405
	linear	dose (ERR and SE per mSv[§])	--	--	-0.001201; 0.001215

Table 6-5. Non-smoking-related cancer risk coefficients (N=34,916 monitored employees with DOB, date of first monitoring and migrant status available) with inclusion of off-site dose.*

Group	Term type	Factor	Number of cases	Number of person-years	Risk estimate (RR or ERR†); 95% CI
Radiogenic cancer, incl. non-underlying, (on-site + off-site dose)	loglinear	SES in 6 classes, migrant, SES·mig Internal	--	--	--
	linear	dose (ERR and 95% CI per mSv)	--	--	**-0.0002299; NB‡, 0.004591**
	loglinear	dose coefficient (at 1 mSv)	--	--	0.9992; 0.9959, 1.002
	loglinear	dose category 0-<1 mSv	127	569,953	Baseline
		dose category 1-<10 mSv	54	131,579	1.019; 0.7241, 1.415
		dose category 10-<50 mSv	40	85,299	1.204; 0.8119, 1.751
		dose category 50-<100 mSv	10	27,497	0.9585; 0.4625, 1.776
		dose category ≥100 mSv	9	24,345	0.8023; 0.3671, 1.555
	linear	dose (ERR and 95% CI per mSv)	--	--	-0.0007546; NB‡, 0.002510
	loglinear	SES in 6 classes, migrant, SES·mig Internal	--	--	--
	linear	dose (ERR and 95% CI per mSv)			**6.754E-06; NB‡, 0.004453**

* All models stratified on age group (14 strata) calendar time (8 strata) duration of employment (<10 years and ≥10 years) and sex. All models included a 10-year lag for internal and external dose. Factors within each model number were adjusted for all other factors in that model. Migrant status refers to those whose SSN was issued outside of ID, MT, WY or UT.
† RR: Relative risk; ERR: Excess relative risk; CI: Confidence interval
‡ NB: lower confidence limit could not be computed, as it was below the boundary (Preston et al. 1993)
§ Standard errors (SE) are reported for the ERR when likelihood-based CIs could not be computed.

Table 6-6. Smoking-related cancer risk coefficients by dose type (N=34,916 monitored employees with DOB, date of first monitoring and migrant status available).*

Dose	Term type	Factor	Number of cases	Number of person-years	Risk estimate (RR or ERR†); 95% CI
On-site with 10-year lag	loglinear	dose coefficient (at 1 mSv)	--	--	0.9971; 0.9952, 0.9989
	loglinear	dose category 0-<1 mSv	448	589,930	Baseline
		dose category 1-<10 mSv	186	12,825	0.9020; 0.7542, 1.075
		dose category 10-<50 mSv	121	77,430	0.8418; 0.6792, 1.036
		dose category 50-<100 mSv	31	24641	0.6842; 0.4625, 0.9754
		dose category ≥100 mSv	25	21,860	0.5090; 0.3272, 0.7569
	linear	dose (ERR and SE per mSv‡)	--	--	-0.001884; 0.0005029
	loglinear	SES in 6 classes, migrant, SES·mig Internal	--	--	--
	linear	**dose (ERR and SE per mSv‡)**			**-0.001825; 0.0005345**
On-site + off-site with 10-year lag	loglinear	dose coefficient (at 1 mSv)	--	--	0.9972; 0.9954, 0.9989
	loglinear	dose category 0-<1 mSv	430	569,953	Baseline
		dose category 1-<10 mSv	191	131,579	0.9107; 0.7620, 1.085
		dose category 10-<50 mSv	131	85,299	0.8757; 0.7108, 1.072
		dose category 50-<100 mSv	30	27,497	0.6141; 0.4123, 0.8804
		dose category ≥100 mSv	29	24,345	0.5373; 0.3561, 0.7805
	linear	dose (ERR and SE per mSv)	--	--	-0.001453; 0.0001201
	loglinear	SES in 6 classes, migrant, SES·mig Internal	--	--	--
	linear	**dose (ERR and SE per mSv‡)**			**-0.001453; 0.1752E-9**

* All models stratified on age group (14 strata) calendar time (8 strata) duration of employment (<10 years and ≥10 years) and sex. Factors within each model number were adjusted for all other factors in that model. Migrant status refers to those whose SSN was issued outside of ID, MT, WY or UT.
† RR: Relative risk; ERR: Excess relative risk; CI: Confidence interval.
‡ Standard errors (SE) are reported for the ERR when likelihood-based CIs could not be computed.

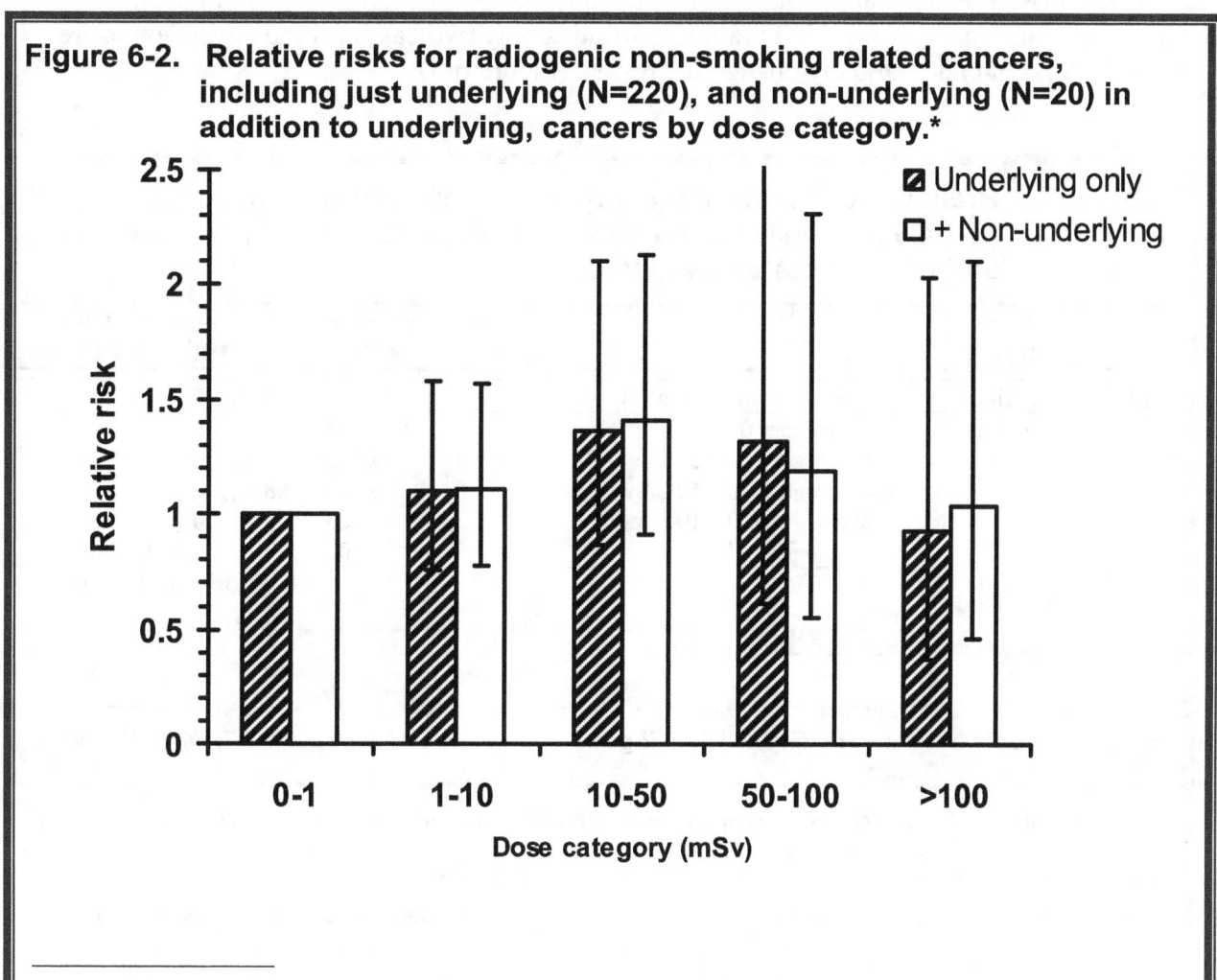

Figure 6-2. Relative risks for radiogenic non-smoking related cancers, including just underlying (N=220), and non-underlying (N=20) in addition to underlying, cancers by dose category.*

* Risk estimates are stratified on sex, age, calendar time and duration of employment (<10 years and ≥10 years) and adjusted in the model for internal exposure category, SES, migrant status, and the interaction of SES and migrant status. Dose is lagged ten years. Error bars represent 95% CIs.

6.5 Brain Tumor Dose-response

A significant positive trend with increasing external dose was observed in the life table analysis for brain cancer with a 20-year lag. This relation was evaluated for all underlying brain tumors combined, while considering other confounders, in Poisson regression analysis. Brain tumor risk was positively associated with dose in a loglinear model, although the lower bound of the CI of the rate ratio was slightly lower than one (Table 6-7). Examination of categorical risk estimates reveals that elevation in risk was observed only in the highest exposure category (≥100 mSv) although CIs were quite wide.

The ERR per mSv was estimated to be positive but with 95% CIs that included zero. After adjustment for confounding by SES, migrant status, their interaction, and internal dose category, the estimated increase per 10 mSv (1 rem) was 8.7% (CI: -0.3%, 33.8%) and the increase in risk remained apparent only in the highest exposure category (Figure 6-3). Inclusion of non-underlying COD brain tumors adds only two cases (to the lowest exposure category) and did not markedly change the results (Figure 6-3).

Table 6-7. Brain tumor (underlying only) risk coefficients by dose type (N=34,916 monitored employees with DOB, date of first monitoring and migrant status available).*

Dose	Term type	Factor	Number of cases	Number of PY	Risk estimate (RR or ERR†); 95% CI
On-site + off-site with 20-year lag	loglinear	dose coefficient (at 1 mSv)	--	--	1.004; 0.999, 1.008
	loglinear	dose category 0-<1 mSv	44	695,104	Baseline
		dose category 1-<10 mSv	9	69,405	0.848; 0.375, 1.731
		dose category 10-<50 mSv	6	46,010	0.852; 0.311, 1.984
		dose category 50-<100 mSv	2	15,063	0.831; 0.132, 2.867
		dose category ≥100 mSv	5	13,102	2.009; 0.639, 5.267
	linear	dose (ERR and 95% CI per mSv)	--	--	0.0055; -0.00101, 0.0204
	loglinear	SES in 6 classes, migrant, SES·mig	--	--	--
Internal					--
	linear	**dose (ERR and 95% CI per mSv)**			**0.0087; -0.00037, 0.0338**

*All models stratified on age group (14 strata) calendar time (8 strata) duration of employment (<10 years and ≥10 years) and sex. Factors within each model number were adjusted for all other factors in that model.
† RR: Relative risk; ERR: Excess relative risk; CI: Confidence interval.

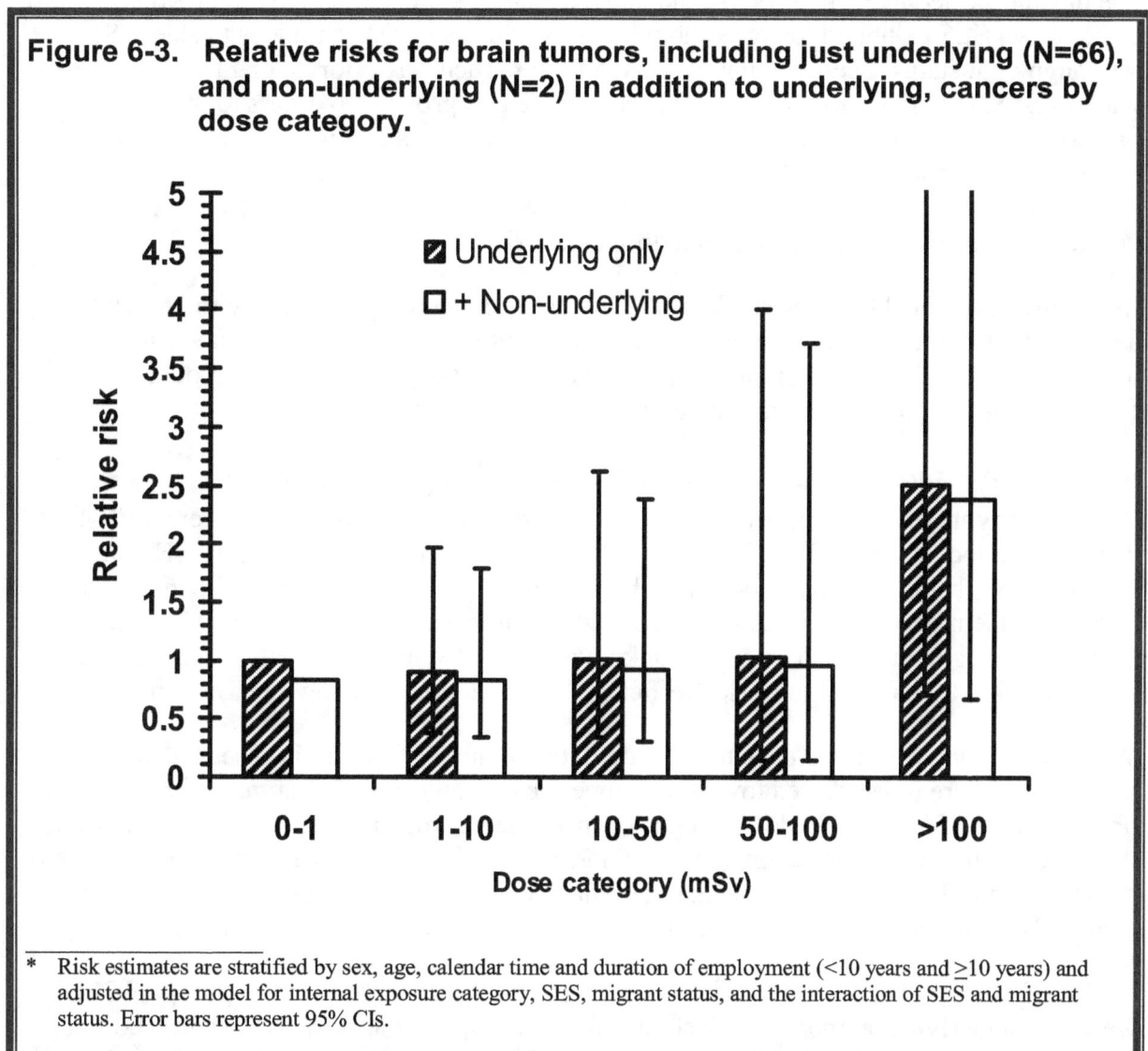

Figure 6-3. Relative risks for brain tumors, including just underlying (N=66), and non-underlying (N=2) in addition to underlying, cancers by dose category.

* Risk estimates are stratified by sex, age, calendar time and duration of employment (<10 years and ≥10 years) and adjusted in the model for internal exposure category, SES, migrant status, and the interaction of SES and migrant status. Error bars represent 95% CIs.

6.6 Leukemias

6.6.1 Baseline Model

A similar evaluation approach was taken for underlying leukemias as for all solid cancers, to develop a baseline model upon which dose-response analyses would be based. SES, migrant status, race and internal dose were not strongly related to leukemia risk. However, it was thought that the SES groupings may have been over-stratified, as the unknown SES group (N=4829) showed extremely imprecise risk estimates (Table 6-8). The interaction between SES and migrant status was also not significant.

Although overstratification of SES may have occurred, there were no clear categories that should have been combined. The most discernible pattern with respect to leukemia mortality risk was the similarity between professional and intermediate workers, between skilled non-manual and skilled manual workers, and between partly skilled and unskilled workers. Those of unknown SES exhibited lower, although highly uncertain, risk estimates (Table 6-8). Although vital status ascertainment may have been particularly poor among those workers without SES, the similarity of all-cancer risk among this group of workers to the professional/skilled non-manual workers (Table 6-3) suggests that underascertainment did not occur. The baseline model, therefore, consisted of SES in 4 categories (Professional/intermediate workers, Skilled workers, Partly skilled/unskilled workers, and workers of unknown SES) and internal dose categories.

In summary, the final baseline model for leukemia included only workers with known DOB and date first monitored. The model stratified on attained age and calendar time (in 5-year intervals), sex, and duration of employment (<10 years and ≥10 years). The model adjusted for SES (in 4 categories) and internal monitoring status (Table 6-9).

6.6.2 Dose-response Analysis with External Ionizing Radiation

When stratifying on sex, age, calendar time and duration of employment, leukemia ERR was positively associated with dose, but the 95% CI included zero. The addition of SES and internal dose changed the point estimates by 11% and 12%, respectively (Table 6-8), with SES adjustment decreasing risk and internal dose adjustment increasing risk. The final baseline model included individuals of unknown migrant status and recategorized SES as described above. The resulting risk coefficients for leukemia are shown in Table 6-9.

With the addition of off-site dose to the model, the number of leukemia cases and amount of person-time were redistributed toward the upper dose categories. The largest effect was observed in the 50-100 mSv dose category, where the RR rose from 1.20 to 1.51; however, the effect on the ERR/mSv estimate was minor, increasing by less than one percent (Table 6-10). Inclusion of non-underlying leukemias slightly reduced the risk estimates, especially among the group most highly exposed to external ionizing radiation (Table 6-10).

Of the fifteen non-underlying leukemia cases, eight were CLL (38% of the group of CLLs were non-underlying, compared to just 12% for other types of leukemia). Examination of radiation-related risk of underlying and non-underlying leukemia deaths by subtype showed that, although based on small numbers of cases, the association of leukemia with external radiation was generally stronger for leukemias that were definitively not chronic lymphocytic (N=52) than for definitive CLL cases (N=21). The risk per unit of dose (on-site and off-site combined) was higher for non-CLLs, and the risks of non-CLLs were particularly elevated in the highest dose categories (Table 6-10). For CLLs, the risks were elevated in two intermediate dose categories, but no cases were observed among the highest exposure category, and the ERR/mSv point estimate was negative. CIs were wide for all analyses by subtype; however, compared to the analysis of underlying and non-underlying leukemias described in the previous paragraph, the ERR/mSv point estimate was higher after removing definitive CLLs and leukemias of ambiguous subtype. For non-CLL leukemias, the ERR/mSv estimate was 0.0054 (or 5.4% per 10 mSv=1 rem) with a likelihood-based 95% CI that included zero (Table 6-11).

6.6.3 Sensitivity Analyses

Alternative dose lags were evaluated, compared to the 5-year lag (Table 6-12) as suggested by the life table trend tests (see §5.4). Increasing the lag to 7 years increased the ERR/mSv estimate for all leukemias, although reducing the lag had the opposite effect. At a lag of 10 years, the ERR/mSv estimate was much reduced, suggesting that a lag of 7 years may be most appropriate.

The assumption of appropriate lag was also evaluated separately for CLL and non-CLL leukemias. For CLL there was very little change in risk coefficients with changing dose lags (either in ERR/mSv or in the RRs per mSv or by category; Table 6-13). The exposed categories exhibited the highest RRs using a 7-year lag. For definitively non-CLL leukemia, the tendencies with lagged dose reflected the patterns observed for all leukemias combined (Table 6-12 and Table 6-13). Maximum risk (both RR at 1 mSv and ERR/mSv) was observed using a 7-year lag period, with a substantial decrease at 10 years and an increase at 20 years (although the estimated ERR/mSv was still below that observed for a 7-year lag). At a 7-year lag, the ERR/mSv estimate for non-CLL leukemia was 0.0054 (CI; -0.0011, 0.024; Table 6-13).

Three additional dose categorization schemes were employed to evaluate the sensitivity of non-CLL leukemia risk estimates to choice of category cutpoints (all employed a 7-year lag and included both on-site and off-site doses):

1. Increasing the number of dose categories to 11 (similar to the categories used in the IARC analyses of Cardis *et al.* 1995) somewhat reduced the ERR estimate for non-CLL leukemias, to 0.0034 (CI: <-0.0010, 0.018) per mSv. The risks generally increased by dose category (diamonds in Figure 6-4); however, CIs at each dose category were quite wide, and there were several dose categories with no cases.

2. Increasing the number of dose categories to 6, with a slight reconfiguration (0-5, 5-20, 20-50, 50-100, 100-200, >200 mSv) also reduced the ERR estimate for non-CLL leukemias, to 0.0031 (CI; <-0.0010, 0.017) per mSv. The risks also tended to increase with dose category, with attenuation in the highest dose category, compared to the 11-category grouping (open squares in Figure 6-4).

3. Expanding the baseline (i.e., lowest-dose) group to include any monitored worker receiving less than 10 mSv (dose categories of 0-10, 10-20, 20-50, 10-100 and ≥100 mSv) decreased the ERR only slightly, to 0.0045 (CI; -0.0014, 0.020). As in the other dose recategorizations, RRs tended to increase with increasing dose category, except at 20-50 mSv (open triangles in Figure 6-4).

Adding a quadratic term for dose to the model using the original dose classification did not improve fit (likelihood ratio test $\chi^2=0.056$, p=0.81). Adding a quadratic term for dose also did not improve model fit under either of the dose recategorization schemes (p=0.85, 0.91 and 0.59, respectively, for the likelihood ratio test of quadratic term in the 11-, 6- and 5-category regroupings described above).

Table 6-8. Comparison of all leukemia (N=69 with 5-year lag) maximum-likelihood risk estimates and likelihood-based CIs produced with (N=34,916) monitored employees with DOB, date first monitored and migrant status available.*

Model number	Term type	Factor	Risk estimate (RR or ERR[†]); 95% CI
1	loglinear	dose coefficient (at 1 mSv)	1.003; 0.9994, 1.0068
2	loglinear	dose category 0-<1 mSv	Baseline
		dose category 1-<10 mSv	1.278; 0.6772, 2.338
		dose category 10-<50 mSv	1.479; 0.6996, 2.955
		dose category 50-<100 mSv	1.192; 0.2789, 3.510
		dose category ≥100 mSv	2.357; 0.8184, 5.930
3	loglinear	SES professional	Baseline
		SES intermediate	0.9682; 0.4123, 2.207
		SES skilled non-manual	1.370; 0.4197, 4.027
		SES skilled manual	1.460; 0.7409, 3.006
		SES partly skilled	1.147; 0.3996, 2.933
		SES unskilled	1.121; 0.3514, 3.048
		SES unknown	0.2112; 0.0115, 1.085
4	loglinear	Local (SSN=ID, UT, MT, WY)	Baseline
		Migrant	0.8762; 0.5342, 1.442
5	loglinear	Internal—not monitored	Baseline
		Internal—monitored, unexposed	1.014; 0.5058, 1.931
		Internal—likely exposed	1.087; 0.6061, 1.953
6	loglinear	Race White	Baseline
		Race Other	1.088; 0.0612, 5.037
7	loglinear	SES professional	Baseline
		SES intermediate	0.9656; 0.4106, 2.204
		SES skilled non-manual	1.357; 0.4102, 4.045
		SES skilled manual	1.449; 0.7233, 3.024
		SES partly skilled	1.133; 0.3857, 2.981
		SES unskilled	1.104; 0.3355, 3.123
		SES unknown	0.2113; 0.0115, 1.086
		Local (SSN=ID, UT, MT, WY)	Baseline
		Migrant	0.9726; 0.5770, 1.642
8	loglinear	Migrant vs local, SES=professional	0.3134
		Migrant vs local, SES=intermediate	0.7187
		Migrant vs local, SES=sk manual	2.222
		Migrant vs local, SES=pt-skilled	0.4770
		Migrant vs local, SES=unskilled	0.7698
		Migrant vs local, SES=unknown	1.0430 (p=0.2943)

Table 6-8. Comparison of all leukemia (N=69 with 5-year lag) maximum-likelihood risk estimates and likelihood-based CIs produced with (N=34,916) monitored employees with DOB, date first monitored and migrant status available.*

Model number	Term type	Factor	Risk estimate (RR or ERR[†]); 95% CI
8	loglinear	SES int vs professional, local	1.727
		SES int vs professional, migrant	0.5789 (p=0.237)
		SES sknm vs professional, local	1.586
		SES sknm vs professional, migrant	1.066 (p>0.5)
		SES skman vs professional, local	1.067
		SES skman vs professional, migrant	1.947 (p>0.5)
		SES ptskill vs professional, local	1.289
		SES ptskill vs professional, migrant	0.6579 (p>0.5)
		SES unsk vs professional, local	1.133
		SES unsk vs professional, migrant	0.9328 (p>0.5)
		SES unkn vs professional, local	0.00002
		SES unkn vs professional, migrant	0.2671 (p>0.5)
8	linear	Dose (ERR and 95% CI per mSv)	0.005349; -0.0006131, 0.01868
9	loglinear	SES in 7 classes	--
	linear	Dose (ERR and 95% CI per mSv)	0.004711; -0.0008474, 0.01717
10	loglinear	SES in 7 classes, Migrant	--
	linear	Dose (ERR and 95% CI per mSv)	0.004717; -0.0008551, 0.01727
11	loglinear	SES in 7 classes, Migrant, Migrant·SES	--
		Dose (ERR and 95% CI per mSv)	--
	linear		0.004144; -0.001053, 0.01585
12	loglinear	SES in 7 classes, Migrant, Migrant·SES, Internal	--
			--
	linear	Dose (ERR and 95% CI per mSv)	0.005134; -0.0009294, 0.02141
13	loglinear	SES in 7 classes, Migrant, Migrant·SES, Internal, Race	--
			--
	linear	Dose (ERR and 95% CI per mSv)	0.005140; -0.0009271, 0.02144
14	loglinear	Internal	--
	linear	Dose (ERR and 95% CI per mSv)	0.006000; -0.0006347, 0.02392
15	loglinear	SES in 7 classes, Internal	--
	linear	Dose (ERR and 95% CI per mSv)	0.005515; -0.0008116, 0.02270

* All models stratified on age group (14 strata) calendar time (8 strata) duration of employment (<10 years and ≥10 years) and sex. Factors within each model number were adjusted for all other factors in that model. Migrant status refers to those whose SSN was issued outside of ID, MT, WY or UT.
† RR: Relative risk; ERR: Excess relative risk; CI: Confidence interval.

Table 6-9. Final baseline model for all leukemia (N=70 with 5-year lag) maximum-likelihood risk estimates and likelihood-based CIs produced with (N=36,169) monitored employees with DOB and date first monitored available.*

Model number	Term type	Factor	Risk estimate (RR or ERR†); 95% CI
1	loglinear	dose coefficient (at 1 mSv)	1.003; 0.9992, 1.007
2	loglinear	dose category 0-<1 mSv dose category 1-<10 mSv dose category 10-<50 mSv dose category 50-<100 mSv dose category ≥100 mSv	Baseline 1.245; 0.6616, 2.266 1.446; 0.6865, 2.874 1.157; 0.2713, 3.389 2.320; 0.8071, 5.814
3	loglinear	SES professional/intermediate SES skilled workers SES partly skilled/unskilled SES unknown	Baseline 1.510; 0.8685, 2.663 1.150; 0.5330, 2.335 0.2083; 0.01155, 1.014 (overall SES p=0.0609)
4	loglinear	Internal—not monitored Internal—monitored, unexposed Internal—likely exposed	Baseline 0.9932; 0.4967, 1.882 1.067; 0.5984, 1.904
5	linear	Dose (ERR and 95% CI per mSv)	0.004971; -0.0007686, 0.01776
6	loglinear linear	SES in 4 classes Dose (ERR and 95% CI per mSv)	-- 0.004310; -0.0009993, 0.01610
7	loglinear linear	Internal Dose (ERR and 95% CI per mSv)	-- 0.005625; -0.0007777, 0.02278
8	loglinear linear	SES in 4 classes, Internal Dose (ERR and 95% CI per mSv)	-- 0.005071; -0.0009588, 0.02114

* All models stratified on age group (14 strata) calendar time (8 strata) duration of employment (<10 years and ≥10 years) and sex. Factors within each model number were adjusted for all other factors in that model. Migrant status refers to those whose SSN was issued outside of ID, MT, WY or UT.
† RR: Relative risk; ERR: Excess relative risk; CI: Confidence interval.

Table 6-10. Leukemia (N=70) risk coefficients (ERR per mSv) by dose type (N=36,169 monitored employees with DOB and date of first monitoring available).*

Dose	Term type	Factor	Leukemia N	Person-Year N	Risk estimate (RR or ERR[†]); 95% CI
On-site with 7-year lag (underlying only)	loglinear	dose coefficient (at 1 mSv)	--	--	1.003; 0.9993, 1.007
	loglinear	dose category 0-<1 mSv	32	558,599	Baseline
		dose category 1-<10 mSv	17	147,469	1.283; 0.6808, 2.338
		dose category 10-<50 mSv	12	90,339	1.491; 0.7079, 2.966
		dose category 50-<100 mSv	3	28,347	1.195; 0.2800, 3.502
		dose category ≥100 mSv	6	25,146	2.400; 0.8347, 6.022
	linear	dose (ERR and 95% CI per mSv)	--	--	0.005256; -0.0006848, 0.01854
	loglinear **linear**	SES in 4 classes, Internal **dose (ERR and 95% CI per mSv)**	--	--	-- **0.005448; -0.0008714, 0.02261**
On-site + off-site with 7-year lag (underlying only)	loglinear	dose coefficient (at 1 mSv)	--	--	1.003; 0.9991, 1.007
	loglinear	dose category 0-<1 mSv	30	536,053	Baseline
		dose category 1-<10 mSv	17	154,538	1.295; 0.6837, 2.378
		dose category 10-<50 mSv	13	99,519	1.595; 0.7727, 3.138
		dose category 50-<100 mSv	4	31,729	1.510; 0.4352, 4.027
		dose category ≥100 mSv	6	28,046	2.236; 0.7794, 5.579
	linear	dose (ERR and 95% CI per mSv)	--	--	0.005050; -0.0007573, 0.01798
	loglinear **linear**	SES in 4 classes, Internal **dose (ERR and 95% CI per mSv)**	--	--	-- **0.005502; -0.0008587, 0.02258**
On-site + off-site with 7-year lag (underlying and non-underlying)	loglinear	dose coefficient (at 1 mSv)	--	--	1.002; 0.9986, 1.006
	loglinear	dose category 0-<1 mSv	37	536,053	Baseline
		dose category 1-<10 mSv	19	154,538	1.101; 0.6076, 1.935
		dose category 10-<50 mSv	17	99,519	1.556; 0.8235, 2.844
		dose category 50-<100 mSv	6	31,729	1.730; 0.6360, 3.994
		dose category ≥100 mSv	6	28,046	1.644; 0.5885, 3.951
	linear	dose (ERR and 95% CI per mSv)	--	--	0.003648; -0.001178, 0.01390
	loglinear **linear**	SES in 4 classes, Internal **dose (ERR and 95% CI per mSv)**	--	--	-- **0.003923; -0.001278, 0.01697**

* All models stratified on age group (14 strata) calendar time (8 strata) duration of employment (<10 years and ≥10 years) and sex. Factors within each model number were adjusted for all other factors in that model.
† RR: Relative risk; ERR: Excess relative risk; CI: Confidence interval.

Table 6-11. Leukemia risk coefficients (ERR per mSv) by dose type (N=36,169 monitored employees with DOB and date of first monitoring available) separated by chronic lymphocytic and non-CLL (both underlying and non-underlying leukemias were included).*

Cause of death	Term type	Factor	Number of cases	Number of person-years	Risk estimate (RR or ERR†); 95% CI
Definitively CLL (ICD-9=204.1)	loglinear	dose coefficient (at 1 mSv)	--	--	0.9977; 0.9834, 1.006
	loglinear	dose category 0-<1 mSv	9	536,053	Baseline
		dose category 1-<10 mSv	2	154,538	0.4464; 0.0670, 1.779
		dose category 10-<50 mSv	8	99,519	2.524; 0.8907, 6.983
		dose category 50-<100 mSv	2	31,729	2.058; 0.3024, 8.507
		dose category ≥100 mSv	0	28,046	No cases
	linear	dose (ERR and SE per mSv)‡	--	--	-0.00145; 0.00201
	loglinear	SES in 4 classes, Internal	--	--	--
	linear	**dose (ERR and SE per mSv)‡**			**-0.00145; 0.280E-7**
Definitively non-CLL (ICD-9= 204-208, excluding 204.1, 204.9, 208.1 and 208.9)	loglinear	dose coefficient (at 1 mSv)	--	--	1.004; 0.9990, 1.007
	loglinear	dose category 0-<1 mSv	22	536,053	Baseline
		dose category 1-<10 mSv	15	154,538	1.538; 0.7619, 3.027
		dose category 10-<50 mSv	6	99,519	1.012; 0.3592, 2.474
		dose category 50-<100 mSv	4	31,729	2.029; 0.5677, 5.697
		dose category ≥100 mSv	5	28,046	2.414; 0.7385, 6.747
	linear	dose (ERR and 95% CI per mSv)	--	--	0.005583; -0.0009145, 0.02127
	loglinear	SES in 4 classes, Internal	--	--	--
	linear	**dose (ERR and 95% CI per mSv)**			**0.005430; -0.001141, 0.02376**

*All models stratified on age group (14 strata) calendar time (8 strata) duration of employment (<10 years and ≥10 years) and sex, included both on-site and off-site doses, and employ a 7-year lag for external dose. Factors within each model number were adjusted for all other factors in that model.
†RR: Relative risk; ERR: Excess relative risk; CI: Confidence interval
‡ Standard errors (SE) are reported for the ERR when likelihood-based CIs could not be computed.

Table 6-12. Effect of varying lags on radiation related leukemia (N=70) risk, for 36,169 monitored employees with DOB and date of first monitoring available.*

Lag	Term type	Factor	Risk estimate (RR or ERR†); 95% CI	
2-year	loglinear	dose coefficient (at 1 mSv)	1.003;	0.9992, 1.007
	loglinear	dose category 0-<1 mSv	Baseline	
		dose category 1-<10 mSv	1.322;	0.7105, 2.393
		dose category 10-<50 mSv	1.466;	0.6946, 2.922
		dose category 50-<100 mSv	1.175;	0.2752, 3.448
		dose category ≥100 mSv	2.357;	0.8191, 5.920
	linear	dose (ERR and 95% CI per mSv)	0.004838; -0.0008165, 0.01746	
	loglinear	SES in 4 classes, Internal	--	
	linear	**dose (ERR and 95% CI per mSv)**	**0.004968; -0.009915, 0.02086**	
5-year	loglinear	dose coefficient (at 1 mSv)	1.003;	0.9992, 1.007
	loglinear	dose category 0-<1 mSv	Baseline	
		dose category 1-<10 mSv	1.245;	0.6616, 2.266
		dose category 10-<50 mSv	1.446;	0.6865, 2.874
		dose category 50-<100 mSv	1.157;	0.2713, 3.389
		dose category ≥100 mSv	2.320;	0.8071, 5.814
	linear	Dose (ERR and 95% CI per mSv)	0.004971; -0.0007686, 0.01776	
	loglinear	SES in 4 classes, Internal	--	
	linear	**Dose (ERR and 95% CI per mSv)**	**0.005071; -0.0009588, 0.02114**	
7-year	loglinear	dose coefficient (at 1 mSv)	1.003;	0.9993, 1.007
	loglinear	dose category 0-<1 mSv	Baseline	
		dose category 1-<10 mSv	1.283;	0.6808, 2.338
		dose category 10-<50 mSv	1.491;	0.7079, 2.966
		dose category 50-<100 mSv	1.195;	0.2800, 3.502
		dose category ≥100 mSv	2.400;	0.8347, 6.022
	linear	dose (ERR and 95% CI per mSv)	0.005256; -0.0006848, 0.01854	
	loglinear	SES in 4 classes, Internal	--	
	linear	**dose (ERR and 95% CI per mSv)**	**0.005448; -0.0008714, 0.02261**	
10-year	loglinear	dose coefficient (at 1 mSv)	1.002;	0.9974, 1.006
	loglinear	dose category 0-<1 mSv	Baseline	
		dose category 1-<10 mSv	0.9940;	0.5156, 1.829
		dose category 10-<50 mSv	1.060;	0.4794, 2.160
		dose category 50-<100 mSv	1.013;	0.2382, 2.946
		dose category ≥100 mSv	1.628;	0.5214, 4.236
	linear	dose (ERR and 95% CI per mSv)	0.002823; -0.001984, 0.01363	
	loglinear	SES in 4 classes, Internal	--	
	linear	**dose (ERR and 95% CI per mSv)**	**0.002864; -0.002121, 0.01604**	

* All models stratified on age group (14 strata) calendar time (8 strata) duration of employment (<10 years and ≥10 years) and sex. Factors within each model number were adjusted for all other factors in that model.
† RR: Relative risk; ERR: Excess relative risk; CI: Confidence interval.

Table 6-13. Effect of varying lags on CLL and non-CLL leukemia risk, for 36,169 monitored employees with DOB and date of first monitoring available.*

Lag	Term type	Factor	Risk estimate (RR or ERR†); 95% CI CLL (N=21)	non-CLL (N=52)
5-year	Loglinear	dose coefficient (at 1 mSv)	0.998; 0.983, 1.006	1.003; 0.999, 1.007
	Loglinear	dose category 0-<1 mSv	Baseline	Baseline
		dose category 1-<10 mSv	0.443; 0.067, 1.765	1.481; 0.735, 2.910
		dose category 10-<50 mSv	2.497; 0.880, 6.916	0.973; 0.345, 2.375
		dose category 50-<100 mSv	2.039; 0.299, 8.435	1.945; 0.545, 5.450
		dose category ≥100 mSv	0; No bounds	2.307; 0.707, 6.433
	Linear	dose (ERR and 95% CI per mSv)	-0.00145; 0.00341	0.00525; -0.0100, 0.0202
	loglinear linear	SES in 4 classes, Internal **dose (ERR and 95% CI per mSv)**	-- **-0.00145; SE=0.00355**	-- **0.0050; -0.00123, 0.0220**
7-year	Loglinear	dose coefficient (at 1 mSv)	0.998; 0.983, 1.006	1.004; 0.999, 1.007
	Loglinear	dose category 0-<1 mSv	Baseline	Baseline
		dose category 1-<10 mSv	0.446; 0.067, 1.779	1.538; 0.762, 3.027
		dose category 10-<50 mSv	2.524; 0.891, 6.983	1.012; 0.359, 2.474
		dose category 50-<100 mSv	2.058; 0.302, 8.507	2.029; 0.568, 5.697
		dose category ≥100 mSv	No cases	2.414; 0.739, 6.747
	Linear	Dose (ERR and 95% CI per mSv)	-0.00145; 0.00201	0.00558; -0.0009, 0.02127
	loglinear linear	SES in 4 classes, Internal **Dose (ERR and 95% CI per mSv)**	-- **-0.00145; SE=0.280E-7**	-- **0.00543; -0.0011, 0.0238**
10-year	Loglinear	dose coefficient (at 1 mSv)	0.997; 0.982, 1.006	1.002; 0.997, 1.006
	Loglinear	dose category 0-<1 mSv	Baseline	Baseline
		dose category 1-<10 mSv	0.400; 0.060, 1.563	1.166; 0.562, 2.307
		dose category 10-<50 mSv	1.976; 0.677, 5.455	0.717; 0.234, 1.813
		dose category 50-<100 mSv	1.830; 0.271, 7.410	1.707; 0.480, 4.728
		dose category ≥100 mSv	0; NB‡, 2.08	1.570; 0.427, 4.600
	Linear	dose (ERR and 95% CI per mSv)	-0.00145; SE=0.00276	0.00278; NB‡, 0.0155
	loglinear linear	SES in 4 classes, Internal **Dose (ERR and 95% CI per mSv)**	-- **-0.00145; SE=0.436E-5**	-- **0.00219; NB‡, 0.0156**
20-year	Loglinear	dose coefficient (at 1 mSv)	0.992; 0.964, 1.005	1.003; 0.997, 1.008
	Loglinear	dose category 0-<1 mSv	Baseline	Baseline
		dose category 1-<10 mSv	0.395; 0.061, 1.48	1.737; 0.832, 3.483
		dose category 10-<50 mSv	1.886; 0.665, 4.95	0.621; 0.143, 1.879
		dose category 50-<100 mSv	0; NB‡, 1.90	1.885; 0.428, 5.848
		dose category ≥100 mSv	0; NB‡, 2.15	2.478; 0.662, 7.478
	Linear	dose (ERR and 95% CI per mSv)	-0.00145; SE=0.00041	0.00501; NB‡, 0.0230
	loglinear linear	SES in 4 classes, Internal **Dose (ERR and 95% CI per mSv)**	-- **-0.00145; SE=0.757E-9**	-- **0.00620; NB‡, 0.0304**

*All models stratified on age group (14 strata) calendar time (8 strata) duration of employment (<10 years and ≥10 years) and sex. Factors within each model number were adjusted for all other factors in that model.
†RR: Relative risk; ERR: Excess relative risk; CI: Confidence interval
‡NB: lower confidence limit could not be computed, as it was below the boundary (Preston et al. 1993)

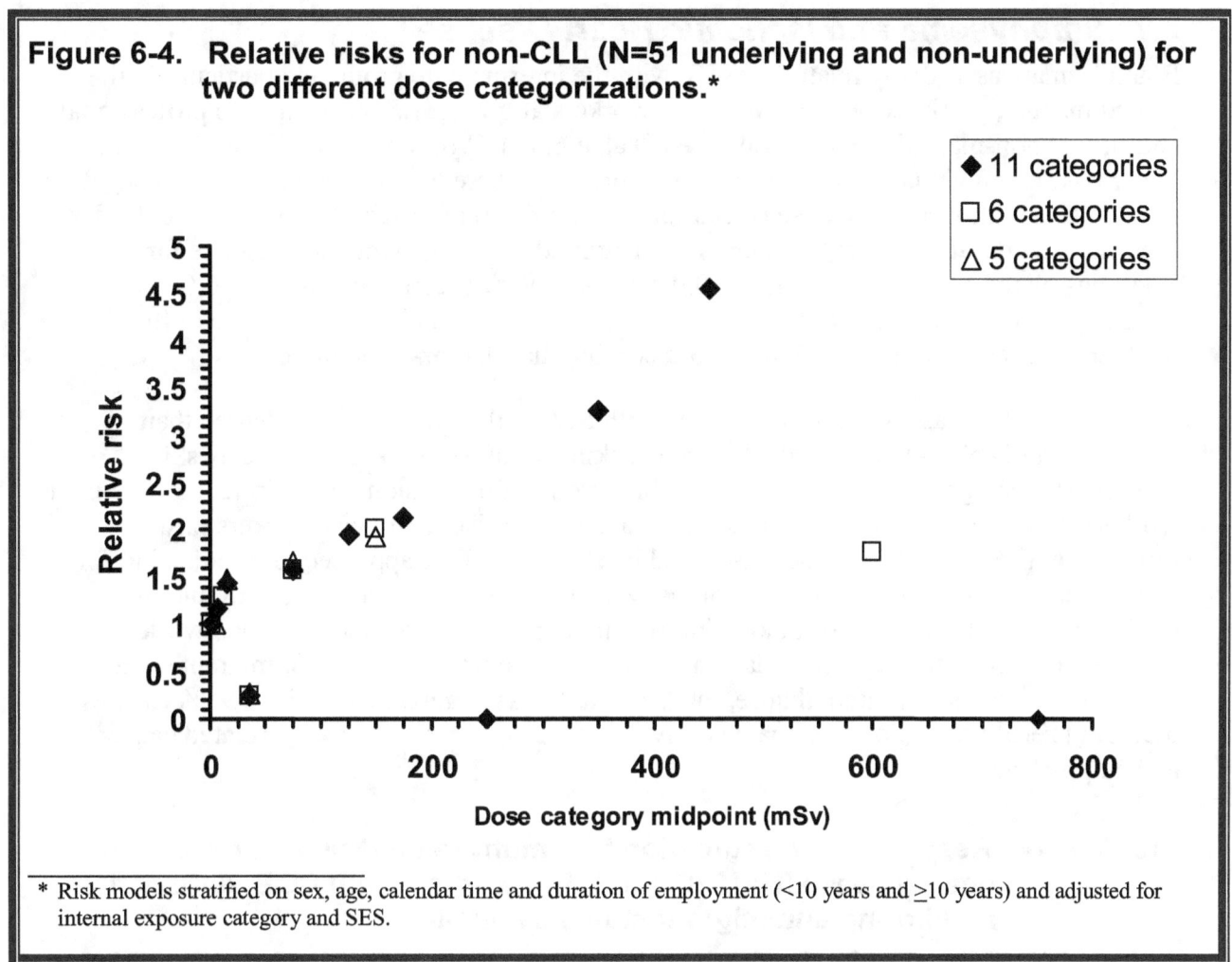

Figure 6-4. Relative risks for non-CLL (N=51 underlying and non-underlying) for two different dose categorizations.*

* Risk models stratified on sex, age, calendar time and duration of employment (<10 years and ≥10 years) and adjusted for internal exposure category and SES.

6.7 Emphysema and Ischemic Heart Disease

Emphysema was strongly related to SES, with the intermediate group and particularly the skilled manual, partly skilled and unskilled workers at greater risk compared to professional and skilled non-manual workers combined (Table 6-14). Workers of unknown SES were at lower risk, although the estimate was very imprecise. There was some evidence of a negative trend in risk with increasing dose (including off-site dose) after stratifying on age, calendar time, sex and duration of employment, of a magnitude similar to the risk estimate for smoking-related cancers (ERR/mSv= -0.001171 for emphysema compared to -0.001453 for smoking-related cancers). However, the CIs were wide for emphysema because of the small number of cases, and the ability to explore confounding factors was limited.

Ischemic heart disease was also associated with SES, although to a lesser degree than emphysema (Table 6-14). Unskilled workers, skilled manual workers and partly skilled workers showed greater risks than the baseline group of professional and skilled non-manual workers. Migrants had substantially lower heart disease risk than local workers (also reflected in the life table analyses discussed in §5.7), which disappeared after adjusting for SES. There was a significant interaction between migrant status and SES, with migrants having higher risks than locals among intermediate, partly skilled and unskilled workers. After adjusting for sex, age, calendar time, duration of employment, SES, migrant status, SES·migrant status and internal dose, the ERR/mSv was negative (but with a 95% CI including zero) with a point estimate nearly five-fold higher than smoking-related cancers and emphysema.

Table 6-14. Results of risk estimation for emphysema (N=69) and ischemic heart disease (N=1296) for 34916 workers with DOB, date of first monitoring and migrant status available.*

Model	Term type	Factor	Risk estimate (RR or ERR[†]); 95% CI Emphysema	Ischemic heart disease
1	loglinear	dose coefficient (at 1 mSv)	0.9985; 0.9919, 1.004	1.000; 0.9989, 1.001
2	loglinear	dose category 0-<1 mSv	Baseline	Baseline
		dose category 1-<10 mSv	0.9913; 0.5298, 1.785	1.077; 0.9345, 1.238
		dose category 10-<50 mSv	0.8312; 0.3634, 1.721	0.9338; 0.7869, 1.104
		dose category 50-<100 mSv	0.8813; 0.2064, 2.575	1.086; 0.8359, 1.390
		dose category ≥100 mSv	0.8046; 0.1845, 2.451	1.043; 0.8040, 1.337
3	loglinear	SES professional & sk-non-man	Baseline	Baseline
		SES intermediate	1.345; 0.5317, 3.355	1.186; 0.9840, 1.428
		SES skilled manual	2.034; 0.9959, 4.514	1.425; 1.219, 1.670
		SES partly skilled	2.666; 1.052, 6.664	1.392; 1.118, 1.725
		SES unskilled	2.445; 0.9131, 6.352	1.743; 1.409, 2.149
		SES unknown	0.4557; 0.0246, 2.437	0.9208; 0.6671, 1.245
4	loglinear	Local (SSN=ID, UT, MT, WY)	Baseline	Baseline
		Migrant	0.7568; 0.4607, 1.245	0.8893; 0.7939, 0.9964
5	loglinear	Internal—not monitored	Baseline	Baseline
		Internal—monitored, unexposed	1.425; 0.7233, 2.685	0.9595; 0.8100, 1.132
		Internal—likely exposed	1.068; 0.5868, 1.941	1.041; 0.9110, 1.189
6	loglinear	Race—White	Baseline	Baseline
		—Other	NA (N=0)	0.9183; 0.4581, 1.621
7	loglinear	SES professional & sk-non-man	Baseline	Baseline
		SES intermediate	1.343; 0.5308, 3.501	1.186; 0.9836, 1.427

Table 6-14. Results of risk estimation for emphysema (N=69) and ischemic heart disease (N=1296) for 34916 workers with DOB, date of first monitoring and migrant status available.*

Model	Term type	Factor	Risk estimate (RR or ERR†); 95% CI	
			Emphysema	Ischemic heart disease
		SES skilled manual	2.006; 0.9718, 4.488	1.423; 1.214, 1.671
		SES partly skilled	2.608; 1.010, 6.641	1.388; 1.110, 1.728
		SES unskilled	2.374; 0.8588, 6.368	1.737; 1.392, 2.158
		SES unknown	0.4577; 0.0248, 2.449	0.9213; 0.6674, 1.246
		Local (SSN=ID, UT, MT, WY)	Baseline	Baseline
		Migrant	0.9408; 0.5590, 1.585	0.9921; 0.8799, 1.119
8	loglinear	Migrant vs local, SES=prof/skn	0.3762	0.7495
		Migrant vs local, SES=int	0.7876	1.290
		Migrant vs local, SES=skman	1.201	0.9359
		Migrant vs local, SES=ptsk	0.5696	1.406
		Migrant vs local, SES=unsk	1.916	1.389
		Migrant vs local, SES=unknown	No cases (p=0.538)	0.4470 (p=0.0013)
8	loglinear	SES int vs prof/sknm, local	0.8550	0.8107
		SES int vs prof/sknm, migrant	1.790 (p>0.5)	1.395 (p=0.009)
		SES skman vs prof/sknm, local	0.9753	1.206
		SES skman vs prof/sknm, migr	3.114 (p>0.5)	1.506 (p=0.164)
		SES ptsk vs prof/sknm, local	1.694	1.015
		SES ptsk vs prof/sknm, migrant	2.564 (p=0.376)	1.904 (p=0.006)
		SES unsk vs prof/sknm, local	1.069	1.310
		SES unsk vs prof/sknm, migrant	1.694 (p>0.5)	2.428 (p=0.007)
		SES unkn vs prof/int, local	No cases	1.401
		SES unkn vs prof/int, migrant	No cases (p>0.5)	0.8358 (p=0.126)
9	linear	Dose (ERR; 95% CI per mSv)	-0.001171; 0.00203	0.0000488; -0.00095; 0.00127
10	loglinear	SES in 6 classes	--	--
	linear	Dose (ERR; 95% CI per mSv)	--	-0.000229; -0.00116, 0.000920
11	loglinear	SES in 6 classes, Migrant	--	--
	linear	Dose (ERR; 95% CI per mSv)	--	-0.000233; -0.00116, 0.000917
12	loglinear	SES in 6 classes, Migrant, SES·migrant	--	--
	linear	Dose (ERR; 95% CI per mSv)	--	-0.000143; -0.00109, 0.00104
13	loglinear	SES in 6 classes, Migrant, SES·migrant, Internal, Race	--	--
	linear	Dose (ERR; 95% CI per mSv)	--	-0.000300; -0.00124, 0.000900
14	loglinear	SES in 6 classes, Migrant, SES·migrant, Internal	--	--
	linear	Dose (ERR; 95% CI per mSv)	--	-0.000299; -0.00124, 0.000902

* All models stratified on age group (14 strata) calendar time (8 strata) duration of employment (<10 years and ≥10 years) and sex. Factors within each model number were adjusted for all other factors in that model. Migrant status refers to those whose SSN was issued outside of ID, MT, WY or UT.

† RR: Relative risk; ERR: Excess relative risk; CI: Confidence interval.

6.8 Non-Hodgkin Lymphoma and Multiple Myeloma

There were strong associations between NHL and both SES and migrant status (Table 6-15). Intermediate (technical/administrative), skilled manual and unskilled workers were at higher risk than professional and skilled non-manual workers. Partly skilled workers exhibited lower risk; however, CIs were quite wide. Migrants were at lower risk than local workers. There appears to have been no substantial interaction between SES and migrant status. No cases of NHL were observed among radiation-monitored non-whites, and internal dose had little association with risk of NHL. The ERR per mSv varied substantially with the adjustment for SES, migrant status and internal dose. The estimated ERR per mSv was approximately 0.002 (2% per 10 mSv) with a CI that included zero. The analysis by dose category, with and without adjustment for these confounders, suggests that any excess in risk was due to exposure in the highest dose category (Table 6-15).

Multiple myeloma was also strongly related to SES (Table 6-15) with partly skilled workers being at highly elevated risk (N=9, or 45% of total cases), compared to professional and skilled non-manual workers. After adjusting for age, sex, calendar year and migrant status, this group of workers had a 6-fold risk of death from multiple myeloma, compared to professional and skilled non-manual workers. Migrants appeared at higher risk, particularly after adjusting for SES; however, there was no evidence of interaction between SES and migrant status. With adjustment for age, calendar year, sex, duration of employment, SES, migrant status and internal dose, the ERR per 10 mSv estimate was approximately 6%, with a CI that included zero. Like NHL, the radiation-related risk of multiple myeloma appeared restricted to the highest-dose category (both with and without adjustment for SES, migrant status and internal exposure); however, the RR and ERR estimates are quite imprecise (Table 6-15).

Table 6-15. Radiation-related risk for non-Hodgkin lymphoma (NHL; N=82) and multiple myeloma (N=20) among 34916 workers with DOB, date of first monitoring and migrant status available.*

Model number	Term type	Factor	Risk estimate (RR or ERR†); 95% CI	
			NHL	Multiple myeloma
1	loglinear	Relative risk at 1 mSv	1.002; 0.9982, 1.005	1.005; 0.9983, 1.010
2	loglinear	dose category 0-<1 mSv	Baseline	Baseline
		dose category 1-<10 mSv	1.283; 0.7353, 2.192	1.088; 0.3587, 3.032
		dose category 10-<50 mSv	0.6117; 0.2572, 1.293	No cases
		dose category 50-<100 mSv	0.6710; 0.158, 1.933	No cases
		dose category ≥100 mSv	1.733; 0.7341, 3.759	2.490; 0.6035, 8.992
3	loglinear	SES prof. & sk-non-man	Baseline	Baseline
		SES intermediate	1.921; 1.017, 3.670	0.7133; 0.1005, 3.422
		SES skilled manual	1.437; 0.7670, 2.754	1.087; 0.2766, 4.335
		SES partly skilled	0.5391; 0.126, 1.606	4.606; 1.337, 16.68
		SES unskilled	2.283; 1.001, 4.926	0.9634; 0.0492, 6.330
		SES unknown	0.8358; 0.234, 2.352	1.083; 0.05526, 7.116
4	loglinear	Local	Baseline	Baseline
		Migrant	0.6438; 0.408, 1.009	1.783; 0.7068, 4.865
5	loglinear	Internal—not monitored	Baseline	Baseline
		Internal—mon, unexposed	1.098; 0.5763, 1.996	0.4686; 0.0712, 1.833
		Internal—likely exposed	1.018; 0.5890, 1.760	1.060; 0.3736, 2.998
6	loglinear	Race—White	Baseline	Baseline
		—Other	No cases	4.343; 0.239, 21.7 (N=1)
7	loglinear	SES prof & sk-non-man	Baseline	Baseline
		SES intermediate	1.877; 0.9925, 3.590	0.7313; 0.1035, 3.473
		SES skilled manual	1.291; 0.6811, 2.503	1.318; 0.3345, 5.221
		SES partly skilled	0.4579; 0.106, 1.386	6.354; 1.781, 23.50
		SES unskilled	1.881; 0.8030, 4.185	1.461; 0.0733, 9.898
		SES unknown	0.8681; 0.242, 2.449	0.9940; 0.0509, 6.494
		Local	Baseline	Baseline
		Migrant	0.6510; 0.404, 1.044	2.507; 0.9489, 7.086
8	loglinear	Mig vs local, SES=prof/sknm	0.6079	2.762
		Migr vs local, SES=int	0.7257	0.6370
		Migr vs local, SES=skman	0.7668	5.174
		Migr vs local, SES=ptsk	No cases in migrants	1.332
		Migr vs local, ES=unsk	0.3373	No cases in locals
		Migr vs local, SES=unkn	0.7738 (p=0.8340)	No cases in locals (p=0.5492)
8	loglinear	SES int vs prof/sknm, local	1.699	1.945
		SES int vs prof/sknm, migr	2.029 (p>0.5)	0.4486 (p>0.5)
		SES skm vs prof/sknm, loc	1.164	0.8177
		SES skm vs prof/sknm, mig	1.469 (p>0.5)	1.532 (p>0.5)
		SES psk vs prof/sknm, local	0.5545	8.846
		SES psk vs prof/sknm, migr	No cases	4.265 (p>0.5)
		SES usk vs prof/sknm, local	1.978	Insufficient cases
		SES usk vs prof/sknm, migr	1.106 (p>0.5)	Insufficient cases
		SES unk vs prof/sknm, loc	0.7351	Insufficient cases
		SES unk vs prof/sknm, mig	0.9359 (p>0.5)	Insufficient cases

Table 6-15. Radiation-related risk for non-Hodgkin lymphoma (NHL; N=82) and multiple myeloma (N=20) among 34916 workers with DOB, date of first monitoring and migrant status available.*

Model number	Term type	Factor	Risk estimate (RR or ERR[†]); 95% CI NHL	Multiple myeloma
9	linear	Dose (ERR; 95% CI per mSv)	0.002303; -0.0015, 0.00953	0.005944; -0.0015, 0.0281
10	loglinear	SES in 6 classes	--	--
	linear	Dose (ERR; 95% CI per mSv)	0.001806; NB[‡], 0.008539	0.005911; -0.0014, 0.0284
11	loglinear	SES in 6 classes, Migrant	--	--
	linear	Dose (ERR; 95% CI per mSv)	0.001522; NB[‡], 0.007999	0.006265; -0.0015, 0.0295
12	loglinear	SES in 6 classes, Migrant, Internal	--	--
	linear	Dose (ERR; 95% CI per mSv)	0.001990; NB[‡], 0.01000	0.006381; NB[‡], 0.0345
13	loglinear	SES in 6 classes, Migrant, Internal	--	--
		dose category 0-<1 mSv	Baseline	Baseline
		dose category 1-<10 mSv	1.287; 0.7033, 2.301	1.002; 0.2981, 3.099
		dose category 10-<50 mSv	0.5934; 0.2340, 1.358	No cases
		dose category 50-<100 mSv	0.6386; 0.1432, 2.019	No cases
		dose category ≥100 mSv	1.544; 0.5874, 3.821	2.051; 0.4058, 9.881

* All models stratified on age group (14 strata) calendar time (8 strata) duration of employment (<10 years and ≥10 years) and sex. A lag of 10 years was used. Factors within each model number were adjusted for all other factors in that model. Migrant status refers to those whose SSN was issued outside of ID, MT, WY or UT.
† RR: Relative risk; ERR: Excess relative risk; CI: Confidence interval.
‡ NB: lower confidence limit could not be computed, as it was below the boundary (Preston *et al.* 1993).

7 Discussion

7.1 Exposures at INEEL

7.1.1 Radiological Exposures

The dose assessment for the INEEL cohort was conducted using dosimetric data obtained from the INEEL as part of an epidemiological cohort study covering the period from 1951-1998. Radioactive material use began at the site in 1951, two years after the site's establishment, at the EBR-I in December and in March 1952 at the Materials Testing Reactor. The types of radioactive material varied by facility with quantities of fission products, neutron activation products, radioactive chemicals, plutonium, americium and assorted calibration sources.

The cohort consisted of 63,561 workers, of whom 35,833 were externally monitored for photon, beta and neutron radiation. A large proportion (27,728 workers) was not monitored. Workers wore film badges before 1966. At that time the site began to use TLDs. Neutron monitoring was also performed but much less frequently. Kodak NTA film was used for this purpose until 1975 when it was replaced by the TLD based albedo dosimeter.

The extent of monitoring for internal radionuclides was less comprehensive than for external dose, and the sampling need was based on job activities and the potential for creation of airborne radioactivity. This difference was true especially in the early years with improvements in the 1970s. Exposure assessment for this epidemiologic analysis was hampered by the lack of available computerized internal dosimetry data before 1986. These data were available as hardcopy images; however, substantial effort would have been required to computerize them for use in an epidemiologic study of a large cohort such as INEEL. This level of detailed information could be achieved much more readily for a case-control study.

The collective, cumulative external dose totals for the site can be summarized as follows:

Penetrating Photon Dose	469.84 Sv
Non-Penetrating Dose	191.83 Sv, excluding the penetrating dose component
Neutron Dose	6.89 Sv

The potential for photon dose under/over estimation was greatest before 1980. Workers had the potential to encounter several types of source terms during a work year. The calibration source energies did not sufficiently match the range of energies found in all source terms. This discrepancy was particularly problematic in the 1950s but extended to the 1960s and 1970s to a lesser degree as well. The number of filter elements, filtration densities and thicknesses played a part in the incorrect assignment of low-energy photon dose to beta dose. This inaccuracy was complicated by the low-energy photon over-response of film. Compensation for the photon incidence angle and backscatter during exposure was also unaccounted for in dose computations. Early problems with temperature stability in the TLD readout system introduced error into the dose computation.

The neutron dose may have been underestimated for several reasons. Quality factor changes were not taken into account, though a factor of ten was used for unknown fast spectrum fields from the early 1960s. Field neutron energy spectra were not known except in very special cases. NTA film responds to fast neutrons, above 500 keV, and TLDs have a large cross section for thermal neutrons. However, the response drops rapidly with increasing neutron energy, although accuracy was greatly increased using TLD albedo dosimetry. There is a missing band of energies that can only be accounted for by using spectral analysis data from the work site, but such data were rarely collected.

Dosimetry distributions are compared between the INEEL cohort and Hanford, Rocky Flats and the Oak Ridge National Laboratory (ORNL) sites included in the most recent combined international nuclear workers study (Cardis *et al*. 1995) in Table 7-1. The Hanford facility in Washington State showed very similar numbers of monitored workers, but its workers exhibited 1.8 times the collective and mean cumulative dose as the INEEL cohort. The ORNL cohort in Tennessee had similar mean dose per worker as in the INEEL cohort, but the number of workers at INEEL was much larger, leading to a larger collective dose. INEEL, Hanford and ORNL all had reactors that contributed to personnel exposure by means of fission and activation products. However, Hanford and Rocky Flats workers were involved with plutonium on a large scale yet INEEL and ORNL workers were exposed to plutonium only in waste processing activities. The mean cumulative dose was lowest for INEEL and was highest at Rocky Flats (Table 7-1).

Table 7-1. Comparison of monitored employees at four DOE facilities. Dose information for Hanford, Rocky Flats and the Oak Ridge National Laboratory (ORNL) facilities is from Cardis *et al.* (1995).

	INEEL	Hanford	Rocky Flats	ORNL
Years of Study (including follow-up)	1951-1998	1944-1986	1951-1979	1943-1984
Number of Workers	35,833	36,925	7495	8313
Collective Dose (Sv)	469.8	884.1	242.5	142.4
Mean Cumulative Dose (mSv)	13.1	23.9	32.4	17.1

The distribution of study cohort members by cumulative dose range showed that over 90% of all workers in each facility received less than 100 mSv total (Table 7-2). INEEL had 12% more workers in the zero to 10 mSv dose range than Hanford and almost half the number of workers in the 10 mSv to 20 mSv range. Eighty-nine percent of Hanford workers had cumulative doses below 50 mSv compared to 93% of INEEL workers.

Table 7-2. Distribution of study cohort members by monitoring status, cumulative dose range and facility. Dose information for Hanford, Rocky Flats and the Oak Ridge National Laboratory (ORNL) facilities is from Cardis *et al.* (1995).

Facility	Total Workers	Not Monitored	Monitored	0 mSv	10 mSv	20 mSv	50 mSv	100 mSv	200 mSv	400+ mSv
INEEL	63,561	27,728	35,833	28,632	2310	2431	1310	749	325	76
Hanford	44,104	7179	36,925	25,278	4131	3587	1684	1112	845	288
Rocky Flats	7571	76	7495	4153	1046	997	627	417	205	50
ORNL	8313	0	8313	6042	884	751	316	187	95	38

Each facility used film badges until the early 1970s, when the TLD replaced it as the preferred monitoring device. Dosimeters were issued on a weekly basis at the four facilities before the mid-1950s, with the LOD ranging between 0.3 and 0.5 mSv (Table 7-3). This combination created a potential missed dose problem at Hanford. Each site reported less than LOD values as zero during these years, and more artificial zeros would have been expected for Hanford because of its 0.5 mSv LOD. Missed dose (from detection limit problems) does not appear to have been a major contributor to the INEEL cohort collective dose, given its small value.

Table 7-3. Dosimeter issuance and LOD by facility, for INEEL and Hanford.

Facility	Years	Number of Dosimeters Issued Annually	LOD (mSv)
Hanford	1944-54	52	0.5
	1955	52	0.3
	1956-57	26	0.3
	1958-63	12	0.3
	1963-71	12	0.2
	1972	12	0.1
INEEL	1951-58	52	0.3
	1958-74	52/12	0.1
	1966-75	4/2/1	0.15
	1973-74	12	0.15
	1974-75	12	0.3
	1974-86	12/4/1	0.15
	1986-89	12/4	0.01
	1989-93	12/4	0.15
	1993	12/4	0.1

7.1.2 Non-radiological Exposures

With more than 60,000 workers in the INEEL cohort who had been employed in the hundreds of buildings at the site, exposure assessment for specific non-radiological hazards was not feasible. In lieu of actual exposure estimates, workers were placed in categories by their job titles or employers to evaluate patterns of cause-specific mortality by subcohorts. The categories were determined after evaluation of work history records plus extensive walk-through evaluations of most active operational areas at the site, as well as consideration of readily available records. Discussions with workers and managers were also sources of information on patterns of potential exposures for this cohort.

With this approach, evaluation of health effects from non-radiological hazards was limited to readily identifiable groups with sufficiently large populations for SMR and SRR analyses. Prime contractors (including DOE employees), subcontractors, construction and service contractors, chemical workers, asbestos workers, drivers, reactor workers, the security force, and painters were the categories selected for the subcohort analysis. These categories were not always mutually exclusive.

Job titles are widely used in the epidemiologic evaluation of occupational risks. Some subcohorts in this study, however, were based on the categorization of employers by type of service or function in addition to job titles, for example construction and service workers. Use of employer industry potentially could have led to a higher degree of misclassification than use of job title. For example, construction firms may have employed workers in different trades, as well as engineers, planners and administrative staff, among others, who experienced different exposures. Beyond the risk of misclassification due to selection of subcohort members, exposure heterogeneity by substance, magnitude and temporality could not be evaluated in this method.

In using either job titles or employers as surrogates for exposures, specific agents responsible for excess health risks cannot be identified except in a few cases (e.g., asbestos or silica). The strategy of subcohort analysis only allowed exploration of differences in health risks and possible or probable exposures may then be proposed.

7.1.2.1 Construction and Service Workers

Construction and service workers were the largest subcohort in this analysis with nearly 50% of the cohort included. Site operation and support activities were housed in more than 500 buildings. These structures were often clustered in industrial complexes such as the NRF, Test Area North, Test Reactor Area, and the ICPP. Many construction workers operated out of facilities based at the Central Facilities Area (CFA) and the site construction headquarters adjoining ICPP. Others were subcontractors for large and small projects who were temporarily employed at the site for the duration of the contracts. Many of these workers were likely to have worked on numerous subcontracts over the decades of construction work at the INEEL.

This group of workers experienced a substantially different mortality pattern than the remainder of the cohort with elevated standardized (mortality) rate ratios for many causes. Similar patterns of excess deaths, particularly for respiratory diseases and some digestive system cancers, have been observed in previous epidemiologic studies of construction workers (Robinson *et al.*1995, Wang *et al.* 1999). Although some of these excesses may have been due to lifestyle-related factors such as smoking and diet among the INEEL cohort, workplace exposures may also have been important.

Classification of the construction trades by the U.S. Census Bureau is shown in Table 7-4. Workers in each of these trades experience potential exposures to myriad substances and, since these trades often work in close proximity during new construction and large refurbishment projects, materials particular to a given set of trades can also be sources of exposures for many others. A recent evaluation of exposures to dusts, fumes, and vapors reported that, even today, some high exposures to many materials continue in the construction industry (Varma 2003). It is noteworthy that many of these trades at the INEEL have been employed in large- and small-scale environmental remediation projects, including decontamination and demolition, with attendant ionizing radiation exposures.

Table 7-4. Construction Worker Classifications from U.S. Census Bureau.

Brickmasons and Stonemasons
Carpenters
Carpet Installers
Concrete and Terrazzo Finishers
Crane and Tower Operators
Drillers
Drywall Installers
Electrical Power Installers
Electricians
Excavating and Loading Machine Operators
Glaziers
Graders, Dozer and Scraper Operators
Helpers, Construction
Inspectors, Construction
Insulation Workers
Laborers, Construction
Operating Engineers
Painters
Paperhangers
Paving, Surfacing And Tamping Equipment Operators
Plasterers
Plumbers, Pipefitters And Steamfitters
Roofers
Sheetmetal Duct Installers
Structural Metal Workers
Tile Setters
Truck Drivers

Malignant neoplasms of the respiratory system in many construction trades have been associated with inhalable and respirable dusts. Non-malignant respiratory diseases such as pneumoconiosis and asthma in construction workers have been shown to result from airborne dusts and fumes (Wang *et al.* 1999). Dust exposures during construction activities included crystalline silica, asbestos and man-made mineral fibers (Varma 2003, Methner 2000). Metals, such as cadmium, chromium, lead, manganese and mercury, are constituents of numerous building materials.

Structural metal workers were likely to have been exposed to these metals in the form of welding fume. Painters may have been exposed to dusts containing these metals and other dusts, particularly silica and asbestos, during surface preparation. Heavy equipment operators are exposed to silica and diesel exhaust during excavation and other earth-moving projects or to asbestos during the demolition of buildings. Roofers experience coal tar pitch fume exposures, including polynuclear aromatic hydrocarbons such benz[a]pyrene and their derivatives. Other associations among construction worker trades, particulate exposures, and health risks have been summarized elsewhere (Robinson *et al.* 1995).

Many of these trades also experience potential exposures to a wide array of organic chemicals. Probable and possible carcinogens in building materials have included benzene, chlorinated hydrocarbons, and aldehydes. Carpet installers, cabinet makers, masons, and insulation workers are exposed to organic compounds that are components of adhesives, solvents and structural materials such as isocyanates and polyurethanes. N-hexane, a compound associated with axonal neuropathies, is present in petroleum and coal derivatives such as low molecular-weight hydrocarbon solvents.

Carbon monoxide has often been present in construction settings because of the use of internal combustion engines. Carbon monoxide, in addition to widely recognized risk for asphyxiation, may also increase the risk for cardiovascular disease and sudden cardiac death (Atkins and Baker 1985). Internal combustion engines also emit particulate, nitrogen oxides and sulphur oxides with attendant risks for inflammation of respiratory diseases such as chronic obstructive pulmonary disease and asthma.

7.1.2.2 Painters

Painters constitute a small subcohort in this analysis with only 690 workers. Although a combination of job titles and employer category was used to select the subcohort, almost 600 were identified by job title. Since most of these subcohort members were selected by job title, the association with possible exposures is stronger than if employer category alone were used. Painters complete tasks beyond the application of paint, such as surface preparation and equipment maintenance, site clean-up and the removal of accidental spills and drips using a variety of solvents.

Painters may have been exposed to inhalable particulates during surface preparations that historically have included blasting as well as scraping. The particulates may include silica from the blasting agents as well as from masonry building materials. Asbestos has been used extensively in building structures such as interior and exterior wall materials that may become airborne during blasting and scraping tasks. Metals such as lead and chromium have been used as pigments in paints. Like asbestos, these hazardous materials become airborne during surface preparations. They also may have been inhaled during spray painting operations.

Painters are also exposed to the solvents in paints as well as those used in adhesives and filler materials (IARC 1989). Historically, the solvents were primarily organic and were derivatives of plant materials such as turpentine. Paints designed for use in industrial settings included epoxies and urethanes. Coal and petroleum-based hydrocarbon solvents such as mineral spirits, naphtha and paint removers were also used by painters. As a result, benzene, methylene chloride and N-hexane exposures were likely. Bystander effects also may contribute to overall exposures for painters since they often work in close proximity to other construction trades such as flooring and insulation installers, electricians, plumbers and drywall installers.

7.1.2.3 Asbestos Workers

The selection of asbestos workers also used a combination of job titles and employer category. Potentially exposed job titles included those that mentioned asbestos specifically, as well as those with recognized exposure risks such as boiler workers, insulation workers and pipefitters. The employer category was also relatively specific for potential asbestos exposures since it was restricted to businesses that were insulation and surface coating firms. Other trades employed at the INEEL but not included in this subcohort, such as carpenters and demolition workers, have reported potential exposures to asbestos (Yeung and Rogers 2001). Seven of the ten asbestosis deaths in the cohort occurred among the subcohort of asbestos workers, and all ten occurred among those identified as construction and service workers.

Insulation workers have been shown in numerous health studies to experience excess risks for respiratory and peritoneal malignancies (Ulvestad *et al.* 2004; Burdorf *et al.* 2003; Menegozzo *et al.* 2002; Jarvholm and Sanden 1998). Asbestos is a known human carcinogen that contributed to these health risks. However, the replacement materials for asbestos have also been associated with elevated risk of respiratory disease (Lockey *et al.* 2002; Drent *et al.* 2000; Lockey *et al.* 1998) although the severity of the disease currently observed in occupational cohorts appears to be much less than that for asbestos.

Insulation materials have changed over the years. Within the timeframe of INEEL construction activities, asbestos has been replaced by man-made mineral fibers as the material of choice (Weeks 1995). The fire-retardant and binding properties of asbestos fibers created tremendous demand for asbestos sheets and cloths for insulation. In reactors, as well as facilities that used boilers for steam and hot water, asbestos was commonly used as lagging on transfer lines. Wall boards and plasters have also contained asbestos either as the primary material or because of its binding properties. The more recently developed insulation materials also include organic chemical foams such as urethanes and isocyanates. The latter of these chemical groups has been associated with respiratory disease and immune-mediated reactions (Tarlo and Liss 2002, Bernstein 1996). Use of these foam insulation materials at INEEL, however, was not discovered during reviews of site records nor was it mentioned by workers and managers at the site.

Recognition of asbestos hazards following publication of epidemiologic study results by public health research organizations had great impact on asbestos-handling work practices. In the 1980s, new application of asbestos containing materials was severely restricted. Remediation projects for friable asbestos were widely targeted and all asbestos-containing materials were identified and labeled. Remediation projects were required to use containment of the sources, and air samples were collected to estimate exposures during removal and to ensure that the building spaces met clearance criteria for subsequent use. These programs substantially reduced potential exposures to asbestos, and thousands of samples were collected at the INEEL to ensure compliance. These data could be used to develop exposure estimates for workers in this subcohort should a public health need for such an evaluation be identified.

7.1.2.4 Chemical Workers

The subcohort of chemical workers included INEEL employees selected by job titles that reflected chemical operation tasks and those who were likely to have been employed at the ICPP (Table 2-10). Because the ICPP was operated by a site-wide contractor for much of the site history, selection of chemical workers based on this latter criterion was not specific. The site-wide contractors were not used for selection since that would have included a large number of workers from all other INEEL facilities in operation during those periods. The exposure profiles for chemical operators in a production facility were also likely to have differed substantially from the exposures experienced by chemical laboratory workers.

Chemical operators tend equipment and make the necessary adjustments to ensure consistent products, usually following written procedures. They are likely to engage in repetitive tasks with little variation from day to day. Alternatively, routine maintenance and overhaul activities are likely to result in higher exposures for chemical operators as equipment containing hazardous materials is evacuated and disassembled (Rich *et al.* 1974).

Chemical operators were responsible for the basic operation at the ICPP, namely, reprocessing spent nuclear reactor fuels to extract enriched uranium. These extractions occurred in a series of compartments within the heavily shielded canyon in Building 601 (Cederberg *et al.* 1974). Other radionuclides were also isolated, such as radioactive barium, for use in nuclear weapons production at another DOE facility, and noble gases for commercial and research purposes. The abundance of these materials in spent fuel depended on the operating characteristics of the reactors and the elapsed time since the core had been removed from the reactor. Some cores used for radioactive barium extraction had been removed from the Materials Test Reactor at the Test Reactor Area only a day or two prior to dissolution. Hence, the fuel elements were very highly radioactive since time had not allowed the abundant short-lived radionuclides, such as iodine-131, to decay (Boardman 1957, Legler 1957).

Raffinates, the aqueous fraction from the extraction process, contained the concentrated nitric acid used for dissolving the spent fuel elements along with all other radionuclides not specifically extracted. These raffinates were stored in a total of 11 underground tanks at the ICPP. Beginning in 1963, the raffinates were converted into solid calcines using fluidized-bed combustion technology for storage in above-ground silos. Chemical sampling and analysis all along the reprocessing stream through the end-stage calcining operation was required for technologic control of the equipment (Dieter et al. 1997).

One of the first steps in spent fuel reprocessing is removal of the cladding material that surrounds each nuclear fuel element. The cladding is engineered to reduce deterioration from the physical and chemical forces in the operating reactor. Different cladding materials were selected for the various nuclear reactor designs. Because of the special materials used for cladding, a number of dissolution 'front-ends' were needed to provide the chemical and physical environment necessary for the various sources of fuels reprocessed at the ICPP. The fuel elements reprocessed at ICPP were clad with zirconium, stainless steel and aluminum, among other materials. The cladding metals were components of the raffinates from the extraction process (Dieter et al. 1997).

The earliest waste calcining facility for raffinate treatment was built as a pilot plant and operated from 1963-1981. Workers reported that the aging facility required increasing maintenance with greater levels of radioactive contamination especially from ruthenium-106. One problem was bursting duct work (Johnson and Bradford 1978). The calcined waste was air-jetted through the ducts and abraded the internal surface, which eventually failed, especially at elbows. According to workers, removal of the radioactive materials required that the floor of the process room be flooded with eight inches of water to shield the widespread contamination prior to workers entering the room for decontamination jobs.

In 1983, the new waste calcining facility was put in operation to continue reducing the inventory of the liquid high level waste in the underground storage tanks. Both calcining facilities used fluidized bed technology to evaporate the liquid component of the waste stream. During this process, the high concentrations of nitric acid are converted to nitrogen oxides, which range in color from transparent to yellow-brown. Yellowish discharge from the stack clearly indicated when calcining operations were underway. The height of the stack allowed the nitrogen oxides and other volatile materials to drift with the prevailing wind to off-site locations and the risk of worker exposure to these gases appears to have been low.

Photographs of workers obtaining process stream samples at the ICPP showed that chemical operators manually removed sample vessels from ports in the exterior walls of the dissolution canyon (Boardman 1957). Use of personal protective equipment other than gloves and safety glasses is not apparent in the photographs. Discussions with employees in the 1990s indicated that physical contact with the sample vessels in Building 601 was reduced as a result of two sequential upgrades to the process stream sampling ports in the 1970s that relied on more remote handling methods. One supervisor reported that the chemical operators responsible for sample vial handling typically received the highest radiation exposures at ICPP.

Workers at the ICPP reported that ground contamination was an increasing problem that was not fully recognized prior to a site-wide survey in the mid-1970s. Following this survey, emissions from various ICPP process streams were additionally controlled, primarily with high-efficiency particulate filters in 1975. Remediation of localized contamination on the ground and building surface (hot-spots) reduced this source of potential exposures for ICPP employees (Rich et al. 1974).

Numerous facilities at the INEEL, including the ICPP, housed chemical laboratory operations, and many workers in these laboratories were included in the chemical worker subcohort. Responsibilities of these laboratories extended from bioassay for internally deposited radionuclides, to environmental samples, to component analysis for various production and experimental operations at the site. Chemical laboratory exposures are particularly difficult to characterize. With routine laboratory operations, such as bioassay of urine and fecal specimens obtained from monitored workers, the inventory of chemicals was more limited than in laboratories designed to have a range of analytical capabilities. Chemical laboratory workers often physically handle chemicals in small vessels as well as bulk quantities. Many if not most operations may have been completed in fume hoods, yet the risk for dermal exposures should not be overlooked.

Many of the materials analyzed in the laboratories contained radioactive components as well as solvents and diluents (Cederberg 1974). Although some organic solvents were used, inorganic compounds, especially radionuclides, have been the analyte of primary interest. For the process chemicals, mass spectrometry and spectrophotometry were used for much of the analysis, and the laboratory preparation of samples would have involved the elimination of the solvents by heating to reduce the dissolved solids to a residue. This material would then have been dissolved in a measured quantity of neat solvent so quantitative estimates of the components could be obtained. A similar sample preparation procedure was used for bioassay samples that could then be quantified by gravimetric, spectrophotometric or radiological methods.

With the increased concerns about environmental contamination, laboratory capacity for organic samples grew at the INEEL and groundwater, soil and air samples were collected and analyzed in increasing amounts. Early waste handling procedures at INEEL had resulted in organic compound contamination of soil and groundwater. The large, modern-day environmental remediation projects at the site substantially increased the number of samples processed. At the same time, enhanced detection technologies in chemical analysis reduced the volume of individual samples so smaller amounts of laboratory chemicals were needed. These reductions, along with improved recognition of health hazards associated with laboratory chemicals, were likely to have lowered exposure potentials for these workers.

Because of the early primary interest in radionuclides, most if not all chemical workers at the INEEL were monitored for external and internal radiological control purposes. More detailed analysis of the radiation data might lead to additional information on exposure levels to specific non-radiological hazardous materials experienced by this subcohort.

7.1.2.5 Security Workers

Workers in this subcohort were selected based on job titles that are contained in the U.S. Bureau of Census codes 803-809. The subcohort was small at 1276 workers with 158 total deaths. Security workers at DOE sites have unique jobs and, for the most part, locating relevant published reports on potential exposures was not successful. These workers were primarily responsible for physical security at the INEEL. DOE security guards are required to maintain proficiency with firearms. Previous studies of firing range exposures have reported relatively high lead exposures (Svensson *et al.* 1992, Goldberg *et al.* 1991, Valway *et al.*

1990). DOE security workers are exposed to internal combustion engine exhaust as a result of vehicle inspections conducted prior to allowing automobiles and trucks to enter sites. INEEL security forces also used helicopters to help maintain site security, which could result in exposures that have not been previously characterized.

Occupational stress is known to be elevated among public safety workers, but extrapolation of this finding to a population of workers with such unique roles is highly uncertain. They staffed the entrance checkpoints for secured facilities throughout the site where metal detectors and x-ray screening devices were used to inspect items brought to the site by workers and visitors. They responded to incidents that involved possible security breaches. The guards also had firing range exposures and completed mock exercises to maintain high skill levels and coordination in preparation for emergencies, and they trained on the use of police tactics to restrain violators of security rules, if necessary.

7.1.2.6 Reactor Workers

The workers in this subcohort, for the most part, had reactor as part of their job title, for example Reactor Engineer. Because of the nature of the facilities at the INEEL, Power Plant and Nuclear Operators were also included in this group. As mentioned in §1.2, a total of 51 nuclear devices were built at the INEEL and 48 of them reached operational status. Some of the reactors, such as the liquid sodium EBR-I and EBR-II, the ship training reactors at the NRF and the three reactors at the TRA, operated for long periods of time. In the early days of reactor safety testing, numerous reactors were run for relatively brief periods of time (See Table 1-4) and then taken through failure exercises to examine the radiation effects and structural damage that occurred in cases of uncontrolled criticalities. Some reactors were built to test a variety of technologies such as the air-cooled reactors for the Aircraft Nuclear Propulsion program at TAN. Yet others were operated on an intermittent basis, used as radiation sources for the conduct of experiments.

Reactor operators worked in control rooms in areas adjacent to the reactor. Much like chemical reactor workers, they were responsible for the monitoring and adjustment of reactor controls to maintain performance parameters, usually following documented procedures. Reactor and nuclear engineers were responsible for the design of the nuclear devices as well as overseeing their construction. Power operators worked with the equipment that generated electrical current from energy generated by the reactors. Other technical staff worked in the vicinity of reactors with sustained operations such as those involved with sample preparation, insertion and removal at the Test Reactor Areas. Job titles for these staff have not been determined.

Reactor workers were likely to have experienced exposures to various forms of high energy ionizing radiation. Systematic exposures to other agents are less clear although a variety of chemical agents were likely to have been used to maintain equipment. In a NIOSH feasibility effort to examine lung cancer risks among DOE reactor workers, asbestos was found to have been a common insulation material in reactor facilities.

7.1.2.7 Drivers

A large number of drivers were identified within the INEEL cohort. Most were truck drivers with a substantial minority of bus drivers. Previous epidemiologic studies of professional drivers have examined health risks related to their occupational exposures, primarily among long-haul truckers and urban bus operators. A hazardous agent of current concern for professional drivers is diesel exhaust (Steenland *et al.* 1998). Local truckers may have been exposed to hazardous materials that they hauled, particularly if they were responsible for loading and unloading the cargo, as many were. For those who operated gasoline powered vehicles, benzene exposures were very likely to occur. In the early decades of site operations, smoking was reported to have been very common on the buses used to transport workers between the site and various cities and towns in southeastern Idaho. If drivers had maintenance responsibilities for their vehicle or were present in shops when maintenance was completed by others, the risk of exposure to a variety of materials, including asbestos from brake linings, would have been higher. Since the late 1980s, most INEEL buses have used liquid natural gas, which should have caused substantial reductions in exposure to diesel exhaust among bus drivers.

7.2 Epidemiological Findings

7.2.1 Life Table Analyses

The INEEL workforce, particularly WM and WF, exhibited higher rates of overall mortality and of cancer mortality than the regional population of Idaho, Montana and Wyoming. Mortality rates for non-whites were, however, lower than expected based on regional rates. This finding could have been partially attributable to the differences in the distribution of racial groups within the INEEL cohort, compared to these regional states.

Many components of the healthy worker effect have been identified in the literature (Arrighi and Hertz-Piccioto 1994), including failure to control for geographic rate differences, initial selection of a healthier workforce into the work environment, and continued selection of healthy individuals who maintain fitness to work. The INEEL cohort showed evidence for a strong healthy worker effect, in which many of these factors were operating. Workers employed longer exhibited much lower mortality rates for most lifestyle-related causes of death. Geographic rate differences were a very important component in the INEEL cohort, as the apparent healthy worker effect disappeared to a great extent, particularly among cancers, when regional mortality rates (the combined states of Idaho, Montana and Wyoming) were used rather than the U.S. rate. This effect may not have been completely controlled-for using this method, however, as a relatively large proportion of the cohort derived from the state of Utah (as evidenced by the state of issue of the SSN), which has lower mortality rates for most cancers (CDC 2004).

The slight elevation in transportation accident mortality rates, evident when the cohort was compared to the general U.S. population, disappeared when using local rates as a comparison group. This finding provides further evidence for the importance of controlling for regional differences among various health outcomes, as does the increase in the homicide SMR and decrease in the suicide SMR when using state rather than national rates.

An additional component of the healthy worker effect has been identified among nuclear workers, sometimes referred to as a "healthy badged worker effect," in which stringent medical evaluations are required before permitting employees to work in radiation-exposed areas (Beral et al. 1988, Smith et al. 1986, Wilkinson et al. 2000, Silver et al. 2004). It was important, therefore, to ascertain the relative health of unbadged workers, and of badged workers who received no measurable dose. The INEEL cohort confirms these findings from other studies, as badged workers exhibited lower mortality on average than unbadged workers, and badged workers receiving a positive dose showed still lower mortality rates, particularly with regard to many lifestyle-related health outcomes, such as lung cancer, emphysema, cardiovascular disease, and cirrhosis of liver. Differential ascertainment success does not account for these differences, as a smaller percentage of badged (4.6%) compared to unbadged (7.7%) workers was lost to follow-up.

A "healthy monitored worker" effect was also apparent with respect to internal radiation monitoring and exposure status. WM monitored for external and internal exposure, and especially those who actually received exposure, were found to exhibit reduced rates of many lifestyle-related causes of death, such as alcoholism, cirrhosis of liver, lung cancer and cardiovascular disease. WF and NWM did not exhibit as strong a "healthy badged worker effect," which may have been due in part to the small numbers of deaths occurring in these groups.

The most noteworthy findings in cancer mortality rates among the full cohort are the elevations in NHL and cancers of "other respiratory" and "other digestive" categories among WM, and in leukemia risk among NWM. The latter all occurred, however, among those who were badged but received no external dose. Two of three bone cancer fatalities in WM occurred among workers with a positive internal dose (but external dose below 10 mSv) leading to a greatly elevated (but highly uncertain) rate ratio. The single case occurring in WF was in a non-monitored individual. Evidence from studies of Russian workers and those medically exposed to internal emitters suggests higher risk of bone cancer following plutonium exposure. Bone cancers have not been found to be elevated, however, in other internally exposed populations of U.S. workers. Better dosimetry information might prove informative on the etiologic nature of bone cancer in this cohort.

NHL was found to be elevated by about 25% for both male and female INEEL workers, compared to the regional population (but not when compared to the U.S. population). Possible reasons for this elevation include the high prevalence of professional employees among WM in the INEEL cohort. WM in the professional SES group showed elevation in NHL compared to the general population and to other SES categories, and other studies (e.g. Seniori Costantini et al. 2001) have found male but not female managers to exhibit elevations in NHL rates. Studies have also associated this cancer with exposure to herbicides and other chemicals, among both men and women (Miligi et al. 2003). NHL risk was elevated among the highest radiation exposure group (those receiving ≥ 100 mSv cumulative dose), although confidence intervals included one.

The observed differences in mortality patterns among employer types were likely due to differences in SES and/or duration of employment (i.e., healthy worker survivor effect) among these groups of workers, as those who worked for prime contractors and especially multiple contractors had much longer durations of employment than employees of subcontractors or unknown contractors (Table 4-6). The SES category differences among employer types are also notable: among males, 72% of subcontractors and 70% of unknown contractors were in the skilled manual, partly skilled or unskilled categories, compared to 49% of the multiple-contractors and just 28% of the prime contractors. It is informative that "other and unspecified" death rates were much lower among multiple-contractor employees. This difference was likely an artifact of the better vital status tracing and percentage of death certificate retrieval for workers with longer duration of employment. Among females, less heterogeneity in SES category was observed: women were most often classified as skilled non-manual employees in each major category; however prime contractor employees were more likely to have been classified as professional or intermediate (i.e., technical and administrative) workers. The findings of elevated breast cancer mortality rate in prime contractor employers and lung cancer mortality rate among subcontractors are likely reflective of SES differences by employer type, as high SES has been associated with increased risk of breast cancer (Yost *et al*. 2001, Dano *et al*. 2003, Gordon 2003) and low SES with increased lung cancer risk (Steenland *et al*. 2002).

This study shows evidence of confounding by approximated religious practices and SES within the INEEL workforce. Workers defined as "local" were more likely to belong to the Church of Jesus Christ of Latter Day Saints (LDS; Burphy *et al*. 2004), which has been associated with lowered mortality from cardiovascular disease and lung cancer (Lyon *et al*. 1978). Local workers also tended to receive higher doses, as they worked for longer periods of time. However, local workers were also more likely to have been of lower SES, which is associated with higher cigarette smoking and subsequent lung cancer and cardiovascular mortality. Therefore, adjustment for migrant status and SES and their interaction was necessary to account for confounding by these two variables. Use of local status (to the INEEL region) as a surrogate of LDS membership may have improved the ability to evaluate the effects of workplace exposure at the INEEL; however, better estimation of smoking and other lifestyle behaviors that affect health are needed to fully understand risk in this cohort.

Construction workers as a group, both males and females, experienced much higher mortality rates for most causes of death than non-construction workers. Among white male construction workers, elevated SRRs were observed for all causes combined, all cancers combined, and cancers of the digestive system, respiratory system, prostate, kidney, and other/unspecified sites. Additionally, elevated rate ratios were observed for ischemic heart disease, other disorders of the circulatory system including cerebrovascular disease, diseases of the respiratory system including emphysema, asthma, asbestosis and other/unspecified respiratory disease, alcoholism, cirrhosis of the liver, diseases of the genitourinary system, accidents particularly transportation and falls, suicide, homicide and other/unspecified causes of death. The SRR analysis for female construction workers was limited due the small number of total deaths (159). Only NHL reached statistical significance; however, ischemic heart disease and all-causes combined were of borderline significance.

These mortality patterns were very similar to the SES results, which is not surprising since most workers identified as construction workers were likely of the skilled manual, partly skilled and unskilled worker category. Construction workers were also less likely to have been migrants than non-construction workers (38% compared to 54%), which complicates interpretation of the mortality findings, because migrants had higher mortality rates for many causes of death, due in part to a higher proportion of LDS membership among locals.

Although much of the observed elevation in mortality rates among construction workers was likely due to lifestyle-related differences in smoking, alcohol consumption, diet and risk-taking behaviors, there is also substantial evidence that work-related exposures may also have contributed to excess mortality risk among construction workers in the INEEL cohort. Rates of asbestosis and cancers of peritoneum and pleura were substantially elevated among construction workers. Renal and pancreatic cancer mortality rates were also elevated among construction workers, and chronic and unspecified nephritis and chronic renal failure were higher among construction workers. Asthma mortality rates were more than tripled among construction workers, and the anemia mortality rate was also substantially elevated compared to other workers. These differences could reflect higher solvent exposure among construction workers, or could be due to the extreme deficit of anemias among the non-construction cohort. Leukemia risk was not elevated among construction workers.

Transportation accident deaths and deaths from accidental falls were also elevated among construction workers. The latter were particularly likely to have been work-related, although transportation accidents were found to be the leading cause of occupational fatalities nationwide during 1992-2001 (Pratt 2004). The actual number of deaths in the INEEL cohort that were work-related is unknown. However, among construction workers 12.5% of accidental deaths from falls occurred within one year of terminating employment, yet 7.3% of accidental deaths from falls among non-construction workers occurred within a year of last employment at the facility.

Bus and truck drivers at the INEEL facility showed elevations in the mortality rate from transportation accidents. Although it is unknown if any of these fatalities were work-related, the observation that rate ratios were not low in the early years after entry in the cohort (as was clearly observed among transportation accidents in non-drivers) suggests workplace factors may have been involved. Drivers were anticipated to show elevation in smoking-related diseases, because of the widespread occurrence of smoking on board the buses, one of the few locations in the workplace in which smoking was permitted. No excess was observed of lung cancer, but emphysema was elevated among the group of drivers. One case of asbestosis was observed among drivers, leading to a high rate ratio compared to both the general population and other workers. This driver was not also identified as an asbestos worker but may have been exposed to asbestos through proximity to the brake linings during vehicle maintenance. The observation of elevated rates of acute glomerulonephritis and acute renal failure was unexpected. It is unknown if these elevations were due to a workplace exposure, to chance, or to some other factor.

Recent studies have reported fairly consistent excess risks for lung cancer among truckers (Steenland *et al.* 1998, Hansen 1993), a finding that has been attributed to diesel exhaust exposures. Urban bus drivers have been reported to have elevated risks for death from myocardial infarctions, mental disorders and malignant neoplasms (Michaels and Zoloth 1991). Excess risk among bus drivers and tramway employees in Denmark for cancers of various types has also been reported (Soll-Johanning *et al.* 1998). Extrapolation of these risks to U.S. bus drivers and, particularly to those who operated the buses in a non-urban area like the INEEL, would be uncertain. Likewise, the population of truck drivers in the INEEL cohort was much more likely to have been short-haul operators where the health risks appear to be lower than those for long-haul truckers (Gustavsson et al. 1996).

Painters were a small subcohort within the INEEL facility, and for the most part exhibited mortality patterns similar to the rest of the cohort. Significant excesses of chronic obstructive pulmonary disease and of NHL were observed among painters, but it is not clear if the excess was related specifically to the workplace. Skilled manual workers, a group into which painters were classified, also showed large excess risk of non-malignant respiratory disease compared to professional workers. Previous studies have found painters to experience elevated risks for lung cancer (Steenland and Palu 1999) and other respiratory diseases (Schwartz 1988). Health risks among painters have been identified through epidemiologic studies using trade union records (Steenland and Palu 1999) and in other cohort studies (Brown et al. 2002, Jensen et al. 1987).

This study is the first report of the mortality experience of security workers at a DOE facility. The DOE security workforce is reported to number about 4000 throughout the U.S. and to be increasing in response to greater concern about terrorist threats in recent years (USDOE 2004). Few studies have evaluated mortality risks in security workers in other industries. One study reported that homicide rates were elevated for public and private security workers (Castillo and Jenkins 1994), a finding not observed in the INEEL cohort. Many of these workers were likely to have physical security responsibilities at facilities with public access. Workers at the highly secure DOE facilities were much less likely to have similar contact with the public. Evaluation of risks among public safety, law enforcement and police officers was not considered relevant.

The excess rates of death from non-transportation accidents (related to both falls and firearms) among security workers at INEEL is of interest in considering potential workplace connections; that is, a much higher percentage of these accidental deaths occurred prior to separation from the site than for non-transportation accidents among the rest of the workforce. Prior to 1999 training of the security workforce occurred within each facility. From 1999 until May 2001 training was centralized at a facility in New Mexico and included courses in the use of weapons in hand-to-hand combat and in firearm usage (USDOE 2004). Since that time, individual DOE sites have resumed at least some of these training activities. A recent report of the DOE's Inspector General was critical of the adequacy of the decentralized training among DOE's high-level security workers, including those at INEEL (USDOE 2004).

Although brain cancer mortality rates were about 12% higher in the overall cohort than in the regional comparison population, the mortality rate from all brain tumors combined was not elevated. However, there was evidence of elevation in combined brain tumor rates for WM who worked between two and ten years, compared to those who worked for less than two years. In addition, brain tumor death rates were more than doubled, but with wide CIs, among both chemical workers and security workers. These elevations were due primarily to tumors that were unspecified as to malignancy (in which a near ten-fold excess rate was observed among chemist and security workers). Higher rates in these groups appear to explain the elevated rates observed among the partly skilled group of workers, into which security guards and chemical operators were classified. The decrease in brain tumor rate ratios (compared to unspecified brain malignancies) suggests that some but not all the elevation may have been due to differential diagnosis among these workers compared to other workers. Reasons for the remaining elevation in rate are not clear, but may include chance, or other workplace exposures such as solvents or electromagnetic radiation (these associations are reviewed in Preston-Martin and Mack 1996). Brain cancer rates have also been observed to be elevated among Mormon men in Utah compared to non-Mormons (Lyon *et al.* 1976). This explanation is less satisfactory, however, as migrants to the INEEL region, who were less likely to be LDS members, showed comparable rates of brain cancer mortality compared to workers who were local to the region.

Previous studies of combined chemical workers and laboratory workers have shown mixed results. Some have found elevated risk for lung and bladder cancers (Greenberg *et al.* 2001) and asthma (Jaakola *et al.* 2003, Fishwick *et al.* 1997) among these workers. Numerous other studies of chemical plant workers where the number of chemicals is much smaller have found excess risks among plant workers who have been exposed to benzene, acrylonitrile, benzidine and other compounds. Dement and Cromer (1992) reviewed previous studies of chemists and laboratory workers. The studies in their review suggest higher mortality risks from malignant lymphoma, leukemia and gastrointestinal system cancers, as well as risk for abnormal pregnancy outcomes that would not have been detected in this cohort mortality study.

Overall findings in the study were similar among males and females, and among whites and non-whites. The female and non-white populations were much younger on average than WM, and future follow-up of these groups is expected to be more informative regarding workplace-related hazards among these groups. Based upon the findings of elevated asbestosis and likely mesothelioma mortality rates among some subgroups of WM, risks associated with asbestos exposure in the INEEL workplace merit further exploration. The observation of excess risk of these diseases among those classified as asbestos workers, combined with boiler operators, pipefitters, applicators and insulation workers, was not surprising, although the magnitude of the risk ratio (a 25-fold excess for asbestosis and 6-fold excess for likely mesotheliomas) is perhaps unexpected. For other groups of workers, such as bus drivers and general construction workers, the specific exposure source leading to excess risk is not clear but may include vehicle maintenance and/or building construction or teardown.

The analysis of asbestos-related disease illustrates some limitations of ICD classifications of mesotheliomas, and of the NIOSH LTAS. LTAS minor categories for "other respiratory cancer" and "peritoneum and other digestive" cancers were a poor substitute for the causes of death that were most likely mesotheliomas: pleura and peritoneal cancer. The creation of special rate files that combine peritoneal and pleura cancer (before ICD-10) with pleura, peritoneal cancer and specified mesotheliomas (in ICD-10) identified excess risks for likely mesothelioma among asbestos workers, which were diluted among the life table categories of "other respiratory" cancers, cancers of peritoneum and "other digestive" and "other and unspecified" cancers. In future, analyses that incorporate ICD-10 coded causes of death should consider creating rate files that combine mesotheliomas with cancers of the pleura and peritoneum to evaluate asbestos-related cancers. Also, in this study a high percentage of "other and unspecified" cancers were actually mesothelioma of unspecified type. The possibility of obtaining death certificates for any "other and unspecified" cancers and evaluating which are mesotheliomas should be explored for any cohort in which asbestos exposures are of interest, if internal cohort comparisons can be made.

7.2.2 Interpretation of Life Table Results

The primary purpose of the life table analysis was to estimate mortality rate ratios, including 95% CIs, for different groups of workers. In some analyses, these groups were defined by categories related to specific workplace exposures (e.g., workers classified into cumulative external dose groups, asbestos workers, drivers, security workers), and in others by their status as potential confounders of the relation of primary interest, between cancers and other causes of death and external radiation exposure. These potential confounders included duration of employment categories, badging status, migrant status to the INEEL region, SES and even internal monitoring status, which were all considered for evaluation as confounders of the multivariate modeling analyses described in §2.6.2.3. These results were interpreted according to the approach put forth by Rothman and Greenland (Modern Epidemiology, 1998, Chapter 12) that, according to modern epidemiologic practice, does not limit the interpretation to significance/hypothesis testing. Rothman and Greenland write that "epidemiologic applications need more than a decision as to whether chance alone could have produced an association. More important is estimation of the magnitude of the association, including an assessment of the precision (or its inverse, the variability) of the estimation method" (pg.183). They further state that "results that are not significant may be compatible with substantial effects. Lack of significance alone provides no evidence against such effects" (pg 192).

Following this approach when evaluating life table results, the authors considered the magnitude of the observed SMR or SRR in relation to the width of the CI. Special consideration was given to those causes of death of *a priori* interest when results were discussed. Some of these point estimates are described as elevated or reduced when the 95% CI includes one. Point estimates and their corresponding CIs are provided in the data tables.

7.2.3 Radiation Dose-response Analyses

In the life table analysis, a significant association was observed between external radiation exposure and breast and digestive (particularly pancreatic) cancers for WF, for lags of 0 to 10 years. WM showed excess leukemia and NHL risk within the highest exposure category. However, the importance of several lifestyle-related factors, such as migrant status, SES and the healthy worker effect (using duration of employment as a surrogate measure) necessitated the use of regression modeling to explore dose-response factors in detail.

In the Poisson regression models, negative trends with increasing external radiation dose were observed for many cancers. Before adjustment for several factors (SES, migrant status as a surrogate for LDS membership, and their interaction, and internal exposure monitoring status), a negative ERR estimate was observed for smoking-related cancers (including lung cancer) and for emphysema. The negative estimates were attenuated somewhat for smoking-related cancers by control for these variables. A reduction remained, however, in the highest dose category even after adjusting for these factors. The preferential assignment of nonsmokers (or other selection among the most highly exposed radiation workers for reduced smoking behavior) to work in the high-dose areas may be the most likely explanation for these findings.

Non-smoking-related cancers showed essentially no association with radiation exposure. However, cancers identified in other studies as radiogenic, including cancers of thyroid, breast, colon, ovary, skin, bone and connective tissue, showed an increasing dose-response trend through the 50-100 mSv category, particularly after adjusting for the confounders described above. It is not clear, if this association is real, why an attenuation of risk was seen at doses greater than 100 mSv.

Although brain cancers were not included in the *a priori* list of radiogenic cancers obtained from Boice *et al.* (1996), other studies of radiation-exposed populations have observed elevated brain cancer rates. In a cohort study of Rocky Flats workers with a total of twelve brain tumor deaths, an elevated rate ratio was seen for brain tumors of unspecified nature both among workers exposed to >2 nCi of plutonium and among workers exposed to $\geq$10 mSv external radiation, but no associations were observed for all brain tumors combined (Wilkinson *et al.* 1987). Several studies of medically exposed populations have also shown associations with ionizing radiation at diagnostic (Preston-Martin *et al.* 1989 a, b) or therapeutic (Tsang *et al.* 1993) levels to adults. Others have shown no risk associated with these exposure types (Ryan *et al.* 1992, Schlehofer *et al.* 1992). The findings presented here indicated no association between internal dose category and brain tumor risk. However, for external dose the RR appeared more than doubled above 100 mSv cumulative dose, compared to those receiving less than 1 mSv. This estimate has wide CIs that include no effect (only five cases were observed in this dose category).

Neoplasms of the lymphatic and hematopoietic system, including leukemia, NHL and multiple myeloma, all showed similar patterns with respect to radiation dose, in which RRs appeared elevated only in the highest dose categories. The ERR per unit dose observed for non-CLL leukemia of 5.4% per 10 mSv (CI, -1.1% to 23.8%) is consistent with that observed in a recent analysis of radiation-exposed workers at the Portsmouth Naval Shipyard (Yiin *et al.* 2004) and with other studies of nuclear workers. The estimated ERR at 10 mSv observed in this study is about two- to four-fold higher than estimated in the Japanese Life Span Study at comparable exposure ages and latencies (i.e., 1.3% to 2.6%, as summarized in Schubauer-Berigan and Wenzl 2001). However, it is important to note that average exposures were much lower in the INEEL cohort compared to the Life Span Study cohort. In addition, the lower CI of the ERR at 10 mSv found in this study overlaps the point estimate for the Life Span Study cohort, indicating that this study did not have sufficient statistical power to detect a difference of this magnitude. CLL did not exhibit a positive association with radiation dose in this cohort at any of the dose lags tested (i.e., between 5 and 20 years). A longer lag was employed for CLL than for other leukemias, based on the longer expected latency and pre-clinical phase for this disease (particularly within a mortality study) compared to other leukemias. CLL did show an increasing trend in risk through the 50-100 mSv dose category, but no cases were observed in the category $\geq$100 mSv.

NHL and multiple myeloma both showed elevated risk in the highest dose category. Multiple myeloma has been found to be associated with radiation exposure in other studies of nuclear workers (Cardis *et al.* 1995), but the interpretation of the role of radiation exposure relative to non-radiological workplace exposures is uncertain. There is evidence in each instance that particular subcohorts at INEEL were at higher risk of these cancers (e.g., migrants and painters for NHL, and partly skilled workers for multiple myeloma). NHL has been found in the literature to be associated with elevated social class, with agricultural occupations (particularly herbicide use; e.g., Miligi *et al.* 2003) with workplace exposures to solvents or metalworking materials (reviewed in Scherr and Mueller 1996) and components (e.g., 2,4-D) of Agent Orange used as a defoliant in the U.S.-Vietnam war (Institute of Medicine 2003). Life table analyses also showed the influence of excluding monitored workers with no dose in the baseline category. The trend of increasing risk with dose was strengthened, and confidence intervals in the high-dose group excluded the null value, by removal of such workers from the baseline group.

The strong confounding apparent from internal exposure category observed in analyses for brain tumors, radiogenic solid cancers, leukemia, lymphoma and multiple myeloma is consistent with other studies that have found lower mortality rates in nuclear workers who were monitored and exposed to plutonium and other radionuclides, likely because of strong selection effects for healthy, stable workers into this job category (Wing *et al.* 2004).

7.3 Strengths and Limitations of Study

Strengths

This large cohort study has many strengths:

- Among U.S. nuclear worker cohorts, the INEEL workforce was similar in size to the Hanford cohort. This large size results in large numbers of observed deaths in which to evaluate dose-response associations for ionizing radiation and other exposures. Unlike some other studies, both males and females, whites and non-whites, were included in this study.

- This study had vital status ascertainment on a large cohort through a fifty-year follow-up period concordant with the entire period of operation of the INEEL facility, although the average length of follow-up was much shorter.

- Data from the site provided high-quality exposure monitoring information for external ionizing radiation. Despite the heterogeneity of activities at the site, radiation monitoring was carried out by just two facilities: the site dosimetry group and the NRF. Neutrons were separated from photon exposures, and off-site doses were maintained separately so that they could be included.

- This cohort is more contemporaneous with current workers than many previous studies of U.S. nuclear workers. The relatively late hire cutoff date of December 31, 1991 and the availability of dosimetry data through 1998 render the study findings highly relevant for current workers.

- The heterogeneity of operations at the INEEL facility made it possible to look at large groups of workers who may not be easily studied in other settings, for example, construction or asbestos workers.

- Findings of disease outcomes that are well-known to be associated with certain occupations (e.g., elevations in the asbestosis mortality rate among asbestos workers and the transportation accident mortality rate among drivers) provide a validation of the work history classifications and subcohort assignments used in this study.

Limitations

There are also several important limitations of the present cohort study of INEEL workers:

- Geographic differences in many causes of death, particularly lifestyle-related causes, are clearly of great importance in understanding the mortality experience of the INEEL cohort, as evidenced by the large increases in most SMRs when state rates were used. The selection of an appropriate comparison population was very difficult. The INEEL site was and is still considered an attractive employer in a very rural region of Idaho, with close proximity to Montana and Utah. About half the cohort was considered "local" (i.e., from Idaho, Utah, Montana and Wyoming). It is not known whether a comparable percentage of Idaho, Montana and Wyoming residents would have been considered local by that measure, nor is it known what percentage of the general population in these states was from Utah, a state with the lowest lung cancer mortality rate in the U.S. (CDC 2004).

- Incomplete ascertainment was a potential problem in this cohort, particularly among early workers. The SSA DMF, the primary source of death information, has very few deaths recorded before 1962. However, use of active vital status tracing procedures minimized the impact of this potential underascertainment.

- About 2% of decedents were missing COD information, because their death certificates could not be found. This 2% was not distributed proportionally into the other specific cause-of-death categories for two reasons: first, they were for the most part individuals identified as deceased before 1979 by the SSA, for whom exhaustive searching did not turn up death certificates. It is possible that some of these individuals were not actually deceased. Second, it did not seem appropriate to presume a random distribution of causes of death across the cohort (since, for example, the years of death for those missing death certificates were earlier than the non-missing). A resulting limitation is that the SMRs may have been underestimated by 2% or more. Internal comparisons using SRRs were less likely to have been affected by this limitation.

- The cohort as analyzed included 487 workers who, it was learned after completing all analyses, did not meet the cohort entry criteria. These workers were DOE employees and contractors whose primary worksite was not the INEEL facility, but who worked temporarily at the site. This error is not likely to have been a substantive limitation of the analysis, as all relevant INEEL doses and work history information were included in the life table and multivariate modeling analyses for these workers, and many workers who met the cohort criteria are also believed to have been employed at other DOE facilities.

- The INEEL cohort is still very young, and it is difficult to extrapolate these findings across the entire lifespan of the cohort members. Although follow-up ended over thirty years after the highest exposures occurred in the cohort, it is possible that risks were insufficiently estimated for diseases of very long latency or that tend to occur later in life (such as CLL and prostate cancer).

- The heterogeneity of cohort makes it difficult to generalize findings of this study. For example, the cohort consisted of a large number of professional workers (particularly scientists and engineers) in addition to the high percentage who were construction and service workers. The subcohort analyses conducted, although highly informative, did not contain information on length of employment in each type of occupation, reducing their utility at the cohort level.

- There is a potential for missed radiation exposure in this study, as dose is accrued only at an annual level before 1986. Estimates of the collective missed external dose are, however, low. Internal radiation exposures and neutron exposures could be only crudely estimated for the cohort, because of limitations in monitoring practices and electronic data availability across the entire cohort.

- Job titles and employer names were used to define subcohorts for evaluation of health effects from non-radiologic exposures. This method is crude and leads to likely misclassification of exposure potentials, particularly when employer-based information was used. For example, the subcohort of construction and service workers included employees who completed both initial construction and installation of building components, as well as their maintenance and replacement. Exposure potentials in new construction are likely to differ substantially from those associated with maintenance and replacement.

- Except in the case of radiation dose-response analysis, few *a priori* hypotheses were evaluated among the subcohorts. Therefore, the potential for false positive findings was exacerbated by lack of control for multiple comparisons. This problem was minimized by the emphasis on estimation rather than hypothesis testing.

- This cohort study shows apparent strong confounding by cigarette smoking has occurred among INEEL workers. Although efforts were somewhat successful in controlling for this confounding through the use of surrogate variables such as SES and migrant status to the INEEL region, its effects remain apparent in the radiation dose-response analyses for smoking-related cancers.

- The cohort definition excluded a large number (over 17,000) of military employees, primarily U.S. Navy personnel on short-term training tours in the NRF. Although the average doses were quite small in this group (Table 2-1), there was a relatively large collective dose among these individuals. Furthermore, their later tours on nuclear submarines could have led to further exposure to ionizing radiation. Any future studies of cancer risk among the Navy submariner population should incorporate doses received in training at the NRF.

8 Conclusions and Recommendations

No excess mortality risk was observed for most cancers following exposure to external ionizing radiation at the INEEL facility. There is evidence that exposure to radiation at cumulative levels above 50 mSv may have been associated with increased risk of leukemia, (particularly non-chronic lymphocytic), brain tumors (above 100 mSv), and NHL at the INEEL facility. In addition, the bone cancer mortality rate was elevated among workers with positive internal exposure; this subcohort could be further followed, with improved dose estimates, to further explore this potential association.

There was evidence of strong negative confounding from smoking, possibly associated with lifestyle factors such as LDS Church membership, among radiation-exposed workers in the INEEL cohort. Smoking-related cancers showed a strong negative dose-response with external radiation exposure. Non-smoking-related cancers (particularly those considered radiogenic in other assessments, such as breast, colon, bone and connective tissue) showed a flat or possibly increasing RR with increasing dose.

There appear to have been other work-related risks associated with the INEEL cohort. This study detected a clear excess of asbestos-related mortality (asbestosis, and pleura and peritoneal cancers) among some groups of INEEL workers. The excess was observed among construction and service workers, and in particular, those who could be defined as asbestos workers. This definition included those so identified by job title, or those who worked in jobs found in other studies to have been associated with asbestos exposure, such as pipefitters, insulation workers and boiler operators. These workers were a fairly small group within the INEEL cohort. Ensuring adequate worker protection during operations, as well as surveillance for asbestos-related disease among this population, could reduce the burden of asbestos-related mortality in the INEEL cohort.

Brain tumor rates were elevated in some groups within the INEEL cohort, particularly among chemical and security workers. It is not clear whether the elevation was related to occupational exposures. Further study of these groups of workers, both at INEEL and across the nuclear worker complex, may be warranted.

The INEEL is a geographically dispersed facility, relying heavily on bus and truck transport to move people and goods long distances across the site. The excess mortality rate of transportation accidents among truck and bus drivers suggests further investigation may be merited to determine whether these accidents have occurred in the workplace in recent years (which was not directly assessed in this study).

Excess mortality rates were observed for deaths from falls and "other accidents," including firearm deaths, among security workers at the INEEL facility. Although this finding was based on a small number of cases (in a small population), an association with workplace exposures cannot be ruled out, and non-fatal accident rates should be evaluated among these workers to determine whether a work-related difference may be occurring among security compared to other workers.

Future efforts within the INEEL cohort may be warranted to explore nested case-control studies for brain tumors, NHL, female breast cancer, and leukemia (including chronic lymphocytic), as well as further examination of the construction and service worker subcohort. Evaluation of ways to obtain information on potential confounders such as LDS membership, or tobacco and alcohol use, is highly recommended, based on strong confounding observed by these factors, which could be only partially controlled through the use of surrogates such as SES and migrant status to the INEEL region.

Several activities designed to further the understanding of hazards encountered in DOE facilities were conducted by researchers at NIOSH. The INEEL study will generate information in support of these other activities, including chemical exposure assessments, codifying institutional memory and documentation of historical health physics measurements. Future research directions for NIOSH may include evaluating health hazards associated with clean-up activities at DOE sites.

Literature Cited

Anderson RN, Minino AM, Hoyert DL, Rosenbert HM. Comparability of cause of death between ICD-9 and ICD-10: preliminary estimates. National Vital Statistics Reports 49(2):1-32, 2001.

Arrighi HM, Hertz-Piccioto I. The evolving concept of the healthy worker survivor effect. Epidemiol 5:189-196, 1994.

Atkins EH, Baker EL. Exacerbation of coronary artery disease by occupational carbon monoxide exposure: a report to two fatalities and a review of the literature. Am J Ind Med 7:73-79, 1985.

Atwood CL, Blackwood LG, Harris GA, Loehr CA. Recommended Methods for Statistical Analysis of Data Containing Less-Than-Detectable Measurements. Idaho National Engineering Laboratory EG&G Idaho, U.S. DOE Idaho Operations Office, 1990.

Bennett WP. Correlation of Instrument and Film Radiation Readings – CPP. Phillips Petroleum Co. Atomic Energy Division, Idaho Operations Office U.S. Atomic Energy Commission, November 1957.

Beral V, Inskip H, Fraser P, Booth M, Brown A, Rose G. Mortality of employees of the United Kingdom Atomic Energy Authority, 1946-1979. Brit Med J 291:440-447, 1985.

Beral V, Fraser P, Carpenter L, Booth M, Brown A, Rose G. Mortality of employees of the Atomic Weapons Establishment, 1951-1982. Br Med J 297:757-770, 1988.

Bernstein JA. Overview of diisocyanate occupational asthma. Toxicology 111: 181-189, 1996.

Boardman BF, The Idaho Chemical Processing Plant (ICPP): A Fact Sheet. US Atomic Energy Commission, National Reactor Testing Station, Idaho Falls, ID. July 1957.

Boice Jr JD, Land CE, Preston DL. Ionizing Radiation. Ch 16 in **Cancer Epidemiology and Prevention, 2nd Ed**. Schottenfeld D, Fraumeni Jr JF, eds. New York: Oxford University Press. 1996, 1521 pp.

Brown LM, Moradi T, Gridley G, Plato N, Dosemeci M, Fraumeni Jr JF. Exposures in the painting trades and paint manufacturing industry and risk of cancer among men and women in Sweden. J Occup Environ Med 44:258-264, 2002.

Burdorf A, Danhan M, Swuste P. Occupational characteristics of cases with asbestos-related diseases in the Netherlands. Ann Occup Hyg 47: 485-492, 2003.

Burphy JS, Schubauer-Berigan MK, Utterback DF. Evaluating the relationship between US state of origin and Mormon religious affiliation in the Idaho National Engineering and Environmental Laboratory cohort. Draft technical report. National Institute for Occupational Safety and Health, Health-Related Energy Research Branch. Cincinnati, OH. 2004, 15 pp.

Cardis E, Gilbert ES, Carpenter L, Howe G, Kato I, Armstrong BK, Beral V, Cowper G, Douglas A, Fix J, Fry SA, Kaldor J, Lave C, Salmon L, Smith PC, Voelz GL, Wiggs LD. Effects of low doses and low dose rates of external ionizing radiation: cancer mortality among nuclear industry workers in three countries. Radiat Res 142:117-132, 1995.

Cardis E, Kato I. International collaborative study of cancer risk among radiation workers in the nuclear industry. III—Procedures document. International Agency for Research on Cancer (IARC) Internal Report 93/003. Lyon: IARC. 1993, 46 pp.

Cassinelli II R, Kock KJ, Steenland K, Spaeth S, Laber P. User Documentation PC LTAS. Life Table Analysis System for Use on the PC. US Dept of Health and Human Services, Centers for Disease Control and Prevention, National Institute for Occupational Safety and Health, Division of Surveillance, Hazard Evaluations, and Field Studies. Cincinnati, OH. Ver. November 25, 2002.

Castillo DN, Jenkins EL. Industries and occupations at high risk for work-related homicide. J Occ Med 36:125-132, 1994.

Cederberg GK, Chamberlain HV, Holcomb WF, Nebeker RL, Palica WP, Smith RR, Summers AD, Idaho Chemical Processing Plant Safety Review Document. Allied Chemical Corporation. March 1974.

Cella, DF, Orav EJ, Kornblith JC, Silberfarb PM, Lee KW, Comis RL, Perry M, Cooper R, Maurer LH, Hoth DF, Perloff M, Bloomfield CD, McIntyre OR, Leone L, Lesnick G, Nissen N, Glicksman A, Henderson E, Barcos M, Crichlow R, Faulkner II CS, Eaton W, North W, Schein PS, Chu F, King G, Chahinian AP. Socioeconomic status and cancer survival. J Clin Oncol 9(8):1500-1509, 1991.

Cember H. **Introduction to Health Physics, 2nd Ed**, Pergamon Press, 1983.

Centers for Disease Control and Prevention (CDC). Wide-ranging OnLine Data for Epidemiologic Research (WONDER). Compressed Mortality Data Request Screen for the years 1979 -1998 with ICD 9 codes. http://wonder.cdc.gov/mortICD9J.html (Accessed August 16, 2004).

Cipperley FV. Personnel Metering Operating Techniques Self-Service System, Effective August 20, 1951 through March 29 1958, United States Atomic Energy Commission, Idaho Operations Office, Idaho Falls, Idaho.

Cipperley FV. Personnel Dosimetry Standard Operating Procedure Effective March 1958, NRTS Idaho Operations Office, United States Atomic Energy Commission, 1958.

Cipperley FV, Gammill WP. Improvements in Personnel Monitoring Procedures at NRTS. Health Physics Society Meeting, Gatlinburg, Tenn, June 1959.

Cusimano JP, Cipperley FV. Personnel dosimetry using thermoluminescent dosimeters. Health Phys 14:339-344, 1968.

Dano H, Andersen O, Ewertz M, Petersen JH, Lynge E. Socioeconomic status and breast cancer in Denmark. Int J Epidemiol 32:218-224, 2003

Dement JM, Cromer JR. Cancer and reproductive risks among chemists and laboratory workers: A review. Appl Occup Environ Hyg 7: 120-126, 1992.

Dieter A, Knecth M, Staiger D, Christian C, Bendixson CL, Hogg GW, Berreth JR. Historical fuel reprocessing and HLW Management in Idaho. Radwaste Magazine, May 1997.

Drent M, Bomans PH, Van Suylen RJ, Lamers RJ, Bast A, Wouters EF. Association of man-made mineral fibre exposure and sarcoidlike granulomas. Respir Med 94: 815-820, 2000.

Fishwick D, Pearce N, D'Souza W, Lewis S, Town I, Armstrong R, Kogevinas M, Crane J. Occupational asthma in New Zealanders: a population based study. Occup Environ Med 54: 301-306, 1997.

Frome EL, Cragle DL, Watkins JP, Wing S, Shy CM, Tankersley WG, West CM. A mortality study of employees of the nuclear industry in Oak Ridge Tennessee. Radiat Res 148:64-80, 1997.

Gilbert ES, Buchanan JA, Holter NA. Description of the process used to create 1992 Hanford mortality study database. PNL-8449, UC-605, Prepared for the U.S. Department of Energy under Contract DE-AC06-76RLO 1830. Hanford, WA, 1992.

Goldberg RL, Hicks AM, O'Leary LM, London S. Lead exposure at uncovered outdoor firing ranges. J Occup Med 33: 718-719, 1991.

Gordon NH. Socioeconomic factors and breast cancer in black and white Americans. Canc Metast Rev 22:55-65, 2003.

Greenberg RS, Mandel JS, Pastides H, Britton NL, Rudenko L, Starr TB. A meta-analysis of cohort studies describing mortality and cancer incidence among chemical workers in the United States and western Europe. Epidemiol 12: 727 -740, 2001.

Gustavsson P, Alfredson L, Brunnberg H, Hammar N, Jakobsson R, Reuterwall C, Ostlin P. Myocardial infarction among male bus, taxi and lorry drivers in middle Sweden, Occup Environ Med, 53:235-40, 1996.

Hancock SL, Tucker MA, Hoppe RT. Factors affecting late mortality from heart disease after treatment of Hodgkin's disease. J Am Med Assoc 270:1949-1955, 1993.

Hansen ES. A follow-up study on the mortality of truck drivers. Am J Indust Med 23:811-821, 1993.

Hauptmann M, Mohan AK, Doody MM, Linet MS, Mabuchi K. Mortality from diseases of the circulatory system in radiologic technologists in the United States. Am J Epidemiol 2003; 157:239-248.

Horan J, Braun JB. IDO-INEL Occupational Radiation Exposure History, 1993.

IARC (International Agency for Research on Cancer). Monographs on the Evaluation of Carcinogenic Risks for Humans. Vol. 47, 1989.

IARC. Monographs on the Evaluation of Carcinogenic Risks for Humans. Tobacco Smoke and Involuntary Smoking, Vol. 83, 2004, 1452 pp.

Idaho Chemical Processing Plant (ICPP) Internal Dose Assessment Manual Radiation Technology, Appendix I, March 1981

IMBA EXPERT, USDOE-EDITION (PHASE I) U.S. Department of Energy, Office of Worker Protection Policy and Programs, 2001.

Institute of Medicine of the National Academies. Veterans and Agent Orange Update 2002. **Committee to review the health effects in Vietnam veterans of exposure to herbicides (Fourth biennial update)**. Washington, DC: National Academies Press. 2003, 619 pp.

International Commission on Radiological Protection (ICRP). Recommendations of the International Commission on Radiological Protection, ICRP Publication 60, Annals of the ICRP, Oxford: Pergamon Press, 1990.

International Commission on Radiological Protection (ICRP). Recommendations of the International Commission on Radiological Protection, ICRP Publication 66, Annals of the ICRP, Oxford: Pergamon Press, 1994.

International Commission on Radiological Protection (ICRP). Recommendations of the International Commission on Radiological Protection, ICRP Publication 68, Annals of the ICRP, Oxford: Pergamon Press, 1995.

Jaakkola JJ, Pilpari R, Jaakkola MS. Occupation and asthma: a population-based incident case-control study. Am J Epidemiol 158: 981-987, 2003.

Jarvholm B, Sanden A. Lung cancer and mesothelioma in the pleura and peritoneum among Swedish insulation workers. Occup Environ Med 55: 766-770, 1998.

Jensen OM, Wahrendorf J, Knudsen JB, Sorenson BL. The Copenhagen case-referent study on bladder cancer. Risks among drivers, painters and certain other occupations. Scand J Work Environ Health 13:129-134, 1987.

Johnson JE, Bradford DJ. Decontamination of the Waste Calcining Facility – A Historical Review. Idaho Chemical Programs. December 1978.

Lantz PM, House JS, Lepkowski JM, Williams DR, Mero RP, Chen J. Socioeconomic factors, health behaviors, and mortality: results from a nationally representative prospective study of US adults. J Am Med Assoc 279:1703-8, 1998.

Legler BM. Preliminary RALA Hot Run Status Report, Phillips Petroleum Co, March 1957.

Lockey JE, LeMasters GK, Levin L, Rice C, Yiin J, Reutman S, Papes D. A longitudinal study of chest radiographic changes of workers in the refractory ceramic fiber industry. Chest 121: 2044-2051, 2002.

Lockey JE, Levin LS, LeMasters GK, McKay RT, Rice CH, Hansen KR, Papes DM, Simpson S, Medvedovic M. Longitudinal estimates of pulmonary function in refractory ceramic fiber manufacturing workers. Am J Respir Crit Care Med 157: 1226-1233, 1998.

Lyon JL, Klauber MR, Gardner JW, Smart CR. Cancer incidence in Mormons and non-Mormons in Utah, 1966-1970. New Engl J Med 294:129-133, 1976.

Lyon JL, Wetzler HP, Gardner JW, Klauber MR, Williams RR. Cardiovascular mortality in Mormons and non-Mormons in Utah, 1969-1971. Am J Epidemiol 108:357-366, 1978.

Marchand LL, Wilkens LR, Kolonel LN, Hankin JH, Lyu L-C. Associations of sedentary lifestyle, obesity, smoking, alcohol use, and diabetes with the risk of colorectal cancer. Cancer Res 57:4787-94, 1997.

McGinnis JM, Foege WH. Actual causes of death in the United States. J Am Med Assoc 270:2207-2212, 1992.

Menegozzo M, Belli S, Borriero S, Bruno C, Carboni M, Grignoli M, Menegozzo S, Olivieri N, Comba P. Mortality study of a cohort of insulation workers. Epidemiol Prev 26: 71-75, 2002.

Methner M. Identification of potential hazards associated with new residential construction. Appl Occup Environ Hyg 15: 189-192, 2000.

Michaels D, Zoloth SR. Mortality among urban bus drivers. Int J Epidemiol 20:399-404, 1991.

Miligi L, Costantini AS, Bolejack V, Veraldi A, Genvenuti A, Nanni O, Ramazzotti V, Tumino R, Stagnaro E, Rodella S, Fontana A, Vindigni C, Vineis P. Non-Hodgkin's lymphoma, leukemia, and exposures in agriculture: results from the Italian multicenter case-control study. Am J Indust Med 44:627-636, 2003.

Moolgavkar SH, Venzon DJ. A method for computing profile likelihood-based confidence bounds. Ann Statist 15:346-359, 1988.

National Council on Radiation Protection and Measurements (NCRP). Use of Personal Monitors to Estimate Effective Dose Equivalent and Effective Dose to Workers for External Exposure to Low-LET Radiation. NCRP Report No. 122, National Council on Radiation Protection and Measurements, 1995.

National Research Council (NRC). **Health Effects of Exposure to Low Levels of Ionizing Radiation. BEIR V.** Committee on the Biological Effects of Ionizing Radiation. National Academy Press. 1990, 421 pp.

Nelson DE, Kirkendall RS, Lawton RL, Chrismon JH, Merritt RK, Arday DA, Giovino GA. Surveillance for smoking-attributable mortality and years of potential life lost, by state—United States, 1990. Mortal Morbid Weekly Report Surveillance Summaries 43(SS-1):1-8, 1994.

Paszat LF, Mackillop WJ, Groome PA, Boyd C, Schulze K, Holowaty E. Mortality from myocardial infarction after adjuvant radiotherapy for breast cancer in the Surveillance, Epidemiology, and End-Results cancer registries. J Clin Oncol 16:2625-2631, 1998.

Perry OR. Idaho National Engineering and Environmental Laboratory (INEEL) Dosimetry Systems History and Minimum Dose Reporting Level (MRL) Information Summary 1951 to March 2002, letter to NIOSH, 2002.

Poulton R, Caspi A, Milne BJ, Thomson WM, Taylor A, Sears MR, Moffitt TE. Association between children's experience of socioeconomic disadvantage and adult health: a life-course study. Lancet 360:1640-1645, 2002.

Pratt S. Work-related roadway crashes—United States, 1992-2002. Mortal Morbid Weekly Report 53(12):260-264, 2004.

Preston DL, Lubin JH, Pierce DA, McConney ME. **Epicure User's Guide**. Seattle: Hirosoft International Corporation. 1993, 330 pp.

Preston-Martin S, Mack WJ. Neoplasms of the nervous system. Chapter 58 in **Cancer Epidemiology and Prevention, 2nd Ed**. Schottenfeld D, Fraumeni Jr JF, eds. New York: Oxford University Press. 1996, 1521 pp.

Preston-Martin S, Mack W, Henderson BE. Risk factors for gliomas and meningiomas in males in Los Angeles County. Cancer Res 49:6137-6143, 1989a.

Preston-Martin S, Thomas DC, Wright WE, Henderson BE. Noise trauma in the aetiology of acoustic neuromas in men in Los Angeles County, 1978-1985. Br J Canc 59:783-786, 1989b.

Puphal K. Work Assessment History of INEEL Bioassay. WESTAT Report, CDC, NIOSH, May 1994.

Puphal K. INEEL External Dosimetry History. WESTAT Report, CDC, NIOSH, June 1996.

Rich BL, Slagle WL, Willis CL, Christiansen HD, Nichols CE, Jenson DK, Wenzel DR, Ison CW, Black DE. Preliminary ICPP Health Physics Upgrade Program, Allied Chemical Corp, October 1974.

Robinson CF, Halperin WE, Alterman T, Braddee RW, Burnett CA, Fosbroke DE, Kisner SA, Lalich NR, Roscoe RJ, Seligman PJ, Sestito JP, Stern FB, Stout NA. Mortality patterns among construction workers in the United States. Occup Med State of the Art Rev 10: 269-283, 1995.

Rothman KJ, Greenland S. Approaches to statistical analysis. Chapter 12 in **Modern Epidemiology, 2nd Ed**. Rothman KJ and Greenland S, eds. Philadelphia: Lippincott-Raven Publishers. 1998, 737 pp.

Ryan P, Lee MW, North B, McMichael AJ. Amalgam fillings, diagnostic dental x-rays and tumors of the brain and meninges. Eur J Canc 28B:91-95, 1992.

SAS Institute Inc., *SAS/STAT® User's Guide, Version 8*, Cary, NC: SAS Institute Inc., 1999. 3884 pp.

Scherr PA, Mueller NE. Non-Hodgkin's lymphomas. Chapter 42 in **Cancer Epidemiology and Prevention, 2nd Ed**. Schottenfeld D, Fraumeni Jr JF, eds. New York: Oxford University Press. 1996, 1521 pp.

Schlehofer B, Blettner M, Becker N, Martinsohn C, Wahrendorf J. Medical risk factors and the development of brain tumors. Cancer 69:2541-2547, 1992.

Schubauer-Berigan MK, Wenzl TB. Leukemia mortality among radiation-exposed workers. Occupat Med State of the Art Rev 16:271-287, 2001.

Schwartz DA, Baker EL. Respiratory Illness in construction industry. Airflow obstruction in painters. Chest 93: 134-137, 1988.

Seniori Costantini A, Miligi L, Kriebel D, Ramazzotti V, Rodella S, Scarpi E, Stagnaro E, Tumino R, Fontana A, Masala G, Vigano C, Vindigni C, Crosigani P, Benvenuti A, Vineis P. A multicenter case control study in Italy on hematolymphopoietic neoplasms and occupation. Epidemiol 12:78-87, 2001.

Shimizu Y, Pierce DA, Preston DL, Mabuchi K. Studies of the mortality of atomic bomb survivors. Report 12, part II. Noncancer mortality: 1950-1990. Radiat Res 152:374-389, 1999.

Silver SR, Daniels RD, Taulbee TD, Zaebst DD, Kinnes GM, Couch JR, Kubale TL, Yiin JH, Schubauer-Berigan MK, Chen P-H. Differences in mortality by radiation monitoring status in an expanded cohort of Portsmouth Naval Shipyard workers. J Occ Environ Med. 46:677-690, 2004.

Smith PG, Douglas AJ. Mortality of workers at the Sellafield plant of British Nuclear Fuels. Br Med J 293:845-854, 1986.

Soll-Johanning H, Bach E, Olsen JH, Tuchsen F. Cancer incidence in urban bus drivers and tramway employees: a retrospective cohort study. Occ Environ Med 55:594-598, 1998.

Steenland K, Beaumont J, Spaeth S, Brown D, Okun A, Jurcenko L, Ryan B, Phillips S, Roscoe R, Stayner L, Morris J. New developments in the Life Table Analysis System of the National Institute for Occupational Safety and Health. J Occup Med 32:1091-1098, 1990.

Steenland K, Deddens J, Stayner L. Diesel exhaust and lung cancer in the trucking industry: exposure-response analyses and risk assessment. Am J Indust Med 34:220-228, 1998.

Steenland K, Henley J, Thun M. All-cause and cause-specific death rates by educational status for two million people in two American Cancer Society cohorts, 1959-1996. Am J Epidemiol 156:11-21, 2002.

Steenland K, Palu S. Cohort mortality study of 57,000 painters and other union members: a 15 year update. Occup Environ Med 56: 315 -321, 1999.

Svensson BG, Schultz A, Nilsson A, Skerving S. Lead exposure in indoor firing ranges. Int Arch Occup Envir Health 64: 219-221, 1992.

Tarlo SM, Liss GM. Diisocyanate-induced asthma: diagnosis, prognosis, and effects of medical surveillance measures. Appl Occup Environ Hyg 17: 902-908, 2002.

Tsang RW, Laperriere NJ, Simpson WJ, Brierley J, Panzarella T, Smyth HS. Glioma arising after radiation therapy for pituitary adenoma: a report of four patients and estimation of risk. Cancer 72:2227-2233, 1993.

Ulvestad B, Kjaerheim K, Martinsen JI, Mowe G, Andersen A. Cancer incidence among members of the Norwegian trade union of insulation workers. J Occup Environ Med 46: 84-89, 2004.

US Department of Energy (USDOE). Department of Energy Standard for the Performance Testing of Personnel Dosimetry Systems, DOE Laboratory Accreditation Program for Personnel Dosimetry Systems, U.S. DOE Assistant Secretary for Environment, Safety and Health, December 1986.

US Department of Energy (USDOE) Office of Inspector General. Audit Report: The Department's Basic Protective Force Training Program. Washington: USDOE Office of Inspector General Office of Audit Services. DOE/IG-0641, March 2004.

Valway SE, Martyny JW, Miller JR, Cook M, Mangione EJ. Lead absorption in indoor firing range users. Am J Public Health 79: 1029-1032, 1989.

Varma DK, Kurtz LA, Sahai D, Finkelstein MM. Current chemical exposures among Ontario construction workers. Appl Occup Environ Hyg 18: 1031-1047, 2003.

Wang E, Dement JM, Lipscomb H. Mortality among North Carolina construction workers. Appl Occup Environ Hyg 14: 45-58, 1999.

Weeks JL. Controlling occupational health hazards in construction. Occup Med State of the Art Rev 10: 407-420, 1995.

Wilkinson GS, Tietjen GL, Wiggs LD, Galke WA, Acquavella JF, Reyes M, Voelz GL, Waxweiler RJ. Mortality among plutonium and other radiation workers at a plutonium weapons facility. Am J Epidemiol 125:231-250, 1987.

Wilkinson GS, Trieff N, Graham R *et al*. Study of Mortality among Female Nuclear Workers. Final report, NIOSH, CDC, 125 pp, 2000.

Wing S, Richardson D, Wolf S, Mihlan G. Plutonium-related work and cause-specific mortality at the United States Department of Energy Hanford site. Am J Indust Med 45:153-164, 2004.

Yeung P, Rogers A. An occupation-industry matrix analysis of mesothelioma cases in Australia 1980-1985. Appl Occup Environ Hyg, 16: 40-44, 2001.

Yiin JH, Schubauer-Berigan MK, Silver SR, Daniels RD, Kinnes GM, Zaebst DD, Couch JR, Kubale TL, Chen P-H. Risk of lung cancer and leukemia from exposure to ionizing radiation and potential confounders among workers at the Portsmouth Naval Shipyard. *(submitted to Radiation Research)* 2004.

Yost K, Perkins C, Cohen R, Morris C, Wright W. Socioeconomic status and breast cancer incidence in California for different race/ethnic groups. Cancer Causes Control 12:703-711, 2001.

Appendix to INEEL final report

A.1 Description of INEEL Facilities A-2
   A.1.1 Central Facilities Area (CFA). A-2
   A.1.2 Power Burst Facility (PBF) A-2
   A.1.3 Test Reactor Area (TRA) A-3
   A.1.4 Test Area North (TAN) A-4
   A.1.5 Idaho Chemical Processing Plant (ICPP) A-5
   A.1.6 Radioactive Waste Management Complex (RWMC) A-7
   A.1.7 Argonne National Laboratory-West (ANL-W) A-7
   A.1.8 Naval Reactors Facility (NRF) A-8
   A.1.9 Auxiliary Reactor Area (ARA). A-8
   A.1.10 Boiling Water Reactor Experiment (BORAX). A-9
   A.1.11 Experimental Breeder Reactor - I (EBR-1). A-9
   A.1.12 Security Training Area (STF). A-9
   A.1.13 Idaho Falls. A-10
A.2 Assignment of Work History Dates for INEEL Cohort Mortality Study A-11
   A.2.1 Assigning hire date A-11
   A.2.2 Assigning termination date A-12
A.3 Imputation of Missing Hire Dates for INEEL Workers A-18
   A.3.1 Background A-18
   A.3.2 Database A-18
   A.3.3 Methods A-18
   A.3.4 Results A-19
A.4 Comparison of INEEL SES categories with those used for the Hanford cohort A-20
A.5 Translating ICD-10 causes of death to ICD-9 for death occurring in 1999 .... A-31
A.6 Additional Life Table Analysis results A-36
A.7 Additional radiological exposure assessment information A-41

A.1 Appendix: Description of INEEL Facilities

A.1.1 Central Facilities Area (CFA).

The CFA is the oldest site at the INEEL complex and was part of the World War II gunnery range used by the United States Navy. Numerous administrative units have been located at the CFA along with environmental and dosimetry laboratories, security and fire operations, medical facilities, communications, warehouses, cafeteria, vehicle and equipment pools including the large bus fleet, and laundry facilities.

The Central Records Facility, a warehouse containing up to 18,000 boxes of records, is located at the CFA. This facility temporarily stores records for the INEEL and disposes of them according to the DOE records retention schedule. All INEEL contractors send inactive records such as personnel, medical, and industrial hygiene to the Central Record Site for temporary retention. Many records with long retention periods are shipped to a Federal Records Center in Seattle, Washington. Records of potential value in epidemiologic studies are covered by a DOE moratorium on destruction of these records and may be at either of these locations. The historical information in these records has been used for various aspects of this study.

A.1.2 Power Burst Facility (PBF).

The PBF was originally designed for testing transient behaviors of nuclear fuels and performing other safety studies of light-water moderated enriched fuel reactor systems. To accomplish this mission, four experimental reactors, known as Special Power Excursion Reactor Tests (SPERT) I-IV, were constructed during the 1950's and early 1960's. The last PBF reactor, SPERT IV, was shut down in 1970. The PBF was divided into five areas: (1) the PBF Control Area, (2) the PBF Reactor Area; (3) Waste Engineering Development Facility, (4) Waste Experimental Reduction Facility, and (5) Mixed Waste Storage Facility.

The PBF Control Area, formerly the Control Area for the SPERT operation, is the facility's administrative center. It houses the PBF reactor controls and instrumentation, administrative offices, instrument and mechanical work areas, raw water supply equipment, and the data acquisition and reduction systems. SPERT-I was constructed below ground level and began testing in 1956. The reactor was decommissioned in 1964, and the pit demolished in 1985.

SPERT-II started operation in 1960, becoming chronologically the third functioning reactor at the PBF. The reactor was designed to study the influence of prompt neutron lifetime on reactor transient behavior by using various moderators or reflectors. Four years later, in 1964, SPERT-II was placed on standby status, and subsequently decommissioned in 1980, when many of its components were removed. The facility was modified in 1986 to provide an area for investigating radioactive and mixed waste treatment technologies and processes which continued until 1990. Subsequently, the PBF was used as a clean lead storage area and as a mechanical craft staging area.

SPERT-III, the second functioning reactor at PBF was designed to study behavior in high-power, high-temperature, heterogeneous light-water reactors. The reactor became operational in 1958, was placed on standby status in 1968, and was decontaminated in 1980. In 1982, the structures were renamed the Waste Experimental Reduction Facility for research on the volumetric reduction of low-level radioactive and mixed waste prior to shipment to a disposal site. Reduction was achieved by incineration, sizing and compacting.

The SPERT-IV reactor was a large pool-type facility, built to extend the range and type of controlled test parameters, and to provide a facility for the kinetic testing of reactor cores of advanced design. Achieving criticality in 1961, the reactor was eventually placed on standby status in 1970. In 1986, the name was changed to the Mixed Waste Storage Facility. These structures served as a Resource Conservation and Recovery Act (RCRA) Storage Facility housing mixed waste, radioactive polychlorinated biphenyl waste, corrosives, and flammables.

The present PBF reactor was built in 1970, north of the SPERT-I reactor and the area was renamed the PBF Reactor Area in 1989. As of 1993, the PBF Reactor was in shutdown status pending decontamination and decommission.

A.1.3 Test Reactor Area (TRA).

Three test reactors have been operated at TRA: Material Test Reactor (MTR); Engineering Test Reactor (ETR), and Advanced Test Reactor (ATR). The area contained 80 buildings and 65 other structures that provided space for the reactors, analytical chemistry and radiation laboratories, office, training, and mechanical craft support services.

The Materials Test Reactor (MTR) was built in 1952, to test structural changes of nuclear fuels and other material samples from irradiation and to provide neutron beam sources for basic physics research. The MTR was permanently shut down in 1970, and many of its buildings have been adapted to support other INEEL operations, including offices, storage, and test areas for the ICPP remote handling equipment.

The Engineering Test Reactor was a gas-cooled reactor system constructed in 1957. Both the reactor and its associated support facilities have been inactive since January 1982.

The major program located at the TRA is the Advanced Test Reactor. Completed in 1965, and started in 1967, the Advanced Test Reactor began operation at full power in 1969. It was originally constructed to continue the irradiation programs being performed by the Engineering Test Reactor, but today it is used for irradiation service for the DOE, Department of Defense, and safety-related programs.

The Advanced Reactivity Measurement Facility (ARMF) consists of two 100 Kw, water-cooled reactors, the ARMF-1 and the Coupled Fast Reactivity Measurement Facility, which share a common pool. The ARMF-1 is a low-power reactor capable of highly precise measurement of reactivity effects produced by small samples. The Coupled Fast Reactivity Measurement Facility is a zoned core reactor with a filtered center measurement position surrounded by a thermal neutron driver core. It additionally houses two other operations, the Pneumatic Facility (also known as the rabbit facility) and the Neutron Radiography Facility.

Additional facilities at TRA include: (1) the Advanced Test Reactor Critical, a low power, full-scale nuclear duplicate of the Advanced Test Reactor utilized to measure the nuclear characteristics of cores that are irradiated in the ATR; (2) the a state-of-the-art Radiation Measurement Laboratory; (3) the Chemistry and Physics Laboratories which complete basic research on nuclear structure and radionuclide metrology; (4) the Radiation Instrumentation Laboratories, a support group for the INEEL, DOE, and Nuclear Regulatory Commission (NRC) in the areas of advanced radiation instrumentation and analysis; (5) the TRA Hot Cells Facility, equipped for the remote handling and metallurgical analysis of radioactive materials; (6) the Test Train Facility which contains nonirradiated fuel storage; (7) the Materials Test Reactor canal used to store irradiated fuel rods, fueled or non-fueled test trains, refurbish irradiated test train hardware, and load fuel; (8) the Warm Waste Treatment Facilities consisting of several facilities designed to reduce radioactive releases to acceptable levels; (9) an office area, and a small machine shop; and (10) additional chemistry, physics, instrumentation, electronic, and computer laboratories which conduct a variety of work for the DOE, NRC, and INEEL.

A.1.4 Test Area North (TAN).

Situated approximately twenty-seven miles from the Central Facilities Area in the northern part of INEEL, TAN was established in the early 1950's to support the Aircraft Nuclear Propulsion program of the U.S. Air Force and the Atomic Energy Commission. Subsequently, four areas at the TAN operated with separate missions: (1) the Initial Engine Test; (2) the Containment Test Facility, formerly known as the Loss-of-Fluid-Test, (3) the Technical Support Facility, and (4) the Water Reactor Research Test Facility.

The Initial Engine Test area was created in 1951, as part of the Aircraft Nuclear Propulsion program to develop and test nuclear-powered jet engines. Three of these engines were built and tested between 1951 and 1961, when the program was terminated.

From 1961 through 1967, the Initial Engine Test area was part of the Space Nuclear Auxiliary Power Transient program. This program evaluated beryllium-reflected reactor performance under: (1) atmospheric conditions; (2) nuclear excursions resulting from the immersion of the reactor in either water or wet earth; and from both (3) non-destructive and (4) destructive reactor tests.

Between 1967 and 1977, the Initial Engine Test area was inactive. In 1977, it became part of a decontamination and decommissioning project for the Hallam Nuclear Power Facility in Lincoln, Nebraska. Reactor components were dismantled and shipped to the INEEL to remove sodium contamination. Since 1978, the Initial Engine Test area has been inactive and much of its equipment has been relocated for use in other areas. Radioactive decontamination and decommissioning and asbestos removal had been scheduled for the facility as of 1993.

Under its original name as the Loss-of-Fluid Test reactor, the Containment Test Facility was established to perform loss-of-coolant experiments under simulated accident conditions. During the experiments, primary attention was directed to an operator's responsibilities, needs and performances, particularly toward recognizing and responding correctly to light water reactor emergency conditions. The Loss-of-Fluid Test reactor has been decommissioned and dismantled. Subsequently, this facility was used for non-radioactive waste storage for the Specific Manufacturing Capabilities (SMC) project. SMC is a "work-for-others" project that began around 1986 and manufactured armor plate from depleted uranium for the U.S. Army. SMC operations occurred in a converted airplane hangar and a similar size building that was more recently constructed.

The Test Support Facility was established as a unique facility to support energy research and defense programs and to maintain specialized facilities for technical engineering and remote handling of radioactive materials. There were six zones at the facility: (1) administrative and support for the Test Support Facility and the rest of TAN; (2) storage, with a liquid waste transfer facility and contaminated-storage building; (3) research and development, with a waste processing facility and a multi-use area for manufacturing and assembly; (4) service and maintenance; (5) warehouse and storage, with storage operations for the Specific Manufacturing Capabilities at the Containment Test Facility; and (6) sanitary waste processing. An earthen berm divides the Test Support Facility along a north-south axis, segregating all radioactive waste handling programs on the west side of the berm.

As of 1994, there were three programs at the Test Support Facility. The Process Experimental Pilot Plant was established to process contact-handled transuranic waste into a permanently disposable form. It developed alternative waste forms, including low-level radioactive wastes, hazardous wastes, and classified materials. The Three Mile Island (TMI) Unit-2 Core Examination project analyzed samples to determine the accident sequence at TMI and to predict nuclear fuel behavior during degraded core cooling situations. The Spent Fuel Programs included methods to transport spent civilian fuel to a Federal facility, testing concrete casks used for dry storage of spent fuel, and monitoring casks that will be used for long-term dry storage.

The Water Reactor Research Test Facility was constructed as an experimental beryllium oxide reactor but was never loaded with nuclear fuel. However, other smaller scale research programs have been located there. These include a testing program, using the quarter scale Separate Effects Test facility, the sodium potassium deactivation project, and the testing of explosives detection systems.

A.1.5 Idaho Chemical Processing Plant (ICPP).

Since it began operations in 1953, the ICPP has been the principal facility for the receipt, interim storage and reprocessing of spent nuclear fuels. The ICPP also manages high-level radioactive solid waste from other DOE facilities and all wastes generated at the ICPP. In 1992, DOE announced that the ICPP would no longer process spent fuel although its interim storage function for radioactive wastes would remain. The facility was also directed to develop remediation technologies for radioactive environmental contamination.

The spent nuclear fuels which are stored at the ICPP have come from the Naval propulsion program and other INEEL research reactors including EBR-II. The fuel was transported to the ICPP by truck where it was placed in storage in either dry containers or in water filled basins. Specific storage facilities at the ICPP include the Graphite Fuel Storage Facility, which received fuel from the ROVER nuclear rocket program, and the Fluorinel Dissolution Process and Fuel Storage Facility.

The fuel re-processing facilities have been an integral component of ICPP operations since 1952. By the mid-1980's, the facility had multiple capabilities for dissolving several types of nuclear fuels with various cladding including aluminum, zirconium, stainless steel, graphite and ceramic. In general, the process involved the solvent extraction of enriched uranium from spent fuels which had been dissolved in inorganic acids. The uranium was then solidified and transported to Oak Ridge National Laboratory. By the end of 1991, the ICPP had processed 760 metric tons of spent fuel and recovered 32 metric tons of enriched uranium.

Prior to 1963, liquid wastes generated at the INEEL were concentrated by evaporation and stored as liquids in eleven large underground stainless steel tanks. The Waste Calcining Facility was constructed in 1963, to convert the high-level radioactive liquid waste to more stable, solid, calcined granules, a process developed at the ICPP. The Waste Calcining Facility was the first fluidized bed for conversion of high-level radioactive liquid waste to a solid calcine that resulted in an eight-fold reduction in volume. The Waste Calcining Facility was shut down in 1981, and subsequently replaced by the New Waste Calcining Facility in 1983. The new process involved spraying liquid waste onto coarse granules in an oven. After evaporation of volatile components, the calcine was transported through ducts by jetted air to storage bins. Between 1963 and 1993, approximately 125,000 cubic feet of solidified high-level radioactive waste was placed in large stainless steel bins which have an expected life span of 500 years. The calcine was scheduled for conversion to glass or glass-ceramic logs followed by permanent storage at the Waste Isolation Pilot Plant, a federal repository in New Mexico.

The ICPP also contains the Rare Gas Plant which recovered krypton-85 from the processing of spent nuclear fuel. The krypton was shipped to the Oak Ridge National Laboratory where it was sold for commercial purposes.

The ICPP has been managed by six contractors, including the current contractor, Westinghouse Idaho Nuclear Co., which assumed its responsibilities in 1984. Among the previous contractors was the Atomic Energy Division, Phillips Petroleum Co., the only company to function simultaneously as both an ICPP contractor and the site contractor.

A.1.6 Radioactive Waste Management Complex (RWMC).

The RWMC mission has three primary components: (1) to provide waste management of transuranic contaminated, solid, and low-level radioactive wastes; (2) to retrieve, examine and certify stored transuranic waste for subsequent shipment to the DOE Waste Isolation Pilot Plant in New Mexico; and (3) to provide research and development, including demonstration projects in waste management. These responsibilities apply not only to wastes generated at the INEEL, but also to the low-level transuranic wastes received from other DOE facilities.

The RWMC was organized into three basic zones: an administrative and support area, the Subsurface Disposal Area and the Transuranic Storage Area. The Subsurface Disposal Area occupied 97 acres. It employed shallow, sub-surface disposal methods for the temporary disposal of low-level beta and gamma emitting wastes. The Transuranic Storage Area was a 57 acre section, used to store transuranic wastes.

The RWMC began operation in 1952, as a 13-acre disposal site for the burial of solid radioactive waste in trenches. At that time, burial use was restricted to beta and gamma emitting radionuclides. In 1954, the RWMC also began accepting transuranic wastes from Rocky Flats. In 1957, the site was enlarged to 88 acres and from 1960 to 1963, the RWMC accepted beta-gamma waste from private sources.

The Transuranic Storage Area was established in 1970 for interim storage of transuranic wastes and was enlarged in 1986. Temporary transuranic element storage was required by AEC policy prior to more permanent storage in a federal repository, at the then-to-be-determined Waste Isolation Pilot Plant. During the last half of the 1970's, research projects were begun to evaluate transuranic waste retrieval methods at both the Subsurface Disposal Area and the Transuranic Storage Area.

During the 1960's, environmental monitoring efforts at RWMC were expanded to include soil and radiation measurements taken around the site perimeter. In 1973, detailed monitoring of the surrounding environment commenced at the RWMC with sampling of soil, surface water and groundwater, and subsequently, both plant and animal life.

A.1.7 Argonne National Laboratory-West (ANL-W).

Construction of the Experimental Breeder Reactor-I began in 1949. ANL-W, the operator of this reactor, was the first contractor at the INEEL. Subsequently, ANL-W operations and facilities were located in the eastern section of the INEEL. These facilities have included the Experimental Breeder Reactor II, Zero Power Physics Reactor which tested reactor fuel configurations, Transient Reactor Test Facility which tested nuclear fuels under simulated reactor accident conditions and the Fuel Manufacturing Facility where uranium-zirconium fuel elements were manufactured for the breeder reactors. DOE oversight for ANL-W is maintained by the Chicago Area Office as part of the Argonne National Laboratory located near Chicago.

The focus of ANL-W operations around 1990 was the Integral Fast Reactor Program, which had a mission to develop technology for on-site re-processing of breeder reactor fuel elements. The Integral Fast Reactor program included: (1) operation of Experimental Breeder Reactor-II, which has been producing electricity since 1961; (2) the Fuel Cycle Facility, which was refurbished to re-process spent breeder reactor fuels into new fuel elements; and (3) the Hot Fuel Examination Facility which performed analysis and re-processing of fuels from other DOE reactors.

A.1.8 Naval Reactors Facility (NRF).

Located near the geographical center of the INEEL, the NRF has been in existence since the laboratory's earliest days. The facility completed research and development for naval propulsion nuclear reactors and served as a training site for naval nuclear reactor operators. The first naval reactor, the Nautilus prototype, became operational in 1953. Three other nuclear power plants for U.S. Navy surface ships and submarines were constructed at the NRF and used to train thousands of sailors. The NRF was also the location of the Expended Core Facility (ECF) which received spent fuel cores from U.S. Navy ships prior to re-processing at ICPP to recover enriched uranium. All employment, and health and safety information for the NRF is maintained independently of other INEEL contract organizations. DOE oversight for the NRF is through the Pittsburgh Naval Reactor Office. Westinghouse Electric Company was the contractor for NRF operations until 1998. Bechtel-Bettis has been the contractor since that time.

A.1.9 Auxiliary Reactor Area (ARA).

Originally known as the Army Reactor Area, the ARA program consisted of four areas, ARA-I through ARA-IV. The program began in 1957 with the development of ARA-I, a compact power reactor capable of being moved with a minimal amount of time between shutdown and startup. The program was phased out in 1965 and all ARA reactors were dismantled or removed, leaving the support buildings vacant. Since 1966, the activities at ARA have included technical support services, and research and development activities at the ARA metallurgy laboratory, the instrument development laboratory and the hot cell facility.

ARA-I functioned as a support facility for ARA II-IV. More recently, it has been used to design, test and treat materials, to measure fatigue on irradiated materials, to study thermonuclear reactor design and to study neutron irradiation effects. ARA-II, which housed the Stationary Low Power Reactor-I (SL-1), was accidentally destroyed during a shutdown in 1961, killing three workers. As a result of the accident, approximately 3000 cubic yards of contaminated soil and reactor hardware have been buried nearby in two pits and a trench, 1600 feet northeast of the ARA-II location.

ARA-III originally housed the Army Gas Cooled Reactor Experiment. ARA-III was a water-moderated, nitrogen-cooled, direct- and close-cycle reactor that generated heat. After becoming critical in 1961, the reactor was placed on standby in 1962, and the program terminated in 1964.

ARA-IV was designed to house the Mobile Low Power Plant-I reactor, a portable gas-cooled, water moderated power reactor. After becoming critical in 1961, the reactor was shut down in 1964, and the program was phased out in 1965.

In 1967, a new program began at ARA-IV with the transfer of the Nuclear Effects Reactor from Nevada. This reactor supplied bursts of high intensity fast neutrons and gamma radiation. The reactor became critical in 1968, but in 1970, was transferred to the Lawrence Livermore National Laboratory. Much of the reactor support equipment has been moved or abandoned, and the facility was decontaminated and decommissioned in 1988. ARA-IV is now known as the Reactive Storage Treatment Area.

A.1.10 Boiling Water Reactor Experiment (BORAX).

The Boiling Water Reactor Experiment included five testing reactors, BORAX I-V, that operated from 1953 through 1964, in an area located five miles to the southwest of the CFA. BORAX I was buried in place and the site abandoned. In 1960, the building used for the BORAX II-IV experiments was removed and replaced by a prefabricated building at the same site. The BORAX V reactor, which used equipment from the previous experiments, was designed to test various nuclear superheating concepts and to advance the boiling water design. BORAX-V achieved criticality in 1962 and, in 1963, produced superheated dry steam for the first time. Since then, the reactor has been dismantled and now awaits disposal.

A.1.11 Experimental Breeder Reactor - I (EBR-1).

EBR-I was the first reactor in the world to generate usable amounts of electricity. With construction beginning in 1949, ANL-W operated EBR-I from 1951 until 1964, when it was taken out of service. Today, it is a National Historic Landmark and is open to the public in an area adjacent to the RWMC. Over the thirteen years of its operation, a total of three core loadings, including a plutonium fuel core, were used in the liquid sodium-potassium cooled reactor.

A.1.12 Security Training Facility (STF).

Originally known as the Experimental Organic Cooled Reactor, the STF was designed and built to advance the organic reactor program. The reactor was scheduled for completion in 1962, but the project was cancelled because a similar reactor had been built in Canada that could provide the desired research information. In 1978 and 1979, the office area was used to support the demolition of the Organic Moderated Reactor Experiment, located directly to the south. In recent years, the STF has been used for material storage, security training maneuvers and the destructive testing of reactor components and other hazardous materials.

A.1.13 Idaho Falls.

At its Idaho Falls, Idaho facilities, the INEEL employed approximately 4000 people (in 1994), in 35 buildings scattered throughout the city's business district. In the early days of the INEEL, the only facilities located in Idaho Falls were those used in support of on-site Atomic Energy Commission activities. However, as the INEEL workforce grew, other activities and offices moved into the city from the remote site, resulting in reduced operating costs and improved productivity. The general mission of the Idaho Falls facilities is "...to provide adequate office, laboratory, computer and storage space for technical, analytical and administrative support activities." These facilities include the INEEL Research Center, the Technical Center, the DOE Idaho Falls Field Office, the INEEL Supercomputing Center and the Willow Creek Building.

The Research Center, which houses 66 laboratories, supports research and development for the DOE and other government agencies. Major programs at the Research Center include investigating strategic and critical materials, researching fossil fuels, engineering, advanced process and industrial research.

The Technical Center includes both the Technical Support Building and the Technical Support Annex. Its programs include nuclear reactor research, new reactor production, special manufacturing capabilities, and facilities management.

The DOE Idaho Falls Field Office activities include administration and operational programs for the INEEL as well as the construction contracts at the site. It provides similar services for facilities in West Valley, New York, and the on-site operation of the Radiological and Environmental Sciences Laboratory at CFA.

The INEEL supercomputing center houses two mainframe computers and numerous smaller computer systems and the necessary hardware and software support. The mainframes are an IBM 3090-300J, used mainly for business purposes, and a Cray X-MP/216, the laboratory's major research computer. The Willow Creek Building is a large office building where employees provide management, technical, scientific and engineering support for the prime contractor of INEEL.

A.2 Assignment of Work History Dates for INEEL Cohort Mortality Study

Background information:

The cohort was identified through linkages of several key employee rosters (through part of 1993), as well as exposure (through part of 1999) and health monitoring files found at the site. This section documents how fields from these files were used to obtain first hire and last termination dates for workers at the INEEL. These decisions resulted from several meetings among the study team. Revision 2 is a modification to allow hire and termination date information from a new source file to be used (NRF5258), and to change the method by which the dose monitoring information is used to assign termination dates. Revision 3 is a modification to allow the incorporation of other NRF doses into the hire date algorithm, when no direct hire date is available. Revision 4 is a correction to adjust for termination dates that appear to be after the date of death. Revision 5 is an adjustment to imputed termination dates: if the termination occurred in the same year as the hire date, and the termination date was imputed from the last dosimetry monitoring year, the termination date was changed to halfway between the hire date and the end of the year of hire, keeping the imputed term date flag = 'Y'.

A.2.1 Assigning hire date

Possible fields for establishing hire date are included in the Roster, SECIMS, NRF, HRS, RDS, MUD and NRF5258 databases (Table A-2). The latter three are dosimetry databases, and include only badging dates, or years of monitoring; therefore, they are less reliable hire date sources. The Roster, SECIMS, NRF, and HRS databases are the primary site rosters that are considered the most accurate sources for hire date. In all cases, the earliest date among fields 1-4 below was used as the hire date. In the absence of a valid entry for the hire date field in one of these files, the badge issue day or first monitoring year from MUD, RDS, NRF or NRF5258 was used to estimate hire date. If no hire date was available from the files, "imputed hire date" (which was estimated from the S-number) was used.

Hierarchy of fields for hire date determination:
1. HIRED in ROSTER
2. HIRE in SECIMS
3. EFFDAT (1st) in NRF
4. HIRE in HRS
5. Year (1st) in NRF5258
6. I_DT in MUD
7. YR (1st) in MUD
8. Begin date HP in NRF
9. ISSUEDAY in RDS
10. PULLDATE (1st) in RDS
11. Imputed hire date

A.2.2 Assigning termination date

The date of last termination could be inferred from most of the same databases (Table A-1). In some databases (e.g., ROSTER, HRS, NRF), the field was thought to be quite accurate; however, information was complete only through the date of receipt of the data base by NIOSH (usually, late 1993). In the SECIMS database, about 25% of the workers who ostensibly terminated employment actually had continuous dose monitoring for more than seven years after the SECIMS date of termination. By contrast, about 90% workers who had both dosimetry and termination dates actually stopped working two years or less after their dose monitoring ceased. Therefore, it appeared that the end of dose monitoring was a better surrogate for employment termination in many cases than the termination date.

To assign a termination date, the fields identified in 1-10 below were checked sequentially, and the latest date used as the termination date (TDATE). To account for inaccurate termination dates, except for workers with an actual termination date between 1995 and 1999, a worker with dose monitoring information in 1996-1999 was considered still employed. These workers would have a "termination censored" field set to 1, and would have termination date set to 12/31/1998 (the date by which the dosimetry is thought to be complete)[1]. A worker with an actual termination date between 1995 and 1999 would retain this termination date.

Other workers with missing termination date had a termination date assigned as either the last pulldate in RDS, or July 1 of the last monitoring year (for the dose files MUD and NRF5258). For other workers with non-missing termination date, if their year of termination (YOT) minus their year of last monitoring was greater than or equal to -1 (i.e. they were monitored no longer than one year after they terminated), the termination date was used. Otherwise (if their YOT minus year of last monitoring < -1), either the last pulldate in RDS, or July 1 of the last monitoring year was used.

The last examination date in OMP (for active employees only), and "active" employee status in the HRS file was used to verify the termination date. If the OMP last exam date or the last EMPSTDT1 in HRS was greater than the final termination date assigned (for active employees), then the termination date was set to 12/31/1998, and the term censoring field set to 1 (i.e., it was assumed the worker is still employed).

The date of termination was corrected if found to occur after the date of death. For workers whose termination date minus DOD was between 1 and 365 days (inclusive), the termination date was reset to the DOD. For workers whose termination date was more than 1 year after their date of death (and for whom re-evaluation confirmed that a correct DC was found), the termination date was either reselected using the next latest termination date available in any of the source files, or was reset to the DOD in the event no other source file was available.

---

[1] Thus, their work duration can be estimated in the analysis, but the fact that the termination date is right-censored will be noted in the study files.

Hierarchy of fields for termination date determination:
1. TERM in ROSTER
2. TERM in SECIMS
3. In NRF, last EFFDAT if LOACODE=T, X, Z, R, or L
4. In HRS, last EMPSTDT1 if EMPSTAT1=T
5. TERM in OMP
6. T_DT in MUD
7. In NRF, last EFFDAT if LOACODE=A, B, D, E, F, K, S, U, or W
8. In HRS, last EMPSTDT1 if EMPSTAT1=L
9. TERMDATE in RDS
10. Apply "Last Dose Year" rule:
    A. If TDATE=. and last dose year > 1996, set TERMCENSOR=1 and TDATE=12/31/1998
    B. Else, if TDATE=., set TDATE= latest of [last pulldate in RDS or last mon. date in NRF or 7/1/(last Yr MUD) or 7/1/(last Yr NRF5258)]
    C. Else, if YOT – last dose year > -1, do not change TDATE
    D. Else, if YOT – last dose year < -1, set TDATE= latest of [last pulldate in RDS or last mon. date in NRF or 7/1/(last Yr MUD) or 7/1/(last Yr NRF5258)]
11. If L_EXMDAT in OMP > termination date, and EMPSTAT=3 (meaning employee is active), set termination date to 12/31/98, and set TERMCENSOR=1.
12. If last EMPSTDT1 in HRS > termination date and EMPSTAT1=A (meaning employee is active), set termination date to 12/31/98 and set TERMCENSOR=1.
13. If 0 < (TDATE – DOD) < 365, set TDATE=DOD
14. If TDATE – DOD >365, and if NIOSH_ID= (on first problematic list), set TDATE = DOD
15. If TDATE – DOD >365, and if NIOSH_ID= (on second problematic list), remove source TDATE and reselect new TDATE from remaining termination date options.
16. If TYEAR = HYEAR and TDATE is imputed as 7/1/(HYEAR), re-impute TDATE as [HDATE + (12/31/(TYEAR) - HDATE)/2], and set TDATE impute code = 'Y'.

Procedure for assigning hire date:

1. Check ROSTER field for date.
   A. If missing, go to step 2.
   B. If filled with VALID date, accept temporarily as HDATE, and go to step 2.
2. Check SECIMS field for date.
   A. If missing, invalid, or > HDATE, go to step 3.
   B. If less than HDATE, replace HDATE with the new date, and go to step 3.
3. Check NRF field for date.
   A. If missing, invalid, or > HDATE, go to step 4.
   B. If less than HDATE, replace HDATE with the new date, and go to step 4.
4. Check HRS field for date.
   A. If missing, invalid, or > HDATE, go to step 5.
   B. If less than HDATE, replace HDATE with the new date, and go to step 5.
5. Check HDATE value.
   A. If missing, go to step 6.
   B. If not missing, go to step 12.
6. Check NRF5258 for first year of dose
   A. If missing, invalid, or > year(HDATE), go to step 7.
   B. If YR less than year(HDATE), replace HDATE with 7/1/YR, set IMPHIRE=Y, and go to step 7.
7. Check MUD for I_DT.
   A. If missing or invalid, go to step 8.
   B. If filled with VALID date, accept temporarily as HDATE, and go to step 8.
8. Check MUD for first YR.
   A. If missing, invalid, or > year(HDATE), go to step 9.
   B. If YR less than year(HDATE), replace HDATE with 7/1/YR, set IMPHIRE=Y, and go to step 9.
9. Check Begin Date in NRF HP file
   A. If missing, invalid, or > year(HDATE), go to step 10.
   B. If YR less than year(HDATE), replace HDATE with 7/1/YR, set IMPHIRE=Y, and go to step 10.
10. Check RDS for ISSUEDAY.
    A. If missing, or invalid, or [ISSUEDAY > HDATE and IMPHIRE=.], or [year(ISSUEDAY) > year(HDATE) and IMPHIRE=Y], go to step 11.
    B. If [year(ISSUEDAY) < year(HDATE) and IMPHIRE=Y], replace HDATE with ISSUEDAY and set IMPHIRE=., and go to step 11.
    C. If [ISSUEDAY<HDATE and IMPHIRE=.], replace HDATE with ISSUEDAY and go to step 11.
11. Check RDS for first PULLDATE
    A. If missing, or invalid, or [PULLDATE > HDATE and IMPHIRE=.], or [year(PULLDATE) > year(HDATE) and IMPHIRE=Y], go to step 12.
    B. If [year(PULLDATE) < year(HDATE) and IMPHIRE=Y], replace HDATE with PULLDATE and set IMPHIRE=., and go to step 12.

C. If [PULLDATE<HDATE and IMPHIRE=.], replace HDATE with PULLDATE and go to step 12.
12. Check HDATE.
   A. If not missing, STOP, accept current value of HDATE (indicate which database it came from).
   B. If missing, use "imputed hire date", which was imputed using S-number.

Procedure for assigning termination date:

1. Check ROSTER field for date.
   A. If missing, go to step 2.
   B. If filled with VALID date, accept temporarily as TDATE, and go to step 2.
2. Check SECIMS field for date.
   A. If missing, invalid, or < TDATE, go to step 3.
   B. If greater than TDATE, replace TDATE with the new date, and go to step 3.
3. Check NRF field for last EFFDAT date.
   A. If missing, invalid, or < TDATE, go to step 4.
   B. If greater than TDATE, and LOACODE=T, X, Z, R, or L, replace TDATE with the new date, and go to step 4.
4. Check last EMPSTDT1 in HRS if EMPSTAT1=T.
   A. If missing, invalid, or < TDATE, go to step 5.
   B. If greater than TDATE, replace TDATE with the new date, and go to step 5.
5. Check OMP field for TERM date.
   A. If missing, invalid, or < TDATE, go to step 6.
   B. If greater than TDATE, replace TDATE with the new date, and go to step 6.
6. Check MUD for T_DT.
   A. If missing, invalid, or < TDATE, go to step 7.
   B. If greater than TDATE, replace TDATE with the new date, and go to step 7.
7. Check last EFFDAT in NRF.
   A. If missing, invalid, or < TDATE, go to step 8.
   B. If greater than TDATE, and LOACODE=A, B, D, E, F, K, S, U, or W, replace TDATE with the new date, and go to step 8.
8. Check last EMPSTDT1 in HRS if EMPSTAT1=L.
   A. If missing, invalid, or < TDATE, go to step 9.
   B. If greater than TDATE, replace TDATE with the new date, and go to step 9.
9. Check RDS for TERMDATE.
   A. If missing, or invalid, or [TERMDATE < TDATE], or [year(TERMDATE) < year(TDATE)], go to step 10.
   B. If [year(TERMDATE) > year(TDATE)], replace TDATE with TERMDATE, and go to step 10.
   C. If [TERMDATE>TDATE and IMPTERM=.], replace TDATE with TERMDATE and go to step 10.
10. Apply "Last Dose Year" rule:
    A. If TDATE=. and last DOSE YEAR > 1996, set TERMCENSOR=1 and

TDATE=12/31/1998
- B. Else, if TDATE=., set TDATE= [last pulldate in RDS (preferred) or 7/1/(last DOSE YEAR)]
- C. Else, if YOT – last DOSE YEAR > -1, do not change TDATE
- D. Else, if YOT – last DOSE YEAR < -1, set TDATE= [last pulldate in RDS (preferred) or 7/1/(last DOSE YEAR)]

11. Verification 1: Check L_EXMDAT in OMP, if EMPSTAT=3.
    - A. If L_EXMDAT > TDATE, set TDATE = 12/31/98, and set TERMCENSOR = 1.
    - B. If missing, invalid, or < TDATE, retain TDATE and go to step 12.
11. Verification 2: Check EMPSTDT1 in HRS if EMPSTAT1=A
    - A. If last EMPSTDT1 > TDATE, set TDATE = 12/31/98 and set TERMCENSOR=1.
    - B. If missing, invalid, or < TDATE, retain TDATE and go to step 13.
13. If TYEAR = HYEAR and TDATE is imputed as 7/1/(HYEAR), re-impute TDATE as [HDATE + (12/31/(TYEAR) - HDATE)/2], and set TDATE impute code = 'Y'.
14. Check T_DATE, and record which file it came from.

Table A-1. Fields for determining hire and termination dates for the INEEL cohort source files.

DATES	ROSTER	SECIMS	NRF	HRS	OMP	MUD	RDS
Begin employment	HIRED1	HIRE1	EFFDAT (1st)2	HIRE1			
End employment	TERMDATE3	TERM3	EFFDAT2 + LOACODE4=T, X, Z, R, L	EMPSTDT2 if EMPSTAT5=T	TERM3	T_DT3	
Issue Badge						I_DT6	ISSUEDAY6
Revoke Badge							TERM_DAY3
Dose reading date			last monitoring date				PULLDAY7
Exam date					L_EXMDAT8		
Chest X-ray date					CHESTXRY8		
Exposure Hx			NRF5258			YR9	

1 Hire date
2 Effective date in personnel file
3 Termination date
4 Leave of absence code
5 Employment status
6 Radiation badge issue date
7 Radiation badge pull date
8 Medical examination dates

A.3 Imputation of Missing Hire Dates for INEEL Workers

A.3.1 Background

One of the critical steps in the INEEL Cohort Mortality Study was the assembly of the complete study roster. Hire date is required both as a person-time begin date in statistical analyses of the full cohort and to compute duration of employment, an eligibility criterion for the IARC Combined International Nuclear Worker study. Inconsistencies in record-keeping practices across time periods and buildings within the facility, lack of complete exposure information, missing work history information and other problems involving the data make this task an extremely challenging and time-consuming process. In order to accurately ascertain the type and duration of a worker's exposure(s), precise hire and termination dates of employment must be determined. A project to impute missing hire dates for several thousand workers was completed in April, 2001. This document contains a description of the statistical analysis and results of this INEEL hire date imputation project.

A.3.2 Database

The INEEL demographic database contains information on all workers who have ever worked at the site for any length of time. In order to update this database in preparation for data analysis, missing hire dates were imputed using linear regression models. Employees in the database were each assigned a unique identification number, or S number, by the INEEL site at the time they were hired. S numbers were assigned sequentially within each of the seven source files that together comprise this database. This imputation process involved five steps: data splitting, model building, model validation, hire date imputation, and model revision.

A.3.3 Methods

The database records were first randomly divided into two approximately equal subsets: a model building subset to create regression models relating S number and hire date, and a test subset to validate these models. The model building subset was first stratified by source file, as the relationship between S number and hire date varied between source files. ROSTER (ROS) and SECIMS (SEC) were the two source files used for model building, since workers with missing hire dates came from these files and there was a piecewise linear relationship between S number and hire date within each file.

Following the creation of linear regression models, these models were applied to the test subset, which contained workers who had both S numbers and hire dates, in order to test the accuracy of the imputed hire dates against true hire dates. Error variables were created, and descriptive statistics were calculated for each error variable separately for ROS and SEC, and for both source files combined. The first error variable was the absolute value of the number of days between the true and imputed hire dates. The second error variable was the true date minus the imputed date, and indicates the direction of the bias (if any) of the imputed dates relative to the true dates.

The models were then revised according to the results of the error analysis in order to maximize the range of S numbers that were included in the models, thereby being able to include as many workers in the imputation as possible based on their S numbers, while minimizing the errors. The final models reflect the most accurate estimators of hire dates.

A.3.4 Results

The demographic database contained 4658 workers with missing hire dates. Linear regression models were able to impute dates for 4028 (86%) of these workers using S numbers. There are 630 (14%) workers whose hire dates remained missing because S numbers for these workers fell in a range in which there was no clear association between S number and hire date.

The median difference between true and imputed hire dates (absolute value) in the test subset is 48 days (Q1=13, Q3=173) in the SEC and 26 days (Q1=12, Q3=55) in the ROS source file. The median difference for all workers in both source files combined is 29 days (Q1=12, Q3=67). The median error (directional bias) for SEC is -10 days, which indicates that the median imputed hire date is 10 days later than the median actual hire date. The median for ROS is +3 days, indicating that the median imputed hire date is 3 days earlier than the median actual hire date. The median among all workers in both source files combined is -1 day, which indicates that the imputed dates do not reflect a substantial bias.

A.4 Comparison of INEEL SES categories with those used for the Hanford cohort

Table A-2. SES classes assigned to 1980 Bureau of Census classifications, compared to assignments used in Hanford cohort development (described in Gilbert et al. 1992).

1980 BOC code	1970 BOC code	1980 Code description	Gilbert SES class	Gilbert SES description	Diff	INEEL SES class	INEEL SES description
003		Legislators	0		*	1	Professional
004	222	Chief execs. & general administrators, public admin.	2	Intermediate	*	1	Professional
005	222	Administrators & officials, public administration	2	Intermediate	*	1	Professional
006	245	Administrators, protective services	2	Intermediate		2	Intermediate
007	201	Financial managers	2	Intermediate	*	1	Professional
007	202	Financial managers	2	Intermediate	*	1	Professional
008	245	Personnel & labor relations managers	2	Intermediate	*	1	Professional
009	245	Purchasing managers	2	Intermediate		2	Intermediate
013	245	Managers, marketing, advertising, & public relations	2	Intermediate		2	Intermediate
014	235	Administrators, education & related fields	2	Intermediate	*	1	Professional
014	240	Administrators, education & related fields	2	Intermediate	*	1	Professional
015	212	Managers, medicine & health	2	Intermediate	*	1	Professional
016	216	Managers, properties & real estate	2	Intermediate		2	Intermediate
017	224	Postmasters & mail superintendents	2	Intermediate		2	Intermediate
018	211	Funeral directors	2	Intermediate		2	Intermediate
019	220	Managers & administrators, n.e.c.	2	Intermediate		2	Intermediate
019	221	Managers & administrators, n.e.c.	2	Intermediate		2	Intermediate
019	245	Managers & administrators, n.e.c.	2	Intermediate		2	Intermediate
023	1	Accountants & auditors	1	Professional		1	Professional
024	202	Underwriters	2	Intermediate		2	Intermediate
025	202	Other financial officers	2	Intermediate		2	Intermediate
026	245	Management analysts	2	Intermediate		2	Intermediate
027	56	Personnel, training, & labor relations specialists	2	Intermediate		2	Intermediate
028	203	Purchasing agents & buyers, farm products	2	Intermediate		2	Intermediate
029	205	Buyers, wholesale & retail trade, except farm products	2	Intermediate		2	Intermediate
033	225	Purchasing agents & buyers, n.e.c.	2	Intermediate		2	Intermediate
034	260	Business & promotion agents	0		*	2	Intermediate
035	213	Construction inspectors	2	Intermediate		2	Intermediate
036	215	Inspectors & compliance officers, exc. Construction	2	Intermediate		2	Intermediate
037	24	Management related occupations, n.e.c.	1	Professional	*	2	Intermediate
037	223	Management related occupations, n.e.c.	2	Intermediate		2	Intermediate
043	2	Architects	1	Professional		1	Professional
044	6	Aerospace engineers	1	Professional		1	Professional
045	15	Metallurgical & materials engineers	1	Professional		1	Professional
046	20	Mining engineers	1	Professional		1	Professional
047	21	Petroleum engineers	1	Professional		1	Professional
048	10	Chemical engineers	1	Professional		1	Professional
049	23	Nuclear engineers	1	Professional		1	Professional
053	11	Civil engineers	1	Professional		1	Professional
054	23	Agricultural engineers	1	Professional		1	Professional
055	12	Electrical & electronic engineers	1	Professional		1	Professional
056	13	Industrial engineers	1	Professional		1	Professional
057	14	Mechanical engineers	1	Professional		1	Professional
058	23	Marine engineers & naval architects	1	Professional	*	2	Intermediate
059	23	Engineers, n.e.c.	1	Professional		1	Professional
063	161	Surveyors & mapping scientists	2	Intermediate		2	Intermediate
064	4	Computer systems analysts & scientists	1	Professional		1	Professional

1980 BOC code	1970 BOC code	1980 Code description	Gilbert SES class	Gilbert SES description	Diff	INEEL SES class	INEEL SES description
065	55	Operations & systems researchers & analysts	1	Professional		1	Professional
066	34	Actuaries	1	Professional		1	Professional
067	36	Statisticians	1	Professional		1	Professional
068	35	Mathematical scientists, n.e.c.	1	Professional		1	Professional
069	53	Physicists & astronomers	1	Professional		1	Professional
073	45	Chemists, except biochemists	1	Professional		1	Professional
074	43	Atmospheric & space scientists	1	Professional		1	Professional
075	51	Geologists & geodesists	1	Professional		1	Professional
076	54	Physical scientists, n.e.c.	1	Professional		1	Professional
077	42	Agricultural & food scientists	1	Professional		1	Professional
078	44	Biological & life scientists	1	Professional		1	Professional
078	52	Biological & life scientists	1	Professional		1	Professional
079	25	Forestry & conservation scientists	1	Professional		1	Professional
083	54	Medical scientists	1	Professional		1	Professional
084	65	Physicians	1	Professional		1	Professional
085	62	Dentists	1	Professional		1	Professional
086	72	Veterinarians	1	Professional		1	Professional
087	63	Optometrists	1	Professional		1	Professional
088	71	Podiatrists	1	Professional		1	Professional
089	61	Health diagnosing practitioners, n.e.c.	1	Professional		1	Professional
089	73	Health diagnosing practitioners, n.e.c.	1	Professional		1	Professional
095	75	Registered nurses	2	Intermediate		2	Intermediate
096	64	Pharmacists	1	Professional		1	Professional
097	74	Dietitians	2	Intermediate		2	Intermediate
098	76	Inhalation therapists	2	Intermediate		2	Intermediate
099	76	Occupational therapists	2	Intermediate		2	Intermediate
103	76	Physical therapists	2	Intermediate		2	Intermediate
104	76	Speech therapists	2	Intermediate		2	Intermediate
105	76	Therapists, n.e.c.	2	Intermediate		2	Intermediate
106	76	Physicians' assistants	2	Intermediate		2	Intermediate
113	103	Earth, environmental, & marine science teachers	2	Intermediate	*	1	Professional
114	104	Biological science teachers	2	Intermediate	*	1	Professional
115	105	Chemistry teachers	2	Intermediate	*	1	Professional
116	110	Physics teachers	2	Intermediate	*	1	Professional
117	135	Natural science teachers, n.e.c.	2	Intermediate	*	1	Professional
118	114	Psychology teachers	2	Intermediate	*	1	Professional
119	116	Economics teachers	2	Intermediate	*	1	Professional
123	120	History teachers	2	Intermediate	*	1	Professional
124	135	Political science teachers	2	Intermediate	*	1	Professional
125	121	Sociology teachers	2	Intermediate	*	1	Professional
126	122	Social science teachers, n.e.c.	2	Intermediate	*	1	Professional
127	111	Engineering teachers	2	Intermediate	*	1	Professional
128	112	Mathematical science teachers	2	Intermediate	*	1	Professional
129	135	Computer science teachers	2	Intermediate	*	1	Professional
133	135	Medical science teachers	2	Intermediate	*	1	Professional
134	113	Health specialties teachers	2	Intermediate	*	1	Professional
135	115	Business, commerce, & marketing teachers	2	Intermediate	*	1	Professional
136	102	Agriculture & forestry teachers	2	Intermediate	*	1	Professional
137	123	Art, drama, & music teachers	2	Intermediate	*	1	Professional
138	124	Physical education teachers	2	Intermediate	*	1	Professional
139	125	Education teachers	2	Intermediate	*	1	Professional
143	126	English teachers	2	Intermediate	*	1	Professional
144	130	Foreign language teachers	2	Intermediate	*	1	Professional
145	132	Law teachers	2	Intermediate	*	1	Professional

1980 BOC code	1970 BOC code	1980 Code description	Gilbert SES class	Gilbert SES description	Diff	INEEL SES class	INEEL SES description
146	122	Social work teachers	2	Intermediate	*	1	Professional
147	133	Theology teachers	2	Intermediate	*	1	Professional
148	134	Trade & industrial teachers	2	Intermediate	*	1	Professional
149	131	Home economics teachers	2	Intermediate	*	1	Professional
153	135	Teachers, postsecondary, n.e.c.	2	Intermediate	*	1	Professional
154	140	Postsecondary teachers, subject not specified	2	Intermediate	*	1	Professional
155	143	Teachers, prekindergarten & kindergarten	2	Intermediate		2	Intermediate
156	142	Teachers, elementary school	2	Intermediate		2	Intermediate
157	144	Teachers, secondary school	2	Intermediate		2	Intermediate
158	145	Teachers, special education	2	Intermediate		2	Intermediate
159	145	Teachers, n.e.c.	2	Intermediate		2	Intermediate
163	174	Counselors, educational & vocational	2	Intermediate		2	Intermediate
164	32	Librarians	2	Intermediate		2	Intermediate
165	33	Archivists & curators	1	Professional		1	Professional
166	91	Economists	2	Intermediate	*	1	Professional
167	93	Psychologists	2	Intermediate	*	1	Professional
168	94	Sociologists	2	Intermediate	*	1	Professional
169	92	Social scientists, n.e.c.	2	Intermediate	*	1	Professional
169	96	Social scientists, n.e.c.	2	Intermediate	*	1	Professional
173	95	Urban planners	2	Intermediate		2	Intermediate
174	100	Social workers	2	Intermediate		2	Intermediate
175	101	Recreation workers	2	Intermediate		2	Intermediate
176	86	Clergy	2	Intermediate		2	Intermediate
177	90	Religious workers, n.e.c.	2	Intermediate		2	Intermediate
178	31	Lawyers	1	Professional		1	Professional
179	30	Judges	1	Professional		1	Professional
183	181	Authors	2	Intermediate		2	Intermediate
184	194	Technical writers	2	Intermediate		2	Intermediate
185	183	Designers	2	Intermediate		2	Intermediate
186	185	Musicians & composers	2	Intermediate		2	Intermediate
187	175	Actors & directors	2	Intermediate		2	Intermediate
188	190	Painters, sculptors, craft-artists, & artist printmakers	2	Intermediate		2	Intermediate
189	191	Photographers	2	Intermediate		2	Intermediate
193	182	Dancers	2	Intermediate		2	Intermediate
194	194	Artists, performers, & related workers, n.e.c.	2	Intermediate		2	Intermediate
195	184	Editors & reporters	2	Intermediate		2	Intermediate
197	192	Public relations specialists	2	Intermediate		2	Intermediate
198	193	Announcers	2	Intermediate		2	Intermediate
199	180	Athletes	2	Intermediate		2	Intermediate
203	80	Clinical laboratory technologists & technicians	2	Intermediate		2	Intermediate
204	81	Dental hygienists	2	Intermediate		2	Intermediate
205	82	Health record technologists & technicians	2	Intermediate		2	Intermediate
206	83	Radiologic technicians	2	Intermediate		2	Intermediate
207	926	Licensed practical nurses	6	Unskilled	*	4	Skilled manual
208	84	Health technologists & technicians, n.e.c.	2	Intermediate		2	Intermediate
208	85	Health technologists & technicians, n.e.c.	2	Intermediate		2	Intermediate
213	153	Electrical & electronic technicians	2	Intermediate		2	Intermediate
214	154	Industrial engineering technicians	2	Intermediate		2	Intermediate
215	155	Mechanical engineering technicians	2	Intermediate		2	Intermediate
216	162	Engineering technicians, n.e.c.	2	Intermediate		2	Intermediate
217	152	Drafting occupations	2	Intermediate		2	Intermediate
218	162	Surveying & mapping technicians	2	Intermediate		2	Intermediate
223	150	Biological technicians	2	Intermediate		2	Intermediate
224	151	Chemical technicians	2	Intermediate		2	Intermediate

1980 BOC code	1970 BOC code	1980 Code description	Gilbert SES class	Gilbert SES description	Diff	INEEL SES class	INEEL SES description
225	162	Science technicians, n.e.c.	2	Intermediate		2	Intermediate
226	163	Airplane pilots & navigators	2	Intermediate		2	Intermediate
226	170	Airplane pilots & navigators	2	Intermediate		2	Intermediate
227	164	Air traffic controllers	2	Intermediate		2	Intermediate
228	171	Broadcast equipment operators	2	Intermediate		2	Intermediate
229	3	Computer programmers	1	Professional	*	2	Intermediate
233	172	Tool programmers, numerical control	2	Intermediate		2	Intermediate
234	173	Legal assistants	2	Intermediate		2	Intermediate
235	156	Technicians, n.e.c.	2	Intermediate		2	Intermediate
235	165	Technicians, n.e.c.	2	Intermediate		2	Intermediate
235	173	Technicians, n.e.c.	2	Intermediate		2	Intermediate
243	231	Supervisors & proprietors, sales occupations	2	Intermediate		2	Intermediate
243	233	Supervisors & proprietors, sales occupations	2	Intermediate		2	Intermediate
253	265	Insurance sales occupations	2	Intermediate	*	3	Skilled non-man
254	270	Real estate sales occupations	0		*	3	Skilled non-man
255	271	Securities & financial services sales occupations	0		*	3	Skilled non-man
256	260	Advertising & related sales occupations	0		*	3	Skilled non-man
257	280	Sales occupations, other business services	3	Skilled non-man		3	Skilled non-man
258	22	Sales engineers	1	Professional	*	2	Skilled non-man
259	281	Sales reps., mining, manufacturing, & wholesale	0		*	3	Skilled non-man
259	282	Sales reps., mining, manufacturing, & wholesale	0		*	3	Skilled non-man
263	283	Sales workers, motor vehicles & boats	0		*	3	Skilled non-man
264	283	Sales workers, apparel	0		*	3	Skilled non-man
265	283	Sales workers, shoes	0		*	3	Skilled non-man
266	283	Sales workers, furniture & home furnishings	0		*	3	Skilled non-man
267	283	Sales workers; radio, television, hi-fi, & appliances	0		*	3	Skilled non-man
268	283	Sales workers, hardware & building supplies	0		*	3	Skilled non-man
269	283	Sales workers, parts	0		*	3	Skilled non-man
274	283	Sales workers, other commodities	0		*	3	Skilled non-man
275	314	Sales counter clerks	3	Skilled non-man		3	Skilled non-man
276	310	Cashiers	3	Skilled non-man		3	Skilled non-man
277	264	Street & door-to door sales workers	0		*	3	Skilled non-man
278	266	News vendors	0		*	3	Skilled non-man
283	262	Demonstrators, promoters & models, sales	0		*	3	Skilled non-man
284	261	Auctioneers	0		*	3	Skilled non-man
285	280	Sales support occupations, n.e.c.	3	Skilled non-man		3	Skilled non-man
303	312	Supervisors, general office	3	Skilled non-man		3	Skilled non-man
304	312	Supervisors, computer equipment operators	3	Skilled non-man		3	Skilled non-man
305	312	Supervisors, financial records processing	3	Skilled non-man		3	Skilled non-man
306	312	Chief communications operators	3	Skilled non-man		3	Skilled non-man
307	312	Supervisors; dist., scheduling, & adjusting clerks	3	Skilled non-man		3	Skilled non-man
308	343	Computer operators	3	Skilled non-man		3	Skilled non-man
309	343	Peripheral equipment operators	3	Skilled non-man		3	Skilled non-man
313	370	Secretaries	3	Skilled non-man		3	Skilled non-man
313	371	Secretaries	3	Skilled non-man		3	Skilled non-man
313	372	Secretaries	3	Skilled non-man		3	Skilled non-man
314	376	Stenographers	3	Skilled non-man		3	Skilled non-man
315	391	Typists	3	Skilled non-man		3	Skilled non-man
316	320	Interviewers	3	Skilled non-man		3	Skilled non-man
317	314	Hotel clerks	3	Skilled non-man		3	Skilled non-man
318	390	Transportation ticket & reservation agents	3	Skilled non-man		3	Skilled non-man
319	364	Receptionists	3	Skilled non-man		3	Skilled non-man
323	394	Information clerks, n.e.c.	3	Skilled non-man		3	Skilled non-man
325	394	Classified-ad clerks	3	Skilled non-man		3	Skilled non-man

1980 BOC code	1970 BOC code	1980 Code description	Gilbert SES class	Gilbert SES description	Diff	INEEL SES class	INEEL SES description
326	394	Correspondence clerks	3	Skilled non-man		3	Skilled non-man
327	394	Order clerks	3	Skilled non-man		3	Skilled non-man
328	394	Personnel clerks, except payroll & timekeeping	3	Skilled non-man		3	Skilled non-man
329	330	Library clerks	3	Skilled non-man		3	Skilled non-man
335	325	File clerks	3	Skilled non-man		3	Skilled non-man
336	394	Records clerks	3	Skilled non-man		3	Skilled non-man
337	305	Bookkeepers, accounting, & auditing clerks	3	Skilled non-man		3	Skilled non-man
338	360	Payroll & timekeeping clerks	3	Skilled non-man		3	Skilled non-man
339	303	Billing clerks	0		*	3	Skilled non-man
343	394	Cost & rate clerks	3	Skilled non-man		3	Skilled non-man
344	341	Billing, posting, & calculating machine operators	3	Skilled non-man		3	Skilled non-man
344	342	Billing, posting, & calculating machine operators	3	Skilled non-man		3	Skilled non-man
345	344	Duplicating machine operators	3	Skilled non-man		3	Skilled non-man
346	355	Mail preparing & paper handling machine operators	3	Skilled non-man		3	Skilled non-man
347	345	Office machine operators, n.e.c.	3	Skilled non-man		3	Skilled non-man
347	350	Office machine operators, n.e.c.	3	Skilled non-man		3	Skilled non-man
347	355	Office machine operators, n.e.c.	3	Skilled non-man		3	Skilled non-man
347	383	Office machine operators, n.e.c.	3	Skilled non-man		3	Skilled non-man
348	385	Telephone operators	3	Skilled non-man		3	Skilled non-man
349	384	Telegraphers	3	Skilled non-man		3	Skilled non-man
353	394	Communications equipment operators, n.e.c.	3	Skilled non-man		3	Skilled non-man
354	361	Postal clerks, exc. mail carriers	3	Skilled non-man		3	Skilled non-man
355	331	Mail carriers, postal service	3	Skilled non-man		3	Skilled non-man
356	332	Mail clerks, exc. postal service	3	Skilled non-man		3	Skilled non-man
357	333	Messengers	3	Skilled non-man		3	Skilled non-man
359	315	Dispatchers	3	Skilled non-man		3	Skilled non-man
363	323	Production coordinators	3	Skilled non-man		3	Skilled non-man
364	374	Traffic, shipping, & receiving clerks	3	Skilled non-man		3	Skilled non-man
365	381	Stock & inventory clerks	3	Skilled non-man		3	Skilled non-man
366	334	Meter readers	3	Skilled non-man		3	Skilled non-man
368	392	Weighers, measurers, & checkers	3	Skilled non-man		3	Skilled non-man
369	323	Samplers	3	Skilled non-man		3	Skilled non-man
373	323	Expediters	3	Skilled non-man		3	Skilled non-man
374	394	Material recording, scheduling, & distrib. clerks, n.e.c.	3	Skilled non-man		3	Skilled non-man
375	326	Insurance adjusters, examiners, & investigators	3	Skilled non-man		3	Skilled non-man
376	321	Investigators & adjusters, except insurance	3	Skilled non-man		3	Skilled non-man
377	311	Eligibility clerks, social welfare	3	Skilled non-man		3	Skilled non-man
378	313	Bill & account collectors	3	Skilled non-man		3	Skilled non-man
379	395	General office clerks	3	Skilled non-man		3	Skilled non-man
383	301	Bank tellers	0		*	3	Skilled non-man
384	362	Proofreaders	3	Skilled non-man		3	Skilled non-man
385	394	Data-entry keyers	3	Skilled non-man		3	Skilled non-man
386	375	Statistical clerks	3	Skilled non-man		3	Skilled non-man
387	382	Teachers' aides	3	Skilled non-man		3	Skilled non-man
389	394	Administrative support occupations, n.e.c.	3	Skilled non-man		3	Skilled non-man
403	983	Launderers & ironers	4	Skilled manual		4	Skilled manual
404	981	Cooks, private household	4	Skilled manual		4	Skilled manual
405	982	Housekeepers & butlers	4	Skilled manual	*	5	Partly skilled
406	980	Child care workers, private household	4	Skilled manual	*	5	Partly skilled
407	984	Private household cleaners & servants	4	Skilled manual	*	6	Unskilled
413	961	Supervisors, firefighting & fire prevention occ's	4	Skilled manual		4	Skilled manual
414	964	Supervisors, police & detectives	4	Skilled manual		4	Skilled manual
415	962	Supervisors, guards	5	Partly skilled		5	Partly skilled
416	961	Fire inspection & fire prevention occupations	4	Skilled manual		4	Skilled manual

Epidemiologic Study of Mortality and Radiation-Related Risk of Cancer Among INEEL Workers

1980 BOC code	1970 BOC code	1980 Code description	Gilbert SES class	Gilbert SES description	Diff	INEEL SES class	INEEL SES description
417	961	Firefighting occupations	4	Skilled manual		4	Skilled manual
418	964	Police & detectives, public service	4	Skilled manual		4	Skilled manual
423	963	Sheriffs, bailiffs, & other law enforcement officers	0		*	4	Skilled manual
423	965	Sheriffs, bailiffs, & other law enforcement officers	4	Skilled manual		4	Skilled manual
424	962	Correctional institution officers	5	Partly skilled		5	Partly skilled
425	960	Crossing guards	0		*	5	Partly skilled
426	962	Guards & police, exc. public service	5	Partly skilled		5	Partly skilled
427	952	Protective service occupations, n.e.c.	6	Unskilled	*	5	Partly skilled
433	230	Supervisors, food preparation & service occupations	2	Intermediate	*	5	Partly skilled
434	910	Bartenders	6	Unskilled	*	5	Partly skilled
435	915	Waiters & waitresses	5	Partly skilled		5	Partly skilled
436	912	Cooks, except short order	5	Partly skilled		5	Partly skilled
437	913	Short-order cooks	5	Partly skilled	*	6	Unskilled
438	914	Food counter, fountain & related occupations	5	Partly skilled	*	6	Unskilled
439	916	Kitchen workers, food preparation	5	Partly skilled	*	6	Unskilled
443	911	Waiters'/Waitresses' assistants	6	Unskilled		6	Unskilled
444	916	Miscellaneous food preparation occupations	5	Partly skilled		5	Partly skilled
445	921	Dental assistants	6	Unskilled	*	5	Partly skilled
446	922	Health aides, except nursing	6	Unskilled	*	5	Partly skilled
446	924	Health aides, except nursing	6	Unskilled	*	5	Partly skilled
447	925	Nursing aides, orderlies, & attendants	6	Unskilled	*	5	Partly skilled
448	903	Supervisors, cleaning & building service workers	6	Unskilled		6	Unskilled
449	901	Maids & housemen	6	Unskilled		6	Unskilled
449	950	Maids & housemen	6	Unskilled		6	Unskilled
453	902	Janitors & cleaners	6	Unskilled		6	Unskilled
453	903	Janitors & cleaners	6	Unskilled		6	Unskilled
454	943	Elevator operators	6	Unskilled		6	Unskilled
455		Pest control occupations	0		*	6	Unskilled
456		Supervisors, personal service occupations	0		*	5	Partly skilled
457	935	Barbers	6	Unskilled	*	5	Partly skilled
458	944	Hairdressers & cosmetologists	6	Unskilled	*	5	Partly skilled
459	932	Attendants, amusement & recreation facilities	6	Unskilled		6	Unskilled
463	933	Guides	6	Unskilled	*	5	Partly skilled
464	953	Ushers	6	Unskilled		6	Unskilled
465	931	Public transportation attendants	6	Unskilled		6	Unskilled
466	934	Baggage porters & bellhops	6	Unskilled		6	Unskilled
467	954	Welfare service aides	6	Unskilled		6	Unskilled
468	942	Child care workers, except private household	6	Unskilled	*	5	Partly skilled
469	933	Personal service occupations, n.e.c.	6	Unskilled		6	Unskilled
469	940	Personal service occupations, n.e.c.	6	Unskilled		6	Unskilled
469	941	Personal service occupations, n.e.c.	6	Unskilled		6	Unskilled
473	801	Farmers, except horticultural	6	Unskilled	*	4	Skilled manual
474	801	Horticultural specialty farmers	6	Unskilled	*	4	Skilled manual
475	802	Managers, farms, except horticultural	6	Unskilled	*	4	Skilled manual
476	802	Managers, horticultural specialty farms	6	Unskilled	*	4	Skilled manual
477	821	Supervisors, farm workers	6	Unskilled		6	Unskilled
479	822	Farm workers	6	Unskilled		6	Unskilled
483	752	Marine life cultivation workers	6	Unskilled		6	Unskilled
484	822	Nursery workers	6	Unskilled		6	Unskilled
485	802	Supervisors, related agricultural occupations	6	Unskilled		6	Unskilled
486	755	Groundskeepers & gardeners, except farm	5	Partly skilled		5	Partly skilled
487	740	Animal caretakers, except farm	6	Unskilled	*	5	Partly skilled
488	625	Graders & sorters, agricultural products	5	Partly skilled		5	Partly skilled
489	215	Inspectors, agricultural products	2	Intermediate		2	Intermediate

1980 BOC code	1970 BOC code	1980 Code description	Gilbert SES class	Gilbert SES description	Diff	INEEL SES class	INEEL SES description
494	821	Supervisors, forestry & logging workers	6	Unskilled		6	Unskilled
495	761	Forestry workers, except logging	6	Unskilled		6	Unskilled
496	761	Timber cutting & logging occupations	6	Unskilled		6	Unskilled
497	221	Captains & other officers, fishing vessels	2	Intermediate	*	4	Skilled manual
498	752	Fishers	6	Unskilled		6	Unskilled
499	780	Hunters & trappers	6	Unskilled		6	Unskilled
503	441	Supervisors, mechanics & repairers	4	Skilled manual		4	Skilled manual
505	473	Automobile mechanics	4	Skilled manual		4	Skilled manual
506	474	Automobile mechanic apprentices	4	Skilled manual		4	Skilled manual
507	492	Bus, truck, & stationary engine mechanics	4	Skilled manual		4	Skilled manual
508	471	Aircraft engine mechanics	4	Skilled manual		4	Skilled manual
509	492	Small engine repairers	4	Skilled manual		4	Skilled manual
514	472	Automobile body & related repairers	4	Skilled manual		4	Skilled manual
515	471	Aircraft mechanics, exc. engine	4	Skilled manual		4	Skilled manual
516	481	Heavy equipment mechanics	4	Skilled manual		4	Skilled manual
517	480	Farm equipment mechanics	4	Skilled manual		4	Skilled manual
518	492	Industrial machinery repairers	4	Skilled manual		4	Skilled manual
519	492	Machinery maintenance occupations	4	Skilled manual		4	Skilled manual
523	492	Electronic repairers, communicat. & indust. equip't	4	Skilled manual		4	Skilled manual
525	475	Data processing equipment repairers	4	Skilled manual		4	Skilled manual
526	482	Household appliance & powertool repairers	4	Skilled manual		4	Skilled manual
527	554	Telephone line installers & repairers	4	Skilled manual		4	Skilled manual
529	552	Telephone installers & repairers	4	Skilled manual		4	Skilled manual
533	485	Miscellaneous electrical & electronic equip't repairers	4	Skilled manual		4	Skilled manual
534	470	Heating, air conditioning, & refrigeration mechanics	4	Skilled manual		4	Skilled manual
535	516	Camera, watch, & musical instrument repairers	4	Skilled manual		4	Skilled manual
536	571	Locksmiths & safe repairers	4	Skilled manual		4	Skilled manual
538	484	Office machine repairers	4	Skilled manual		4	Skilled manual
539	571	Mechanical controls & valve repairers	4	Skilled manual		4	Skilled manual
543	571	Elevator installers & repairers	4	Skilled manual		4	Skilled manual
544	502	Millwrights	4	Skilled manual		4	Skilled manual
547	483	Specified mechanics & repairers, n.e.c.	4	Skilled manual		4	Skilled manual
547	486	Specified mechanics & repairers, n.e.c.	4	Skilled manual		4	Skilled manual
547	492	Specified mechanics & repairers, n.e.c.	4	Skilled manual		4	Skilled manual
549	491	Not specified mechanics & repairers	4	Skilled manual		4	Skilled manual
549	495	Not specified mechanics & repairers	4	Skilled manual		4	Skilled manual
553	441	Supervisors; brickmasons, stonemasons, & tile setters	4	Skilled manual		4	Skilled manual
554	441	Supervisors, carpenters & related workers	4	Skilled manual		4	Skilled manual
555	441	Supervisors, electricians & power transmis. installers	4	Skilled manual		4	Skilled manual
556	441	Supervisors; painters, paperhangers, & plasterers	4	Skilled manual		4	Skilled manual
557	441	Supervisors; plumbers, pipefitters, & steamfitters	4	Skilled manual		4	Skilled manual
558	441	Supervisors, n.e.c.	4	Skilled manual		4	Skilled manual
563	410	Brickmasons & stonemasons	4	Skilled manual		4	Skilled manual
564	411	Brickmason & stonemason apprentices	4	Skilled manual		4	Skilled manual
565	440	Tile setters, hard & soft	4	Skilled manual		4	Skilled manual
565	560	Tile setters, hard & soft	4	Skilled manual		4	Skilled manual
566	420	Carpet installers	4	Skilled manual		4	Skilled manual
567	415	Carpenters	4	Skilled manual		4	Skilled manual
569	416	Carpenter apprentices	4	Skilled manual		4	Skilled manual
573	615	Drywall installers	5	Partly skilled	*	4	Skilled manual
575	430	Electricians	4	Skilled manual		4	Skilled manual
576	431	Electrician apprentices	4	Skilled manual		4	Skilled manual
577	433	Electrical power installers & repairers	4	Skilled manual		4	Skilled manual
579	510	Painters, construction & maintenance	4	Skilled manual		4	Skilled manual

1980 BOC code	1970 BOC code	1980 Code description	Gilbert SES class	Gilbert SES description	Diff	INEEL SES class	INEEL SES description
583	512	Paperhangers	4	Skilled manual		4	Skilled manual
584	520	Plasterers	4	Skilled manual		4	Skilled manual
585	522	Plumbers, pipefitters, & steamfitters	4	Skilled manual		4	Skilled manual
587	523	Plumber, pipefitter, & steamfitter apprentices	4	Skilled manual		4	Skilled manual
588	421	Concrete & terrazzo finishers	4	Skilled manual		4	Skilled manual
589	445	Glaziers	4	Skilled manual		4	Skilled manual
593	601	Insulation workers	5	Partly skilled	*	4	Skilled manual
594	436	Paving, surfacing, & tamping equipment operators	4	Skilled manual		4	Skilled manual
595	534	Roofers	4	Skilled manual		4	Skilled manual
596	575	Sheetmetal duct installers	4	Skilled manual		4	Skilled manual
597	550	Structural metal workers	4	Skilled manual		4	Skilled manual
598	614	Drillers, earth	5	Partly skilled	*	4	Skilled manual
599	575	Construction trades, n.e.c.	4	Skilled manual		4	Skilled manual
613	441	Supervisors, extractive occupations	4	Skilled manual		4	Skilled manual
614	694	Drillers, oil well	5	Partly skilled	*	4	Skilled manual
615	603	Explosives workers	5	Partly skilled	*	4	Skilled manual
616	640	Mining machine operators	5	Partly skilled	*	4	Skilled manual
617	640	Mining occupations, n.e.c.	5	Partly skilled	*	4	Skilled manual
633	441	Supervisors, production occupations	4	Skilled manual		4	Skilled manual
634	561	Tool & die makers	4	Skilled manual		4	Skilled manual
635	562	Tool & die maker apprentices	4	Skilled manual		4	Skilled manual
636	575	Precision assemblers, metal	4	Skilled manual		4	Skilled manual
637	461	Machinists	4	Skilled manual		4	Skilled manual
639	462	Machinist apprentices	4	Skilled manual		4	Skilled manual
643	404	Boilermakers	4	Skilled manual		4	Skilled manual
644	540	Precision grinders, fitters, & tool sharpeners	4	Skilled manual		4	Skilled manual
645	514	Patternmakers & model makers, metal	4	Skilled manual		4	Skilled manual
646	454	Lay-out workers	4	Skilled manual		4	Skilled manual
647	453	Precious stones & metals workers	4	Skilled manual		4	Skilled manual
649	435	Engravers, metal	4	Skilled manual		4	Skilled manual
653	535	Sheet metal workers	4	Skilled manual		4	Skilled manual
654	536	Sheet metal worker apprentices	4	Skilled manual		4	Skilled manual
655	403	Miscellaneous precision metal workers	4	Skilled manual		4	Skilled manual
655	442	Miscellaneous precision metal workers	4	Skilled manual		4	Skilled manual
655	446	Miscellaneous precision metal workers	4	Skilled manual		4	Skilled manual
655	533	Miscellaneous precision metal workers	4	Skilled manual		4	Skilled manual
656	514	Patternmakers & model makers, wood	4	Skilled manual		4	Skilled manual
657	413	Cabinet makers & bench carpenters	4	Skilled manual		4	Skilled manual
658	443	Furniture & wood finishers	4	Skilled manual		4	Skilled manual
659	575	Miscellaneous precision woodworkers	4	Skilled manual		4	Skilled manual
666	613	Dressmakers	5	Partly skilled	*	4	Skilled manual
667	551	Tailors	4	Skilled manual		4	Skilled manual
668	563	Upholsterers	4	Skilled manual		4	Skilled manual
669	542	Shoe repairers	4	Skilled manual		4	Skilled manual
673	514	Apparel & fabric patternmakers	4	Skilled manual		4	Skilled manual
674	575	Miscellaneous precision apparel & fabric workers	4	Skilled manual		4	Skilled manual
675	546	Hand molders & shapers, except jewelers	4	Skilled manual		4	Skilled manual
676	514	Patternmakers, lay-out workers, & cutters	4	Skilled manual		4	Skilled manual
677	506	Optical goods workers	4	Skilled manual		4	Skilled manual
678	426	Dental laboratory & medical appliance technicians	4	Skilled manual		4	Skilled manual
679	405	Bookbinders	4	Skilled manual		4	Skilled manual
683	675	Electrical & electronic equipment assemblers	5	Partly skilled		5	Partly skilled
684	444	Miscellaneous precision workers, n.e.c.	4	Skilled manual		4	Skilled manual
684	530	Miscellaneous precision workers, n.e.c.	4	Skilled manual		4	Skilled manual

1980 BOC code	1970 BOC code	1980 Code description	Gilbert SES class	Gilbert SES description	Diff	INEEL SES class	INEEL SES description
684	575	Miscellaneous precision workers, n.e.c.	4	Skilled manual		4	Skilled manual
684	580	Miscellaneous precision workers, n.e.c.	0		*	4	Skilled manual
686	631	Butchers & meat cutters	5	Partly skilled	*	4	Skilled manual
686	633	Butchers & meat cutters	5	Partly skilled	*	4	Skilled manual
687	402	Bakers	4	Skilled manual		4	Skilled manual
688	501	Food batchmakers	4	Skilled manual		4	Skilled manual
689	450	Inspectors, testers, & graders	4	Skilled manual		4	Skilled manual
689	452	Inspectors, testers, & graders	4	Skilled manual		4	Skilled manual
693	475	Adjusters & calibrators	4	Skilled manual		4	Skilled manual
694	575	Water & sewage treatment plant operators	4	Skilled manual		4	Skilled manual
695	525	Power plant operators	4	Skilled manual		4	Skilled manual
696	545	Stationary engineers	4	Skilled manual		4	Skilled manual
696	666	Stationary engineers	5	Partly skilled	*	4	Skilled manual
699	575	Miscellaneous plant & system operators	4	Skilled manual		4	Skilled manual
703	653	Lathe & turning machine set-up operators	5	Partly skilled		5	Partly skilled
704	652	Lathe & turning machine operators	5	Partly skilled		5	Partly skilled
705	653	Milling & planing machine operators	5	Partly skilled		5	Partly skilled
706	656	Punching & stamping press machine operators	5	Partly skilled		5	Partly skilled
707	690	Rolling machine operators	5	Partly skilled		5	Partly skilled
708	650	Drilling & boring machine operators	5	Partly skilled		5	Partly skilled
709	621	Grinding, abrading, buffing, & polishing mach. op.	5	Partly skilled		5	Partly skilled
709	651	Grinding, abrading, buffing, & polishing mach. op.	5	Partly skilled		5	Partly skilled
713	692	Forging machine operators	5	Partly skilled		5	Partly skilled
714	692	Numerical control machine operators	5	Partly skilled		5	Partly skilled
715	692	Misc. metal, plastic, stone, & glass working mach. op.	5	Partly skilled		5	Partly skilled
717	660	Fabricating machine operators, n.e.c.	5	Partly skilled		5	Partly skilled
719	503	Molding & casting machine operators	4	Skilled manual	*	5	Partly skilled
723	635	Metal plating machine operators	5	Partly skilled		5	Partly skilled
724	626	Heat treating equipment operators	5	Partly skilled		5	Partly skilled
725	690	Misc. metal & plastic processing mach. operators	5	Partly skilled		5	Partly skilled
726	652	Wood lathe, routing, & planing machine operators	5	Partly skilled		5	Partly skilled
727	662	Sawing machine operators	5	Partly skilled		5	Partly skilled
728	690	Shaping & joining machine operators	5	Partly skilled		5	Partly skilled
729	660	Nailing & tacking machine operators	5	Partly skilled		5	Partly skilled
733	690	Miscellaneous woodworking machine operators	5	Partly skilled		5	Partly skilled
734	530	Printing machine operators	4	Skilled manual	*	5	Partly skilled
735	515	Photoengravers & lithographers	4	Skilled manual	*	5	Partly skilled
736	422	Typesetters & compositors	4	Skilled manual	*	5	Partly skilled
737	530	Miscellaneous printing machine operators	4	Skilled manual	*	5	Partly skilled
738	672	Winding & twisting machine operators	5	Partly skilled		5	Partly skilled
739	671	Knitting, looping, taping, & weaving mach. operators	5	Partly skilled		5	Partly skilled
739	673	Knitting, looping, taping, & weaving mach. operators	5	Partly skilled		5	Partly skilled
743	612	Textile cutting machine operators	5	Partly skilled		5	Partly skilled
744	663	Textile sewing machine operators	5	Partly skilled		5	Partly skilled
745	664	Shoe machine operators	5	Partly skilled		5	Partly skilled
747	611	Pressing machine operators	5	Partly skilled		5	Partly skilled
748	630	Laundering & dry cleaning machine operators	5	Partly skilled		5	Partly skilled
749	670	Miscellaneous textile machine operators	5	Partly skilled		5	Partly skilled
749	674	Miscellaneous textile machine operators	5	Partly skilled		5	Partly skilled
753	690	Cementing & gluing machine operators	5	Partly skilled		5	Partly skilled
754	604	Packaging & filling machine operators	5	Partly skilled		5	Partly skilled
754	643	Packaging & filling machine operators	5	Partly skilled		5	Partly skilled
755	690	Extruding & forming machine operators	5	Partly skilled		5	Partly skilled
756	641	Mixing & blending machine operators	5	Partly skilled		5	Partly skilled

1980 BOC code	1970 BOC code	1980 Code description	Gilbert SES class	Gilbert SES description	Diff	INEEL SES class	INEEL SES description
757	690	Separating, filtering, & clarifying machine operators	5	Partly skilled		5	Partly skilled
758	690	Compressing & compacting machine operators	5	Partly skilled		5	Partly skilled
759	644	Painting & paint spraying machine operators	5	Partly skilled		5	Partly skilled
763	690	Roasting & baking machine operators, food	5	Partly skilled		5	Partly skilled
764	690	Washing, cleaning, & pickling machine operators	5	Partly skilled		5	Partly skilled
765	690	Folding machine operators	5	Partly skilled		5	Partly skilled
766	622	Furnace, kiln, & oven operators, exc. food	5	Partly skilled		5	Partly skilled
768	690	Crushing & grinding machine operators	5	Partly skilled		5	Partly skilled
769	612	Slicing & cutting machine operators	5	Partly skilled		5	Partly skilled
773	505	Motion picture projectionists	4	Skilled manual	*	5	Partly skilled
774	645	Photographic process machine operators	5	Partly skilled		5	Partly skilled
777	653	Miscellaneous machine operators, n.e.c.	5	Partly skilled		5	Partly skilled
777	690	Miscellaneous machine operators, n.e.c.	5	Partly skilled		5	Partly skilled
779	692	Machine operators, not specified	5	Partly skilled		5	Partly skilled
779	695	Machine operators, not specified	5	Partly skilled		5	Partly skilled
783	680	Welders & cutters	5	Partly skilled		5	Partly skilled
784	665	Solderers & brazers	5	Partly skilled		5	Partly skilled
785	602	Assemblers	5	Partly skilled		5	Partly skilled
786	612	Hand cutting & trimming occupations	5	Partly skilled		5	Partly skilled
787	694	Hand molding, casting, & forming occupations	5	Partly skilled		5	Partly skilled
789	425	Hand painting, coating, & decorating occupations	4	Skilled manual	*	5	Partly skilled
789	543	Hand painting, coating, & decorating occupations	4	Skilled manual	*	5	Partly skilled
793	435	Hand engraving & printing occupations	4	Skilled manual	*	5	Partly skilled
794	694	Hand grinding & polishing occupations	5	Partly skilled		5	Partly skilled
795	636	Miscellaneous hand working occupations	5	Partly skilled		5	Partly skilled
795	694	Miscellaneous hand working occupations	5	Partly skilled		5	Partly skilled
796	610	Production inspectors, checkers, & examiners	5	Partly skilled		5	Partly skilled
797	610	Production testers	5	Partly skilled		5	Partly skilled
798	392	Production samplers & weighers	3	Skilled non-man	*	5	Partly skilled
799	624	Graders & sorters, except agricultural	5	Partly skilled		5	Partly skilled
803	441	Supervisors, motor vehicle operators	4	Skilled manual	*	5	Partly skilled
804	715	Truck drivers, heavy	5	Partly skilled		5	Partly skilled
805	715	Truck drivers, light	5	Partly skilled		5	Partly skilled
805	763	Truck drivers, light	6	Unskilled	*	5	Partly skilled
806	705	Driver-sales workers	5	Partly skilled		5	Partly skilled
808	703	Bus drivers	5	Partly skilled		5	Partly skilled
809	714	Taxicab drivers & chauffeurs	5	Partly skilled		5	Partly skilled
813	711	Parking lot attendants	5	Partly skilled		5	Partly skilled
814	710	Motor transportation occupations, n.e.c.	5	Partly skilled		5	Partly skilled
823	226	Railroad conductors & yardmasters	2	Intermediate	*	5	Partly skilled
823	704	Railroad conductors & yardmasters	5	Partly skilled		5	Partly skilled
824	455	Locomotive operating occupations	4	Skilled manual		4	Skilled manual
825	712	Railroad brake, signal, & switch operators	5	Partly skilled		5	Partly skilled
825	713	Railroad brake, signal, & switch operators	5	Partly skilled		5	Partly skilled
826	456	Rail vehicle operators, n.e.c.	4	Skilled manual	*	5	Partly skilled
828	701	Shop captains & mates, except fishing boats	5	Partly skilled		5	Partly skilled
829	661	Sailors & deckhands	5	Partly skilled		5	Partly skilled
833	23	Marine engineers & naval architects	1		*	2	Intermediate
834	694	Bridge, lock, & lighthouse tenders	5	Partly skilled		5	Partly skilled
843	441	Supervisors, material moving equipment operators	4	Skilled manual		4	Skilled manual
844	844	Operating engineers	6	Unskilled	*	5	Partly skilled
845	760	Longshore equipment operators	6	Unskilled	*	5	Partly skilled
848	424	Hoist & winch operators	4	Skilled manual	*	5	Partly skilled
848	706	Hoist & winch operators	5	Partly skilled		5	Partly skilled

1980 BOC code	1970 BOC code	1980 Code description	Gilbert SES class	Gilbert SES description	Diff	INEEL SES class	INEEL SES description
849	424	Crane & tower operators	4	Skilled manual		4	Skilled manual
853	436	Excavating & loading machine operators	4	Skilled manual		4	Skilled manual
855	412	Graders, dozer, & scraper operators	4	Skilled manual		4	Skilled manual
855	436	Graders, dozer, & scraper operators	4	Skilled manual		4	Skilled manual
856	706	Industrial truck & tractor equipment operators	5	Partly skilled		5	Partly skilled
859	694	Miscellaneous material moving equipment operators	5	Partly skilled		5	Partly skilled
863	441	Supervisors; handlers, equip't cleaners, & laborers	4	Skilled manual	*	5	Partly skilled
864	780	Helpers, mechanics & repairers	6	Unskilled		6	Unskilled
865	750	Helpers, construction trades	6	Unskilled		6	Unskilled
866	605	Helpers, surveyor	5	Partly skilled	*	6	Unskilled
867	780	Helpers, extractive occupations	6	Unskilled		6	Unskilled
869	751	Construction laborers	6	Unskilled		6	Unskilled
873	780	Production helpers	6	Unskilled		6	Unskilled
875	754	Garbage collectors	6	Unskilled		6	Unskilled
876	760	Stevedores	6	Unskilled		6	Unskilled
877	762	Stock handlers & baggers	6	Unskilled		6	Unskilled
878	762	Machine feeders & offbearers	6	Unskilled		6	Unskilled
883	753	Freight, stock, & material handlers, n.e.c.	6	Unskilled		6	Unskilled
883	770	Freight, stock, & material handlers, n.e.c.	6	Unskilled		6	Unskilled
885	623	Garage & service station related occupations	5	Partly skilled	*	6	Unskilled
887	764	Vehicle washers & equipment cleaners	6	Unskilled		6	Unskilled
888	634	Hand packers & packagers	5	Partly skilled	*	6	Unskilled
889	780	Laborers, except construction	6	Unskilled		6	Unskilled
889	785	Laborers, except construction	6	Unskilled		6	Unskilled

A.5 Translating ICD-10 causes of death to ICD-9 for death occurring in 1999

The results of the translation crosswalk that resulted from duplicate ICD revision coding for deaths occurring in 1996 (Anderson et al. 2001) were initially used to create a translation table from ICD-10 to ICD-9. A number of the ICD-10 codes were linked to multiple ICD-9 codes (these are identified in Table A-3 with the field DUP=*). In most such cases, the ICD-10 codes overwhelmingly were associated with one code in ICD-9. The ICD-9 and ICD-10 code books were reviewed for each of these ICD-10 codes, to determine if a literal translation existed that could favor a single code ICD-9. In all but three cases, this approach was successful in ascertaining which ICD-9 code should be matched to the ICD-10 code. The three difficult-to-translate cases were C509 (which maps to ICD-9 1749 if female, and 175 if male), C80 (which maps to either 1990 or 1991 in ICD-9), and M480 (which maps to 7230 and 7240 in ICD-9). For the latter two ICD-10 causes, a case-by-case decision was made.

In Table A-3, the decimal has been omitted for the ICD-9 revision, and that the "nature of injury" causes in 9 are preceded by the letter N, and "external causes" are preceded by an E.

One critical limitation of these translations is that the coding rules have changed in the new revisions. That is, the decision tree for selecting an underlying cause-of-death usually changes in new revisions. This has not been taken into account. To be strictly correct in translating these codes, each cause of death in the NDI file should be translated from ICD-10 to ICD-9, and the set of coding rules rerun, using rules for ICD-9 to select an underlying cause of death. We decided not to do this, because the rules changed very little for cancers, the cause of death of greatest interest in this study. These changes in coding rules would primarily affect influenza, dementia or Alzheimer's disease rates, as the coding rules changed dramatically between revisions 9 and 10.

Table A-3. Cross-walk used between ICD-10 and ICD-9 codes for INEEL deaths occurring in 1999.

ICD10	ICD9	DUP1	USE_CODE1	ICD10	ICD9	DUP	USE_CODE
A047	0084			C793	1983	*	
A403	0382			C795	1985		
A410	0381			C798	1988	*	
A415	0384			C80	1990	*	
A419	0389	*		C80	1991	*	
B171	0705	*		C833	2000	*	
B238	2795			C835	2001	*	
B24	2795	*		C845	2020	*	
B377	1128			C851	2028		
B49	1179	*		C859	2028		
B948	1398			C900	2030		
C139	1489			C911	2041		
C159	1509			C920	2050		
C169	1519			C930	2060		
C189	1539			C959	2089		
C220	1550			C97	1990		
C221	1551			D381	2357	*	
C259	1579			D410	2369	*	
C269	1599			D469	2387	*	
C329	1619			D509	280		
C343	1625			D619	2849	*	
C349	1629	*		D649	2859		
C439	1729			D696	2875		
C449	1739			D70	2880	*	
C450	1639			E039	2449		
C457	1958	*		E109	2500		
C459	1991			E119	2500		
C499	1719	*		E142	2503		
C509	1749	*	IF SEX=F	E144	2505		
C509	175	*	IF SEX=M	E145	2506		
C539	1809			E146	2507	*	
C56	1830	*		E149	2500		
C61	185			E348	2598		
C64	1890			E41	261	*	
C679	1889			E43	262	*	
C710	1910			E46	2639	*	
C719	1919			E662	2788		
C760	1950			E668	2780	*	
C779	1969			E743	2713		
C780	1970			E780	2720		
C782	1972			E785	2724		

C785	1975			E789	2729	*
C786	1976			E835	2754	*
C787	1977			E859	2773	
C790	1980			E86	2765	
C791	1981			E870	2760	
E871	2761	*		I38	4249	
E872	2762			I420	4254	
E875	2767			I429	4259	*
E878	2769			I442	4260	
E880	2738	*		I454	4265	*
E889	2779			I461	4299	
F03	2909	*		I469	4275	*
F069	2949			I472	4271	
F100	3050	*		I48	4273	
F101	3050			I490	4274	
F102	303			I495	4278	
F179	2929			I499	4279	*
F329	311	*		I500	4280	
F340	3011	*		I501	4281	
G122	3352	*		I509	4289	*
G20	3320	*		I516	4292	
G301	3310	*		I519	4299	
G309	3310	*		I619	431	
G35	340	*		I633	4340	
G473	7805	*		I634	4341	
G610	3570			I639	4349	
G629	3569	*		I64	436	
G711	3592			I672	4370	*
G909	3379			I679	4379	*
G912	3313	*		I694	438	
G919	3314			I698	438	
G931	3481			I702	4402	
G935	3484	*		I709	4409	
G939	3489	*		I710	4410	
I080	396			I713	4413	
I10	4019	*		I714	4414	
I110	4020	*		I719	4416	
I120	4030	*		I739	4439	
I131	4040	*		I743	4442	
I209	413			I802	4511	*
I219	410			I859	4561	
I248	411			I959	4589	
I249	411	*		J110	4870	
I250	4292			J111	4871	
I251	4140	*		J129	4809	
I255	4148			J152	4824	

I259	4149		J154	4823	
I269	4151		J180	485	*
I270	4160		J188	486	
I279	4169		J189	486	
I313	4239		J439	492	
I330	4210		J448	4918	*
I340	4240		J449	496	
I350	4241		J47	494	
I359	4241		J61	501	
J679	4959		M069	7140	
J690	5070		M199	7159	*
J80	5185		M332	7104	
J81	5184	*	M349	7101	
J841	515	*	M358	7108	
J849	5169	*	M45	7200	*
J90	5119	*	M480	7230	*
J958	5185	*	M480	7240	*
J960	7991		M513	7225	*
J961	7991		N179	5849	
J969	7991		N180	585	
J980	5191	*	N189	585	*
J984	5188	*	N19	586	*
J988	5198	*	N319	5965	*
K219	5308	*	N390	5990	
K221	5302	*	N40	600	
K228	5308		Q447	7516	
K254	5314		Q899	7599	
K264	5324		R001	4278	
K279	5378	*	R060	7860	
K297	5355		R064	7860	
K403	5501		R090	7990	
K529	558	*	R092	7991	
K550	5570		R13	7872	*
K559	5579		R18	7895	*
K567	5601		R402	7800	
K578	5621		R471	7845	
K579	5621		R53	7807	*
K631	5698	*	R54	797	
K650	5679	*	R568	7803	
K659	5679	*	R570	7855	
K703	5712		R579	7855	*
K704	5728		R58	4590	*
K709	5713		R628	7834	*
K729	5738	*	R64	7994	*
K741	5719		R99	7999	*
K746	5715	*	S019	N8739	*

K760	5718	*		S029	N8033	*
K766	5723	*		S062	N8540	*
K767	5724			S065	N8522	*
K769	5719	*		S069	N8540	*
K810	5750	*		S099	N8540	*
K831	5762	*		S199	N9590	*
K901	5791			S224	N8070	*
K922	5789	*		S269	N8610	*
L032	6820			S271	N8602	*
L039	6829			S280	N8628	*
L984	7079	*		S299	N9591	
L988	7098	*		S328	N8054	*
M051	7142	*		S720	N8208	*
S727	N8210	*		V899	E848	
T07	N9598			V953	E8413	*
T099	N9591			W01	E885	
T141	N8798	*		W10	E880	
T149	N9599			W17	E884	
T178	N9348			W18	E885	
T179	N9349	*		W19	E888	
T189	N938			W20	E916	
T300	N9490	*		W30	E9190	
T406	N9650	*		W31	E9198	
T509	N9779	*		W69	E910	
T58	N986	*		W78	E911	
T751	N9941			W80	E912	
T794	N9584			X00	E8909	
T798	N9588			X44	E947	
T813	N9983			X47	E8609	
T818	N9988	*		X59	E9043	
T828	N9967			X62	E9505	
V030	E8227			X72	E9550	
V031	E8147			X73	E9551	
V436	E8121	*		X74	E9559	
V485	E8160	*		Y434	E9331	
V595	E8121			Y832	E8782	
V800	E8272	*		Y834	E8784	
V877	E8109			Y836	E8786	
V892	E8199			Y839	E8789	

[1]DUP indicates a duplicate ICD-9 code for the given ICD-10 code. Only the "preferred" ICD-9 code is provided
USE_CODE identifies the situation in which a given ICD-9 code would be selected for that ICD-10 code.

A.6 Additional Life Table Analysis results

Table A-4. Standardized mortality ratios (SMRs) and standardized rate ratios (SRRs) for white male workers. Badged compared to unbadged workers as persons. Comparison population for SMR analysis is combined Idaho, Montana and Wyoming 1960-1999 (from INEEL05.doc).

Cause of death	SMR Unbadged (# observed)	SMR Badged (# observed)	SRR badged/ unbadged (95% CI)
MN Buccal cavity	0.93 (18)	0.72 (22)	0.76 (0.41-1.42)
MN Pharynx	0.78 (7)	0.63 (9)	0.85 (0.32-2.28)
MN Digestive	1.07 (245)	1.05 (375)	0.96 (0.82-1.13)
MN Esophagus	1.28 (31)	0.90 (35)	0.69 (0.43-1.12)
MN Stomach	1.02 (32)	1.06 (51)	1.00 (0.64-1.56)
MN Intestine	1.00 (79)	1.07 (134)	1.07 (0.81-1.41)
MN Rectum	0.86 (16)	0.90 (26)	1.03 (0.55-1.93)
MN Liver & Gall Bladder	1.22 (18)	1.12 (26)	0.91 (0.50-1.66)
MN Liver unspecified	0.92 (5)	0.95 (8)	1.14 (0.37-3.49)
MN Pancreas	1.09 (56)	1.12 (90)	1.01 (0.72-1.41)
MN Peritoneum & other	1.97 (8)	0.80 (5)	0.39 (0.13-1.19)
MN Respiratory	1.26** (392) CI 1.13-1.39	1.03 (509)	0.81 (0.71-0.92)
MN Larynx	0.92 (9)	1.03 (16)	1.12 (0.49-2.54)
MN Trachea, Bronchus, Lung	1.26** (376) CI 1.13-1.39	1.01 (478)	0.79 (0.69-0.91)
MN Other Respiratory	1.94 (7)	2.59** (15)	1.34 (0.54-3.30)
MN Breast	1.86 (2)	1.77 (3)	0.83 (0.14-4.96)
MN Male Genital	1.20 (122)	1.02 (157)	0.85 (0.67-1.08)
MN Prostate	1.23* (121) CI 1.01-1.46	1.05 (156)	0.85 (0.67-1.08)
MN Testes	0.34 (1)	0.20 (1)	0.58 (0.04-9.25)
MN Urinary Organs	1.17 (61)	1.08 (88)	0.90 (0.65-1.25)
MN Kidney	1.14 (30)	1.12 (47)	0.98 (0.62-1.55)
MN Bladder	1.20 (31)	1.04 (41)	0.83 (0.52-1.33)
MN Other & Unspecified Sites	1.08 (144)	1.12 (237)	1.02 (0.83-1.26)
MN Skin Melanoma	1.21 (21)	1.10 (31)	0.91 (0.52-1.59)
MN Brain & Other Nervous System	1.13 (35)	1.05 (53)	0.94 (0.61-1.44)
MN Thyroid	0.64 (1)	1.60 (4)	2.87 (0.32-25.7)
MN Bone	0.34 (1)	0.44 (2)	1.19 (0.11-13.2)
MN Connective Tissue	0.50 (3)	1.16 (11)	2.21 (0.62-7.91)
MN Other & Unspec	1.11 (77)	1.20* (132)	1.07 (0.81-1.42)
MN Lymphatic & Hematopoietic	1.06 (107)	1.05 (167)	0.98 (0.77-1.25)
Non-Hodgkin's Lymphoma	1.15 (42)	1.36* (79)	1.18 (0.81-1.71)
Hodgkin's Disease	0.33 (2)	0.80 (8)	2.51 (0.53-11.8)
Leukemia	1.09 (44)	0.99 (62)	0.89 (0.61-1.31)
Myeloma	1.04 (19)	0.59 (18)	0.59 (0.31-1.13)
Benign & Unspec. Neoplasms	1.46 (19)	0.89 (18)	0.57 (0.30-1.10)

Cause of death	SMR Unbadged (# observed)	SMR Badged (# observed)	SRR badged/ unbadged (95% CI)
All Cancers	1.15**(1091) CI 1.08-1.22	1.04 (1558) CI 0.99-1.10	--
Diabetes mellitus	0.92 (63)	1.14 (122)	1.22 (0.90-1.65)
Diseases of Blood & Blood-Forming Organs	1.02 (13)	0.66 (13)	0.60 (0.28-1.30)
Non-pernicious & unspecified anemias	1.39 (6)	0.78 (5)	0.50 (0.15-1.63)
Alcoholism	0.78 (17)	0.56** (20)	0.72 (0.38-1.38)
Other Mental Disorders	1.17 (28)	1.09 (40)	0.93 (0.57-1.51)
Diseases of Nervous System & Sense Organs	1.29* (87)	1.03 (109)	0.79 (0.60-1.05)
Diseases of Heart	1.04 (1410)	0.84**(1741)	0.80 (0.74-0.86)
Ischemic Heart Disease	1.08**(1192)	0.84**(1421)	0.77 (0.71-0.83)
Hypertension with Heart Dis.	1.67* (27)	1.79** (42)	0.99 (0.61-1.61)
Other Diseases of Circulatory System	1.01 (340)	0.86** (425)	0.83 (0.72-0.96)
Hypertension w/o Heart Dis.	1.13 (12)	1.06 (17)	0.93 (0.44-1.95)
Cerebrovascular Disease	1.04 (213)	0.83** (246)	0.77 (0.64-0.94)
Disease of Respiratory System	1.20** (445) CI 1.09-1.32	0.88** (495)	0.72 (0.63-0.82)
Pneumonia	0.97 (86)	0.88 (115)	0.90 (0.68-1.19)
Chronic & Unspec. Bronchitis	0.94 (10)	0.84 (13)	0.86 (0.38-1.97)
Emphysema	1.39** (85)	0.79 (71)	0.56 (0.40-0.76)
Asthma	0.95 (9)	0.69 (10)	0.75 (0.30-1.86)
Asbestosis	3.33 (4)	3.21* (6)	0.96 (0.27-3.43)
Silicosis	0.69 (1)	0.51 (1)	0.59 (0.04-9.45)
Other Respiratory Disease	1.29** (245)	0.92 (273)	0.70 (0.59-0.83)
Diseases of Digestive System	1.17* (186) CI 1.00-1.34	0.76** (191)	0.65 (0.53-0.79)
Cirrhosis of Liver	1.31** (92) CI 1.05-1.60	0.66** (76)	0.51 (0.38-0.69)
Diseases of Genito-Urinary System	1.13 (50)	0.81 (54)	0.70 (0.47-1.03)
Acute Glomerulo-nephritis & Acute Renal Failure	1.37 (6)	0.29 (2)	0.21 (0.04-1.02)
Chronic & Unspec. Nephritis, Renal Failure	1.25 (28)	1.17 (40)	0.90 (0.56-1.47)
Diseases of skin	0.87 (2)	1.76 (6)	1.96 (0.39-9.71)
Diseases of musculoskeletal & connective tissue	0.56 (6)	0.54 (9)	0.89 (0.32-2.51)
Accidents	0.94 (340)	0.62** (361)	0.67 (0.57-0.77)
Transportation accidents	0.89 (185)	0.64** (214)	0.72 (0.59-0.87)
Accidental falls	1.30 (41)	0.77 (37)	0.60 (0.38-0.93)
Suicide	0.83 (121)	0.73** (168)	0.88 (0.70-1.12)
Homicide	0.95 (24)	0.49** (20)	0.53 (0.29-0.96)
HIV-related	1.53 (13)	0.66 (8)	0.42 (0.17-1.02)
Other & unspecified	2.49** (169)	2.04** (217)	0.84 (0.68-1.04)
All deaths	1.09** (4464)	0.89** (5617)	--

Table A-5. SRRs for badged with zero dose and badged with positive dose, compared to unbadged workers, as persons. White males only.

Cause of death	SMR Badged-Zero dose (# obs)	SMR Badged-positive dose (# observed)	SRR badged-zero/unbadged (95% CI)	SRR badged-pos/unbadged (95% CI)
MN Buccal cavity	1.17 (10)	0.54* (12)	1.22 (0.56-2.66)	0.57 (0.28-1.20)
MN Pharynx	1.26 (5)	0.38* (4)	1.59 (0.50-5.03)	0.57 (0.17-1.94)
MN Digestive	1.20* (122)	0.98 (253)	1.13 (0.91-1.40)	0.90 (0.75-1.07)
MN Esophagus	1.22 (13)	0.78 (22)	0.94 (0.49-1.80)	0.58 (0.34-1.01)
MN Stomach	1.30 (18)	0.96 (33)	1.25 (0.70-2.24)	0.91 (0.56-1.48)
MN Intestine	1.08 (38)	1.08 (96)	1.10 (0.75-1.62)	1.07 (0.79-1.44)
MN Rectum	1.33 (11)	0.72 (15)	1.61 (0.74-3.48)	0.83 (0.41-1.69)
MN Liver & Gall Bladder	1.38 (9)	1.02 (17)	1.12 (0.50-2.51)	0.80 (0.41-1.56)
MN Liver unspec.	1.66 (4)	0.67 (4)	2.11 (0.57-7.88)	0.80 (0.21-2.97)
MN Pancreas	1.28 (29)	1.05 (61)	1.15 (0.73-1.80)	0.94 (0.65-1.35)
MN Peritoneum & other	0 (1.81 exp)	1.12 (5)	--	0.54 (0.17-1.64)
MN Respiratory	1.18* (162)	0.97 (347)	0.94 (0.79-1.13)	0.76 (0.66-0.88)
MN Larynx	1.17 (5)	0.97 (11)	1.35 (0.45-4.04)	1.09 (0.45-2.64)
MN Trachea, Bronchus & Lung	1.14 (150)	0.95 (328)	0.91 (0.75-1.10)	0.75 (0.65-0.87)
MN Other Resp.	4.29** (7)	1.92 (8)	2.23 (0.78-6.38)	0.99 (0.36-2.73)
MN Breast	0 (0.47 exp)	2.46 (3)	--	1.15 (0.19-6.89)
MN Male Genital	1.10 (50)	0.98 (107)	0.95 (0.68-1.32)	0.83 (0.64-1.08)
MN Prostate	1.14 (50)	1.01 (106)	0.96 (0.69-1.34)	0.83 (0.64-1.08)
MN Testes	0 (1.44 exp)	0.27 (1)	--	0.76 (0.05-12.2)
MN Urinary Organs	0.91 (21)	1.15 (67)	0.81 (0.49-1.33)	0.97 (0.69-1.38)
MN Kidney	1.03 (12)	1.16 (35)	0.98 (0.50-1.91)	1.01 (0.62-1.65)
MN Bladder	0.78 (9)	1.14 (32)	0.65 (0.31-1.37)	0.94 (0.57-1.55)
MN Other & Unspecified Sites	1.15 (68)	1.10 (169)	1.05 (0.78-1.40)	1.01 (0.81-1.27)
MN Skin Melanoma	0.77 (6)	1.23 (25)	0.65 (0.26-1.62)	1.04 (0.58-1.87)
MN Brain & Other Nervous System	1.30 (18)	0.96 (35)	1.18 (0.67-2.09)	0.85 (0.53-1.36)
MN Thyroid	2.92 (2)	1.10 (2)	4.76 (0.43-52.5)	2.04 (0.18-22.5)
MN Bone	0.75 (1)	0.31 (1)	1.47 (0.09-23.5)	0.79 (0.05-12.6)
MN Connective Tissue	0.76 (2)	1.31 (9)	1.28 (0.21-7.64)	2.42 (0.66-8.97)
MN Other & Unspec	1.24 (38)	1.19 (94)	1.09 (0.74-1.61)	1.06 (0.78-1.43)
MN Lymphatic & Hematopoietic	1.04 (47)	1.05 (120)	0.95 (0.67-1.34)	1.00 (0.76-1.33)
Non-Hodgkin's Lymphoma	1.60* (26)	1.27 (53)	1.32 (0.81-2.16)	1.05 (0.70-1.58)
Hodgkin's Disease	0.35 (1)	0.97 (7)	1.00 (0.09-11.0)	3.08 (0.64-14.9)
Leukemia	1.00 (18)	0.98 (44)	0.89 (0.51-1.55)	0.98 (0.60-1.60)
Myeloma	0.25* (2)	0.77 (16)	0.26 (0.06-1.14)	0.73 (0.38-1.43)
Benign & Unspec. Nature Neoplasms	0.68 (4)	0.97 (14)	0.42 (0.14-1.26)	0.61 (0.31-1.23)

Cause of death	SMR Badged-Zero dose (# obs)	SMR Badged-positive dose (# observed)	SRR badged-zero/unbadged (95% CI)	SRR badged-pos/unbadged (95% CI)
Benign of nervous system	0 (0.59 exp)	0 (1.51 exp)	--	--
Unspecified of nervous system	0.37 (1)	0.74 (5)	0.39 (0.04-3.30)	0.88 (0.25-3.07)
Other benign & unspecified	1.17 (3)	1.46 (9)	0.52 (0.15-1.86)	0.61 (0.26-1.46)
All Cancers	1.14** (480)	1.00 (1078)	--	--
Diabetes mellitus	1.28 (39)	1.09 (83)	1.39 (0.93-2.07)	1.14 (0.82-1.58)
Diseases of Blood & Blood-Forming Organs	0.69 (4)	0.65 (9)	0.62 (0.20-1.90)	0.56 (0.24-1.31)
Non-pernicious anemia	1.02 (2)	0.67 (3)	0.66 (0.13-3.28)	0.40 (0.10-1.60)
Alcoholism	0.20** (2)	0.69 (18)	0.26 (0.06-1.13)	0.91 (0.47-1.76)
Other Mental Disorders	1.10 (12)	1.08 (28)	0.92 (0.47-1.81)	0.91 (0.54-1.54)
Diseases of Nervous System & Sense Organs	1.32 (40)	0.92 (69)	1.02 (0.70-1.48)	0.71 (0.51-0.97)
Diseases of Heart	0.87** (523)	0.83** (1218)	0.83 (0.75-0.91)	0.78 (0.73-0.85)
Ischemic Heart Disease	0.84** (411)	0.85** (1010)	0.77 (0.69-0.86)	0.77 (0.71-0.84)
Hypertension with Heart Disease	1.83* (13)	1.78** (29)	1.10 (0.56-2.14)	0.99 (0.58-1.68)
Other Diseases of Circulatory System	0.96 (147)	0.81** (278)	0.96 (0.79-1.17)	0.78 (0.66-0.91)
Hypertension w/o Heart Disease	1.28 (6)	0.96 (11)	1.23 (0.46-3.29)	0.84 (0.37-1.92)
Cerebrovascular Disease	0.94 (86)	0.78** (160)	0.88 (0.68-1.13)	0.71 (0.58-0.87)
Disease of Respiratory Syst.	1.05 (173)	0.81** (322)	0.87 (0.73-1.04)	0.67 (0.58-0.77)
Pneumonia	1.00 (40)	0.83 (75)	1.02 (0.70-1.49)	0.85 (0.62-1.16)
Chronic & Unspec. Bronchitis	0.42 (2)	1.03 (11)	0.45 (0.10-2.10)	1.06 (0.45-2.50)
Emphysema	0.70 (19)	0.83 (52)	0.50 (0.30-0.83)	0.59 (0.42-0.84)
Asthma	0.72 (3)	0.67 (7)	0.68 (0.18-2.52)	0.76 (0.28-2.06)
Asbestosis	1.87 (1)	3.75* (5)	0.48 (0.05-4.30)	1.19 (0.32-4.49)
Silicosis	1.57 (1)	0 (1.34 exp)	2.46 (0.15-39.3)	--
Other Resp. Disease	1.27* (107)	0.78** (166)	0.98 (0.78-1.23)	0.60 (0.49-0.73)
Diseases of Digestive System	0.95 (68)	0.69** (123)	0.82 (0.62-1.09)	0.58 (0.46-0.73)
Cirrhosis of Liver	0.86 (27)	0.59** (49)	0.68 (0.44-1.04)	0.45 (0.32-0.64)
Diseases of Genito-Urinary System	0.85 (17)	0.79 (37)	0.75 (0.43-1.30)	0.70 (0.46-1.08)
Acute Glomerulo-nephritis & Acute Renal Failure	0.51 (1)	0.20 (1)	0.44 (0.05-3.63)	0.15 (0.02-1.22)

Cause of death	SMR Badged-Zero dose (# obs)	SMR Badged-positive dose (# observed)	SRR badged-zero/unbadged (95% CI)	SRR badged-pos/unbadged (95% CI)
Chronic & Unspec. Nephritis, Renal Failure & Other Renal Sclerosis	1.00 (10)	1.24 (30)	0.77 (0.37-1.59)	0.98 (0.58-1.65)
Diseases of skin	0.98 (1)	2.09 (5)	1.10 (0.10-12.2)	2.26 (0.43-11.6)
Diseases of musculoskeletal	0.21 (1)	0.68 (8)	0.34 (0.04-2.83)	1.11 (0.38-3.22)
Symptoms & Ill-Defined Conditions	0.60 (12)	0.94 (44)	0.68 (0.35-1.31)	1.03 (0.67-1.59)
Accidents	0.68** (114)	0.59** (247)	0.73 (0.59-0.91)	0.64 (0.55-0.76)
Transportation accidents	0.77** (74)	0.59** (140)	0.87 (0.66-1.14)	0.66 (0.53-0.83)
Accidental falls	0.84 (12)	0.74 (25)	0.64 (0.34-1.22)	0.58 (0.35-0.95)
Suicide	0.74* (49)	0.72** (119)	0.89 (0.69-1.15)	0.88 (0.70-1.12)
Homicide	0.25** (3)	0.58* (17)	0.30 (0.09-0.99)	0.63 (0.33-1.17)
HIV-related	0.50 (2)	0.73 (6)	0.36 (0.08-1.62)	0.50 (0.19-1.35)
Other & unspecified	2.25** (69)	1.96** (148)	0.91 (0.69-1.21)	0.84 (0.68-1.04)
All deaths	0.96 (1761)	0.86** (3872)	--	--

A.7 Additional radiological exposure assessment information

INTERNAL DOSE FILE CREATION PROCEDURE

There are on record for the site a total of 306 persons who have a known positive internal dose (PID). These records are on hard copy and have been coded.

Source files:

I	BIOASSAY
II	WBSAMPPRM
III	WBNUCLRM
IV	EXP_HIST
V	FINAL_RESULTS_LUNG_BIO_BY_NIOSHID
VI	306 CODED = KNOWN INTERNAL DOSE FROM HARD COPY

Creation of file VII (monitored but zero)

1. Remove subjects of file V from files I, II and III
2. Remove subjects of file VI from file I, II, and III
3. Variables in file should be ID and year of first sample, i.e., if a person has several years of samples use on the earliest date for year

Creation of file VIII (monitored but non-zero)

1. Remove all persons from file IV who have only zero for WBC, SWBC, SUA, T, ST in all years
2. Remove subjects of file VI from file IV
3. Remove all persons who have a total SUA sample frequency <3 summed over all years recorded
4. Variables in file should be NIOSH ID and year of sample. Use only one year value if multiple samples left in that year and include all years

Creation of file IX (monitored but zero)

1. Remove all persons from file IV who have only zero for WBC, SWBC, SUA, T, ST in all years
2. Remove subjects of file VI from file IV
3. Remove all persons who have a total SUA sample frequency > or = 3 summed over all years recorded
4. Variables in file should be NIOSH ID and year of first sample, i.e., if a person has several years of samples use on the earliest date for year.

Creation of three final internal dose files

A - MONITORED WITH POSITIVE INTERNAL DOSE = V + VI + VIII

Note: File variables are NIOSH ID and each year of sample with multiple samples within a year being covered by that single year value.

B - MONITORED WITH NO INTERNAL DOSE = VII + IX

Note: Variables in file should be NIOSH ID and year of first sample, i.e., if a person has several years of samples use on the earliest date for year.

C - NOT MONITORED = NIOSH ID'S FOR COHORT – NIOSH ID'S OF MONITORED FILES (V+VI+VII+VIII+IX)

Note: Variables should be NIOSH ID only.

www.ingramcontent.com/pod-product-compliance
Lightning Source LLC
Chambersburg PA
CBHW081720170526
45167CB00009B/3641